Introductory Mathematical Analysis

Introductory Mathematical Analysis

Said Taan El-Hajjar

Revised by

Faisal Al Showeikh

Copyright © 2011 by Said Taan El-Hajjar.

Library of Congress Control Number: 2011909854

ISBN:	Hardcover	978-1-4628-8788-0
	Softcover	978-1-4628-8789-7
	Ebook	978-1-4628-8790-3

All rights reserved. No part of this book may be reproduced or transmitted in any form or by any means, electronic or mechanical, including photocopying, recording, or by any information storage and retrieval system, without permission in writing from the copyright owner.

This book was printed in the United States of America.

To order additional copies of this book, contact:
Xlibris Corporation
1-888-795-4274
www.Xlibris.com
Orders@Xlibris.com
101029

For my Children,

Reine and Taan

And

For my Wife,

Hanadi

Contents

PREFACE

During the past few years, the teaching of mathematics underwent a considerable change in most countries. At the secondary level, this change is marked by the introduction of notions about the set theory, linear programming and algebraic structures. The geometry, which used to constitute the principal part of the program, is treated today in a narrow connection with the structure of vector spaces.

This book aims to emphasize on the importance of mathematics in real life by showing the utility of this theoretical major considered unfairly unpleasant. To reach this goal we proposed to learners problems related to other fields: Physics, Biology, and Demography…

The scientific content, simplified but rigorous, work to reach the targets, controlled and evaluated.

Far from being a dry account of facts, this work is a tool that permits and promotes the active participation of learners. We wanted to create an instructive product suitable for an education based on the balanced and well defined division of labor between three parts: the student – the instructor – the administrator. The instructor is not only the possessor of the knowledge, but is more the animator of the class, he/she guides it in its choice of orientations, maintains it interest, promotes research among its members, and helps find results while indicating their applications.

At the heart of the educative system is the student, a person with the right to an education worthy of a free individual, independent and seeking knowledge. It is the student who, without feeling forced to put up the knowledge, he/she must learn, must feel constantly faced with situations that simulate him/ her desire to know more.

The proofs are often omitted. The objective is being to initiate the learner for reasoning and for preparing him/ her to approach the exercises easily.

The exercises meet the pedagogical objectives and acquire knowledge.

While expecting a radical reform of the mathematics programs, we hope to supply our students with a series of problems which allow them a better profit from their actual program, and hence a better success in their examinations.

Finally, any remarks, criticisms, or suggestions that colleagues would like to make will be received with attention.

INTRODUCUION

Our objective remains the same to offer to the pupils a big variety of exercises in complement of the manual that they use in progress, and to help the pupils to learn strictness and logic.

We have resort to exercises- type proposed in an increasing order of difficulty, that will allow the pupils, if they solve them all, to make a systematic enrichment of the program. We take the liberty to remind them to the pupils that it would be vain to think than one can learn to make mathematics. We hope that this work will be efficient help for our colleagues and for the candidates official examinations, and that it will contribute besides to make the pupils like mathematics.

The goal of this book is to show both sides – the beauty of limits, derivatives, integrals and their values. The effort is not all concentrated on theorems and proofs, although the mathematics is there. The emphasis is less on rigor, and much more on understanding. We tried to explain rather than to deduce. In the book, and also in class, ideas come with examples.

Once you work with limits and integrals, you understand them. The ability to reason mathematically will develop, if it is given enough to do.

And the essential ideas of calculus are not too hard.

We would like to say clearly that this is a book about calculus .It is not so totally watered down that all the purpose is drained out. We do not believe that students or instructors want an empty course; three hours a week can achieve something worthwhile, provided the textbook helps. I hope and believe that you will see , behind the informal and personal style of this book , that it is written to teach real calculus.

In closing, this is an opportunity for me to say thank you. I am extremely grateful to readers who have read and commented on the book , and have seen what I stands for.

Chapter 1 **Equations and Inequalities**

Introduction

Mathematics has often been criticized because it has emphasized logic and structure that seemingly have no physical representation or practical application. This is unfortunate because probably no other subject has greater application than mathematics. Equations and inequalities are the prime instruments for understanding and exploring our scientific, economic, and social world. Since early antiquity, man has been interested in finding solutions to quadratic equations. In early Babylonian manuscripts certain types of these equations are detected inspired by daily life problems. Greek mathematicians followed by Arab mathematicians worked on these equations. It was the Arab mathematician Al Khawarizmi who first indexed every type of these equations and provided their solutions in his book "Al Jabr Wal Mouqabalah". Today, more than ever before, all fields of knowledge are dependent on equations for solving problems, stating theories, and predicting outcomes. It is an indispensable tool in creating new knowledge.

Why?

Solving equations is a widespread activity. We encounter them in Physics, Chemistry, Economy, and in Operational research.... This chapter aims at initiating the student to general methods of solving such equations and inequalities. It provides students with an analytical way to form an equation or an inequality for a certain given problem solving. It also supplies students with significant strategies in finding an appropriate response to situation which is unique and novel to the problem solver.

Objectives

❖ After studying the material in this chapter, you should be able to:

- Evaluate an expression with one variable
- Solve a linear equation with one unknown
- Solve a rational equation with one unknown
- Solve an irrational equation with one unknown
- Solve an absolute value equation
- Solve an inequality with one unknown

Section 1.1

Evaluating an Expression

Evaluate an expression with one variable

A mathematical expression can have a variable as part of the expression. If $x = 2$, the expression $5x + 4$ becomes $5 \times 2 + 4$ which is equal to $10 + 4$ or 14. To evaluate an expression with a variable, simply substitute the value of the variable into the expression and simplify.

Example 1.1-1 Evaluate the following expressions:

a. $5x + 10$ if $x = 3$

b. $7x + 5$ if $x = 9$

Solution

a. The expression $5x + 10$ if $x = 3$ is equal to 25

b. The expression $7x + 5$ if $x = 9$ is equal to 68

Evaluate an expression with two variables

A mathematical expression can have variables as part of the expression. If $x = 7$ and $y = 2$, the expression $3x + y - 4$ becomes $3 \times 7 + 2 - 4$ which is equal to $21 + 2 - 4$ or 19. To evaluate an expression with two or more variables, substitute the value of the variables into the expression and simplify.

Example 1.1-2 Evaluate the following expressions:

a. $4x + 5y + 4$ if $x = 7$ and $y = 0$

b. $8x + 5y + 8$ if $x = 9$ and $y = 9$

Solution

a. The expression $4x + 5y + 4$ if $x = 7$ and $y = 0$ is equal to 32

b. The expression $8x + 5y + 8$ if $x = 9$ and $y = 9$ is equal to 125

Section 1.2

Linear Equations

Definition 1.2 -1 Equation

An equation is a mathematical statement that has two expressions separated by an equal sign. The expression on the left side of the equal sign has the same value as the expression on the right side. One or both of the expressions may contain variables.

Example 1.2-1

The following statements are equations

$$x = 4 + 8$$

$$4x - 2x^3 = x - 1$$

$$\frac{x+3}{x-1} = 7x + 2$$

$$\sqrt{x-2} = \frac{3}{2}x - 5$$

Definition 1.2 -2

Any equation that can be written in the form $ax + b = 0$ where $a \in \Re - \{0\}$ and $b \in \Re$, is said to be a linear equation of the first degree in one unknown .

Example 1.2-2

$$3x + 5 = 0$$

$$6x - 4 = x + 3$$

Definition 1.2 - 3 Solving an Equation

Solving an equation means manipulating the expressions and finding the value of the variables that satisfy the given equation.

Example 1.2-3

Solve the equation $x - 2 = 6$

Solution

To keep an equation equal, we must do exactly the same thing to each side of the equation. If we add (or subtract) a quantity from one side, we must add (or subtract) that same quantity to the other side. To solve this equation we would add 2 to both sides. The equation would become $x - 2 + 2 = 6 + 2$. This becomes $x = 6 + 2$ or $x = 8$.

Example 1.2-4

Solve the equation $5x = 25$

Solution

To keep both sides of an equation equal, we must do exactly the same thing to each side of the equation. If we multiply (or divide) one side by a quantity, we must multiply (or divide) the other side by that same quantity. In order to solve this equation we would divide both sides by 5.The equation would become $5x / 5 = 25 / 5$. When simplified, this would become $x = 25 / 5$ or $x = 5$. It is possible to substitute the value of x back into the original equation $5 \times 5 = 25$.

Example 1.2-5

Solve the equation $4x-1=15$

Solution

To keep both sides of an equation equal, we must do exactly the same thing to each side of the equation. First, add one to each side of the equation so that $4x-1+1=15+1$ or $4x = 16$. If we multiply (or divide) one side by a quantity, we must multiply (or divide) the other side by that same quantity. In order to solve this equation we would divide both sides by 4. The equation would become $4x/4 = 16/4$. When simplified, this would become $x = 16/4$ or $x = 4$. It is possible to substitute the value of x back into the original equation $4 \times 4 -1=15$.

Example 1.2-6

Solve the equation $x \div 4 = 6$

Solution

The solution of an equation is finding the value of the unknown x. Use the multiplication property of equations to find the value of x. The multiplication property of equations states that the two sides of an equation remain equal if both sides are multiplied by the same number.

$x \div 4 = 6$

$x \div 4 \times 4 = 6 \times 4$

$x \div 1 = 24$

$x = 24$

Check the answer by substituting the answer (24) back into the equation.

$24 \div 4 = 6$

Section 1.3

Rational Equations

Definition 1.3 -1

Any equation that can be written in the form of $\frac{P(x)}{Q(x)} = 0$, where P(x) is a polynomial function with one variable x and Q(x) is also a polynomial function with one variable x such that Q(x) is different from zero is called rational equation with one variable x.

Example 1.3-1

$$\frac{8x-4}{x} = 0 \quad with\ x \neq 0$$

$$\frac{7x^3}{3x+4} = 0 \quad with\ x \neq \frac{-4}{3}$$

Example 1.3-2

Solve the following rational equations:

a. $\frac{x-2}{2x-1} = 0$

b. $\frac{x+1}{2x-3} = \frac{x}{2x-4}$

c. $\frac{x^2-1}{x-1} = 1$

Solution

a. $\frac{x-2}{2x-1} = 0$ is defined if and only if $2x - 1 \neq 0$, then $x \neq \frac{1}{2}$.

By cross multiplication, we get $x - 2 = 0$ and then $x = 2 \neq \frac{1}{2}$; so this solution is acceptable.

b. $\frac{x+1}{2x-3} = \frac{x}{2x-4}$ is defined if and only if $2x - 3 \neq 0$ and $2x - 4 \neq 0$, then

$x \neq \frac{3}{2}$ and $x \neq 2$. By cross multiplication, we get

$(x + 1)(2x - 4) = x(2x - 3)$ which implies that $x = 4$. The obtained value of

x is different from $\frac{3}{2}$ and 2, then it is acceptable.

c. $\frac{x \ \ -1}{x^2-1} = 1$ is defined if and only if $x^2 - 1 \neq 0$, then $x \neq 1$ and $x \neq -1$.

By cross multiplication, we get $(x^2 - 1) = x - 1$ which implies that

$(x - 1)(x + 1) - (x - 1) = 0$.

Factorize by $(x - 1)$ to get $(x - 1)(x + 1 - 1) = 0$ implies that $x(x - 1) = 0$.

Hence, either $x = 0$ or $x = 1$. But x should be different from 1, then the only accepted value for x is 0.

Example 1.3-3

Manal wants to invest a sum of \$ 10 000 000 during 10 years, she has the chance to choose between two offers:

Offer I

Places the sum at a bank paying an annual interest rate of 8 % compounded semiannually.

Offer II

Places this sum in another bank paying compound interest at an annual interest rate of 8 % compounded monthly.

Which of the two offers is more advantageous for Manal?

Solution

The future value attained after 10 years if Manal chooses the first offer is :

$$FV = PV(1 + i)^{n} = 10\,000\,000\left(1 + \frac{0.08}{2}\right)^{2 \times 10} = \$\,219\,112\,31.43.$$

The future value attained after 10 years if Manal chooses the second offer is :

$$FV = PV(1 + i)^{n} = 10\,000\,000\left(1 + \frac{0.08}{12}\right)^{12 \times 10} = \$\,22196402.35.$$

Therefore, offer II is more advantageous for Manal.

Example 1.3-4

Taan wants to buy a farm for \$ 50 000 000. He borrows the sum from a bank paying an annual interest rate of 6 % compounded monthly. Taan wants to repay this loan with equal monthly payments during 12 years.

a. Calculate the value of each payment.
b. What will the total interest paid by Taan be ?

Solution

a. It is an annuity with 50 000 000 as the present value , $k = 12$,

   $n = 12(12) = 144$ and $i = \dfrac{0.06}{12}$.

Then $A = R \times \frac{1-(1+i)^{-n}}{i}$ which gives $50\,000\,000 = R \times \frac{1-(1+\frac{0.06}{12})^{-144}}{\frac{0.06}{12}}$,

then $50\,000\,000 = R \times 102.4747$. Consequently, $R = \frac{50\,000\,000}{102.4747} \approx \$\,487925.31$.

The value of the periodical payment is \$ 487925.31 .

b. The sum paid by Taan is $487925.31 \times 12 \times 12 = \$\,70\,261\,244.64$.

   The total interest values paid by Taan is 70 261 244.64 – 50 000 000 = \$ 20 261 244.64.

Section 1.4

Simple Irrational Equations

Definition 1.4 -1

Any equation that can be written in the form of $\sqrt{P(x)} = 0$, where P(x) is a positive polynomial function with one variable x (i.e. $P(x) \geq 0$), is called simple irrational equation with one variable x.

Example 1.4-1

$\sqrt{x-2} = 0$

$\sqrt{3x^3 - 5x} + 6x = 0$

$\sqrt{x+5} = \sqrt{5x^2 - 5x^6}$

Example 1.4-2

Solve the following irrational equations:

a. $\sqrt{x-3} = 0$

b. $\sqrt{-2x-3} = 5$

c. $\sqrt{x+4} - \sqrt{2x+1} = 0$

Solution

a. $\sqrt{x-3} = 0$ is defined if $x - 3 \geq 0$, then $x \geq 3$. So an acceptable solution of this equation is a value greater than or equal to 3; otherwise, the given equation has no solution. To solve this equation, we must get rid of the square root. The simplest way is to square both sides. By squaring both sides, we get:

$x - 3 = 0$ which implies that $x = 3$. The value 3 is greater than or equal to 3, then it is acceptable.

Remark 1.4-1

We can replace x by its value 3 in the given equation to check the validity of the obtained answer.

b. $\sqrt{-2x-3} = 5$ is defined if $-2x - 3 \geq 0$, then $x \leq -\frac{3}{2}$. So an acceptable solution of this equation is a value less than or equal to $-\frac{3}{2}$; otherwise, the given equation has no solution. By squaring both sides, we get:

$-2x - 3 = 25$ which implies that $x = -14$. The value -14 is less than or equal to $-\frac{3}{2}$, then it is acceptable .

c. $\sqrt{x+4} - \sqrt{2x+1} = 0$ is defined if $x + 4 \geq 0$ and $2x + 1 \geq 0$, then $x \geq -4$ and $x \geq -\frac{1}{2}$. It is obvious that a common interval for these two conditions is $\left[-\frac{1}{2}, +\infty\right)$. So an acceptable solution of this equation is a value greater than or equal to $-\frac{1}{2}$; otherwise, the given equation has no solution . To solve this equation, it is preferable first to take one of the two given square roots to the other side of the equation and then squaring both sides ; otherwise, the square root remains after squaring, due to the fact of the identity $(a - b)^2 = a^2 - 2ab + b^2$. In this case $2ab = 2\sqrt{(x+4)(2x+1)}$ and the square root does not vanish .

Therefore, we write $\sqrt{x+4} = \sqrt{2x+1}$. By squaring both sides, we get:

$x + 4 = 2x + 1$ which implies that $x = 3$. The value 3 belongs to the interval

$\left[-\frac{1}{2}, +\infty\right)$, then it is acceptable.

Section 1.5

Absolute Value Equations

Definition 1.5 -1

Any equation that can be written in the form of $|P(x)| = k$, where P(x) is a function with one variable x and k is any real number, is called simple absolute value equation with one variable x.

Example 1.5 -1

$$|7x^4 + 3x - 4| = 2$$

$$|x-5| = |2x - x^2|$$

$$\left|\frac{x+3x^2}{x+1}\right| = |2-x|$$

Property 1.5-1

For any real number a, we have $|a| = \begin{cases} +a \text{ if } a \geq 0 \\ \text{or} \\ -a \text{ if } a \leq 0 \end{cases}$

Example 1.5 -2

$|13| = 13$

$|-13| = -(-13) = 13$

$$\left| -\frac{2}{3} \right| = -\left(-\frac{2}{3}\right) = \frac{2}{3}$$

$$\left| \sqrt{5} - 1 \right| = \sqrt{5} - 1$$

$$\left| 1 - \sqrt{5} \right| = -(1-\sqrt{5}) = \sqrt{5} - 1$$

Property 1.5-2

For any real number x and any positive real number k , we have

$$|x| = k \Leftrightarrow x = \begin{cases} +k \ \ if \ x \geq 0 \\ \quad or \\ -k \ \ if \ x \leq 0 \end{cases}$$

Proof

$$|x| = k \Leftrightarrow \begin{cases} +x = k \ \ if \ x \geq 0 \\ \quad or \\ -x = k \ \ if \ x \leq 0 \end{cases} \Leftrightarrow \begin{cases} x = k \ \ if \ x \geq 0 \\ \quad or \\ x = -k \ \ if \ x \leq 0 \end{cases}$$

Example 1.5 -3

Solve the following equations:

a. $|2x+4| = 6$

b. $|x+3| + |x-2| = 7$

Solution

a. $|2x+4| = 6 \Rightarrow \begin{cases} 2x+4 = 6 \\ 2x+4 = -6 \end{cases}$

$$\Rightarrow \begin{cases} 2x = 6-4 \\ 2x = -6-4 \end{cases}$$

$$\Rightarrow \begin{cases} x = 1 \\ x = -5 \end{cases}$$

b. $|x+3| + |x-2| = 7$

The expression without absolute value of $|x+3| + |x-2|$ is defined by the following table:

x	$-\infty$	-3	2	$+\infty$
$\lvert x-2 \rvert$	2 - x	2 - x	x - 2	
$\lvert x+3 \rvert$	- x - 3	x + 3	x + 3	
$\lvert x-2 \rvert + \lvert x+3 \rvert$	- 2x - 1 5	5	5 2x + 1	

For $x \in \left]-\infty, -3\right[$, we have : - 2 x – 1 = 7 , then x = - 4 acceptable .

For $x \in [-3, 2]$, we have: 5 = 7 which is impossible and the equation has no solutions.

For $x \in \left]2, +\infty\right[$, we have : 2 x + 1 = 7 , then x = 3 acceptable .

Hence the solution is x = - 4 or x = 3.

Section 1.6

Quadratic Equations

Definition 1.6 -1

Any equation that can be written in the form of $a x^2 + b x + c = 0$,where a is a real number different from zero, b, and c are any real numbers , is called quadratic equation with one variable x .

Remark 1.6 -1

If $a x^2 + b x + c = 0$ is a quadratic equation, then a, b and c are called coefficients and x is the unknown.

Example 1.6 -1

The following equations are quadratic equations with one variable x :

$2 x^2 - 7 x + 4 = 0$

$- 8 x^2 + 3 x = 0$

$x^2 - 3 = 0$

$x^2 = 0$

Property 1.6-1

To solve a quadratic equation of the form $a x^2 + b x + c = 0$, we need to find the discriminant $\Delta = b^2 - 4ac$ of the given equation such that:

a. If $\Delta < 0$, then the equation has no roots.

b. If $\Delta = 0$, then the equation has two equal roots or a double root :

$$x_1 = x_2 = \frac{-b}{2a}.$$

c. If $\Delta > 0$, then the equation has two distinct roots :

$$\begin{cases} x_1 = \dfrac{-b+\sqrt{\Delta}}{2a} \\ x_2 = \dfrac{-b-\sqrt{\Delta}}{2a} \end{cases}$$

Proof

$$a x^2 + b x + c = 0 \Rightarrow a[(x^2 + \frac{b}{a} x) + \frac{c}{a}] = 0$$

$$\Rightarrow a[(x^2 + \frac{b}{a} x + \frac{b^2}{4a^2}) - \frac{b^2}{4a^2} + \frac{c}{a}] = 0$$

$$\Rightarrow (x + \frac{b}{2a})^2 - \frac{b^2}{4a^2} + \frac{c}{a}] = 0 \text{ since } a \neq 0$$

$$\Rightarrow (x + \frac{b}{2a})^2 - \frac{b^2 - 4ac}{4a^2} = 0$$

$$\Rightarrow (x + \frac{b}{2a})^2 - \frac{\Delta}{4a^2} = 0 \text{, where } \Delta = b^2 - 4ac$$

$$\Rightarrow (x + \frac{b}{2a})^2 = \frac{\Delta}{4a^2}$$

$4a^2$ is always positive, then $(x + \frac{b}{2a})^2$ admits the sign of Δ.

a. If $\Delta < 0$, then $(x + \frac{b}{2a})^2$ is equal to a negative number which is impossible, and then the equation has no roots.

b. If $\Delta = 0$, then $(x + \frac{b}{2a})^2 = 0$, and then the equation has a double root equals to $-\frac{b}{2a}$.

c. If $\Delta > 0$, then $x + \frac{b}{2a} = \pm\sqrt{\frac{\Delta}{4a^2}} \Rightarrow x = \pm\frac{\sqrt{\Delta}}{2a} - \frac{b}{2a}$, and then the equation has two distinct roots: $\frac{-b+\sqrt{\Delta}}{2a}$ or $\frac{-b-\sqrt{\Delta}}{2a}$.

Example 1.6 -2

Solve the following quadratic equations:

a. $2x^2 + x + 1 = 0$

b. $x^2 + 4x + 4 = 0$

c. $-x^2 - 5x + 6 = 0$

Solution

a. $\Delta = b^2 - 4ac$

$= 1^2 - 4(2)(1)$

$= -7.$

Then $\Delta < 0$, and the equation has no roots.

b. $\Delta = b^2 - 4ac$

$= 4^2 - 4(1)(4)$

$= 0.$

Then $\Delta = 0$, and the equation has two equal roots or a double root :

$$x_1 = x_2 = \frac{-b}{2a} = \frac{-4}{2(1)} = -2 .$$

c. $\Delta = b^2 - 4ac$

$= (-5)^2 - 4(-1)(6)$

$= 25 + 24$

$= 49$

Then $\Delta > 0$, and the equation has two distinct roots :

$$\begin{cases} x_1 = \dfrac{-b+\sqrt{\Delta}}{2a} = \dfrac{-(-5)+\sqrt{49}}{2(-1)} \\ \quad = \dfrac{5+7}{-2} = -6. \\ x_2 = \dfrac{-b-\sqrt{\Delta}}{2a} = \dfrac{-(-5)-\sqrt{49}}{2(-1)} \\ \quad = \dfrac{5-7}{-2} = 1. \end{cases}$$

Property 1.6-2

To solve a quadratic equation of the form $a x^2 + b x + c = 0$, where b is even, then let $b' = \dfrac{b}{2}$ and use $\Delta' = b'^2 - a c$ to reduce the calculation.

a. If $\Delta' < 0$, then the equation has no roots.

b. If $\Delta' = 0$, then the equation has two equal roots or a double root :

$$x_1 = x_2 = \frac{-b'}{a}.$$

c. $\Delta' > 0$, then the equation has two distinct roots :

$$\begin{cases} x_1 = \dfrac{-b' + \sqrt{\Delta}}{a} \\ x_2 = \dfrac{-b' - \sqrt{\Delta}}{a} \end{cases}$$

Remark 1.6 -1

If x_1 and x_2 are roots of $a x^2 + b x + c = 0$, then

$a x^2 + b x + c = a(x - x_1)(x - x_2)$

Example 1.6 -3

Solve the following quadratic equations:

a. $x^2 + 4x + 5 = 0$.

b. $x^2 + 8x + 16 = 0$.

c. $x^2 - 8x + 7 = 0$.

Solution

a. $\Delta' = b'^2 - ac$

$= 2^2 - (1)(5)$

$= -1$.

Then $\Delta' < 0$, and the equation has no roots.

b. $\Delta' = b'^2 - ac$

$= 4^2 - (1)(16)$

$= 0$.

Then $\Delta' = 0$, and the equation has two equal roots or a double root :

$$x_1 = x_2 = \frac{-b'}{a} = \frac{-4}{1} = -4.$$

c. $\Delta' = b'^2 - ac$

$= (-4)^2 - (1)(7)$

$= 16 - 7$

$= 9$

Then $\Delta' > 0$, and the equation has two distinct roots :

$$x_1 = \frac{-b' + \sqrt{\Delta'}}{a} = \frac{-(-4) + \sqrt{9}}{1} = 7 \text{ or } x_1 = \frac{-b' - \sqrt{\Delta'}}{a} = \frac{-(-4) - \sqrt{9}}{1} = 1.$$

Remark 1.6 -2

$x^2 + 8x + 16 = 1(x - (-4))(x - (-4)) = (x + 4)^2$

$x^2 - 8x + 7 = 1(x - 7)(x - 1) = (x - 7)(x - 1)$

Remark 1.6 -3

To solve a quadratic equation of the form $ax^2 + bx + c = 0$, where $a + b + c = 0$, then the equation has two roots $x_1 = 1$ and $x_2 = \frac{c}{a}$.

Example 1.6 – 4

Solve the quadratic equation $x^2 - 9x + 8 = 0$.

Solution

$a + b + c = 1 - 9 + 8 = 0$. Then the equation has two roots $x_1 = 1$ and

$$x_2 = \frac{c}{a} = \frac{8}{1} = 8.$$

Remark 1.6 -2

To solve a quadratic equation of the form $ax^2 + bx + c = 0$, where $a - b + c = 0$, then the equation has two roots $x_1 = -1$ and $x_2 = \frac{-c}{a}$.

Example 1.6 – 5

Solve the quadratic equation $x^2 + 9x + 8 = 0$.

Solution

$a - b + c = 1 - 9 + 8 = 0$. Then the equation has two roots $x_1 = -1$ and

$$x_2 = -\frac{c}{a} = \frac{-8}{1} = -8.$$

Section 1.7

Inequalities

An inequality is similar to an equation. There are two expressions separated by a symbol that indicates how one expression is related to the other. In an equation such as $7x = 49$, the " = " sign indicates that the expressions are equivalent. In an inequality, such as $7x > 49$, the " > " sign indicates that the left side is larger than the right side.

To solve the inequality $7x > 49$, we follow the same rules that we did for equations. In this case, divide both sides by 7 so that $x > 7$. This means that x is a value and it is always larger than 7, and never equal to or less than 7.

The "less than" symbol (<) may also be seen in inequalities.

Property 1.7-1

To solve the inequality of the form $a x > b$ where a is a negative number, then when we divide both sides by a, we must change the sign " > " to " < " and vice versa .

Example 1.7– 1

Solve the inequality $-4x > 8$.

Solution

Divide both sides by - 4, we get: $x < \frac{8}{-4}$, then $x < -2$.

Property 1.7-2

To solve an equality of the form $a x^2 + b x + c < 0$, or $a x^2 + b x + c > 0$, first solve the quadratic equation $a x^2 + b x + c = 0$. Second study the sign of the polynomial $a x^2 + b x + c$ as follows :

a. If it has no roots, then $a x^2 + b x + c$ admits the sign of a.

b. If it has a double root $x_1 = x_2 = \frac{-b}{2a}$, then

x	$-\infty$	$-b/2a$	$+\infty$
$ax^2 + bx + c$	same sign as a	0	same sign as a

c. If it has two distinct roots $\begin{cases} x_1 = \frac{-b+\sqrt{\Delta}}{2a} \\ x_2 = \frac{-b-\sqrt{\Delta}}{2a} \end{cases}$, then if $x_1 < x_2$, we have :

x	$-\infty$	x_1		x_2	$+\infty$
$ax^2 + bx + c$	same sign as a	0	opposite sign of a	0	same sign as a

Finally, choose the interval that satisfies the given inequality.

Example 1.7– 2

Solve in $\Re$ the following inequalities:

a. $x^2 + x + 1 > 0$

b. $x^2 - 2x + 1 < 0$

c. $-2x^2 + 5x - 3 \geq 0$

Solution

a. Consider the equation $x^2 + x + 1 = 0$

$\Delta = b^2 - 4ac$

$= (1)^2 - 4(1)(1)$

$= -3$

Then $\Delta < 0$ and $x^2 + x + 1$ admits the sign of $a = +1 > 0$ for every x. Hence

$x^2 + x + 1 > 0$ for every $x \in \Re$.

b. Consider the equation $x^2 - 2x + 1 = 0$.

$\Delta' = b'^2 - ac$

$= (-1)^2 - (1)(1)$

$= 0.$

Then $\Delta' = 0$, and the equation has two equal roots or a double root :

$$x_1 = x_2 = \frac{-b'}{a} = \frac{1}{1} = 1 .$$

x	$-\infty$		1		$+\infty$
$x^2 - 2x + 1$		+	0	+	

Hence, $x^2 - 2x + 1 < 0$ has no solution.

c. Consider the equation $-2x^2 + 5x - 3 = 0$.

$a + b + c = -2 + 5 - 3 = 0$, then one of the root $x_1 = 1$ and the other

root $x_2 = \dfrac{c}{a} = \dfrac{-3}{-2} = \dfrac{3}{2}$.

x	$-\infty$		1		3/2		$+\infty$
$-2x^2+5x-3$		—	0	+	0	—	

Hence, $-2x^2 + 5x - 3 \geq 0$ for $x \in \left[1, \frac{3}{2}\right]$.

Property 1.7-3

For any positive real number k, and any unknown x, we have $|x| \leq k$ if and only if $-k \leq x \leq k$.

Proof

$$|x| \leq k \Leftrightarrow \begin{cases} x \leq k & for\ x \geq 0 \\ -x \leq k & for\ x \leq 0 \end{cases}$$

$$\Leftrightarrow \begin{cases} x \leq k & for\ x \geq 0 \\ x \geq -k & for\ x \leq 0 \end{cases}$$

$$\Leftrightarrow -k \leq x \leq k$$

Property 1.7 - 4

For any positive real number k, and any unknown x, we have

$|x| \geq k$ if and only if $\begin{cases} x \leq -k \\ x \geq k \end{cases}$.

Proof

$$|x| \geq k \Leftrightarrow \begin{cases} x \geq k & for\ x \geq 0 \\ -x \geq k & for\ x \leq 0 \end{cases}$$

$$\Leftrightarrow \begin{cases} x \geq k & for\ x \geq 0 \\ x \leq -k & for\ x \leq 0 \end{cases}$$

$$\Leftrightarrow \begin{cases} x \leq -k \\ x \geq k \end{cases}$$

Example 1.7– 3

Solve the following inequalities:

a. $|\ x-3\ | \leq 4$

b. $|\ 2x+4\ | \succ 1$

Solution

a. $|\ x-3\ | \leq 4 \Rightarrow -4 \leq x-3 \leq 4$

$$\Rightarrow -4+3 \leq x \leq 4+3$$

$$\Rightarrow -1 \leq x \leq 7$$

That is $x \in [-1, 7]$.

b. $|\ 2x+4\ | > 1 \Rightarrow \begin{cases} 2x+4 < -1 \\ 2x+4 > 1 \end{cases}$

$$\Rightarrow \begin{cases} 2x < -1-4 \\ 2x > 1-4 \end{cases}$$

$$\Rightarrow \begin{cases} 2x < -5 \\ 2x > -3 \end{cases}$$

$$\Rightarrow \begin{cases} x < \dfrac{-5}{2} \\ x > \dfrac{-3}{2} \end{cases}$$

That is $x \in \left]-\infty, \dfrac{-5}{2}\right[\cup \left]\dfrac{-3}{2}, +\infty\right[$.

Example 1.7– 4

Solve the following system of inequalities:

$$\begin{cases} 3x-1 > x+3 \\ 4x \leq 24 \\ -x < 1 \end{cases}$$

Solution

$$\begin{cases} 3x - x > 1 + 3 \\ 4x \le 24 \\ -x < 1 \end{cases} \Rightarrow \begin{cases} x > 2 \\ x \le 6 \\ x > -1 \end{cases}$$

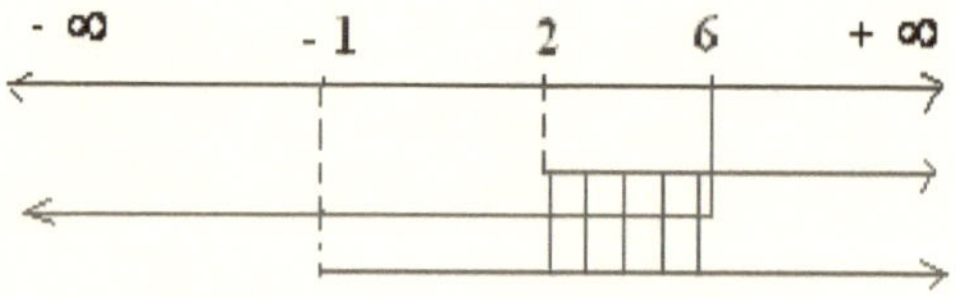

- ∞ - 1 2 6 + ∞

The solution set is $x \in (2, 6]$.

Summary

☒ For any real number a, we have $|a| = \begin{cases} +a \text{ if } a \ge 0 \\ \text{or} \\ -a \text{ if } a \le 0 \end{cases}$

☒ For any real number x and any positive real number k, we have

$$|x| = k \Leftrightarrow x = \begin{cases} +k \text{ if } x \ge 0 \\ \text{or} \\ -k \text{ if } x \le 0 \end{cases}$$

☒ $\Delta = b^2 - 4ac$ of the given equation such that :

a. If $\Delta < 0$, then the equation has no roots.

b. If $\Delta = 0$, then the equation has two equal roots or a double root:

$$x_1 = x_2 = \frac{-b}{2a}.$$

c. If $\Delta > 0$, then the equation has two distinct roots:

$$\begin{cases} x_1 = \dfrac{-b + \sqrt{\Delta}}{2a} \\ x_2 = \dfrac{-b - \sqrt{\Delta}}{2a} \end{cases}$$

☒ For any positive real number k, and any unknown x, we have

a. $|x| \le k$ if and only if $-k \le x \le k$

b. $|x| \ge k$ if and only if $\begin{cases} x \le -k \\ x \ge k \end{cases}$

EXERCISES

In Problems 1.1 - 1.4, evaluate the expressions for the given value of x.

1.1. $f(x) = 4x^2 + 3x - 2$ for $x = 2$

1.2. $f(x) = 2x(x-3)$ for $x = \sqrt{2}$

1.3. $f(x) = x^4(x^3 - 3x)$ for $x = -2$

1.4. $f(x) = (x^2 - x)(3x + 2)$ for $x = \frac{1}{2}$

In Problems 1.5 -1.8, determine the value of the functions for the given value of x.

1.5. $f(x) = \frac{x^3 - 2x^2 + x}{x+4}$ for $x = -1$

1.6. $f(x) = \frac{x}{x+1} + x + 4$ for $x = 1$

1.7. $f(x) = \frac{x-3}{2x-1} + \frac{2}{x^2}$ for $x = 2$

1.8. $f(x) = (\frac{x^2 - 1}{x+7})(\frac{1}{3x^2})$ for $x = -2$

In Problems 1.9 - 1.12, calculate f (x) for the given value of x.

1.9. $f(x) = \sqrt{x+9}$ for $x = 7$

1.10. $f(x) = 2x - \sqrt{2x+9}$ for $x = 20$

1.11. $f(x) = \sqrt{-x+10} + \sqrt{-x-1}$ for $x = -26$

1.12. $f(x) = (3x+8)\sqrt{2x+9}$ for $x = -4$

In Problems 1.13 - 1.16, evaluate the expressions for the given value of x.

1.13. $f(x) = |-x-5|$ for $x = 10$

1.14. $f(x) = 3x^3 + 8x - |5x-2|$ for $x = -3$

1.15. $f(x) = |12x - 10x^4| - |x-1|$ for $x = 1$

1.16. $f(x) = |x^2+2x-3|\ |-4x+3|$ for $x = -2$

In Problems 1.17 - 1.19, solve in $\Re$ the given equations.

1.17. $\dfrac{2x-3}{3} + \dfrac{x-1}{2} = x - \dfrac{7}{6}$

1.18. $\dfrac{5x+2}{3} = x + \dfrac{2x}{3} + \dfrac{2}{3}$

1.19. $\dfrac{2}{x} + \dfrac{3}{x} = \dfrac{1}{2} + \dfrac{1}{3}$

In Problems 1.20 - 1.22, find x.

1.20. $\dfrac{2x-3}{x+1} = 3$

1.21. $\dfrac{3x+1}{3x} = \dfrac{2x-4}{2x}$

1.22. $\dfrac{2-x}{x} + \dfrac{x}{x+1} = 0$

In Problems 1.23 - 1.25, solve in $\Re$ the given equations.

1.23. $\sqrt{x^2-3} = 3 - x$

1.24. $\sqrt{x+4} = \sqrt{3x-2}$

1.25. $\sqrt{x^2-1} + 1 + x = 0$

In Problems 1.26 - 1.28, use the discriminant Δ to solve each quadratic equation, then factorize when possible.

1.26. $x^2 + \frac{1}{2}x + 2 = 0$

1.27. $-x^2 + 5x + 3 = 0$

1.28. $9x^2 - 12x + 4 = 0$

In Problems 1.29 - 1.31, solve in $\Re$ the given absolute value equations.

1.29. $|8x+7| = 15$

1.30. $|x-5| = 2|x|$

1.31. $|-3x+2| + |2x-3| = 5$

In Problems 1.32 - 1.35, solve in $\Re$ the given inequalities.

1.32. $3x + 4 \le 2(x - 1)$

1.33. $7x + 13 > 12x - 7$

1.34. $2x - 1 + \frac{x}{2} \ge \frac{3x+1}{2} + x$

1.35. $\frac{x}{2} - \frac{3}{4} < \frac{x}{2} + \frac{1}{4}$

In Problems 1.36 - 1.39, solve the given absolute value inequalities.

1.36. $|2x-1| \le 3$

1.37. $|3x-2| < 7$

1.38. $|3x-1| > 2$

1.39. $|10x-1| \ge 4$

In Problems 1.40 - 1.43, determine in $\Re$ the set of values of x for the given inequalities.

1.40. $2x^2 - 4x < 0$

1.41. $2(x+1)^2 - 4(x+1) \geq 0$

1.42. $(x^2 - x)^2 - x(x-1)^2 \leq 0$

1.43. $\dfrac{4x^2 - 25 - 3(2x-5)}{x+4} > 0$

1.44. A B C D is a rectangular land with sides A B = 200 meters and A D = 150 meters. Locate a point M on D C such that the area of the triangle A D M is equal to 3 / 7 of the area of the trapezium A B C M.

1.45. Determine three consecutive odd integers whose sum is equal to 81.

1.46. Find two consecutive integers such that the difference of their square is equal to 555.

1.47. A customer bought 10 copybooks of the same format. With a discount of 10 %, he paid ¢ 29250 .What is the price of one copybook before the reduction?

1.48. One quarter of a herd of camels was seen at an oasis. Twice the square root of the number of camels went into the desert, and three times five camels stayed . How many camels did the herd have? (Old Hindu problem)

1.49. A father, of 42 years old, has a son of 15 years old. After how many years the age of the father will be double the age of his son?

1.50. Consider a semi-circle with diameter A B = 6 cm and of center O. Let P be the midpoint of O B . Through a point M of (A B), draw the tangent M T to the semi-circle. Determine the point M so that: $MT^2 = \frac{5}{4} MP^2$.

1.51. Consider four squares of respective sides a, a + 1, a + 2, and a + 3. Can we find a so that the area of the largest square is equal to the sum of the areas of the others?

1.52. A user charges compound interest for the money he lends. Find the interest rate that should be charged so that the sum doubles in a 2 year period.

1.53. On a road of length equal to 480 km, two motorcyclists started simultaneously. The second, who achieved 2 km per hour less than the first, arrived the final line in one hour late . What is the velocity of the first?

In Problems 1.54 - 1.56, solve the given biquadrate equations .

1.54. $x^4 - 5x^2 - 6 = 0$

1.55. $3x^4 + 2x^2 - 5 = 0$

1.56. $x^4 - 4x^2 + 4 = 0$

In Problems 1.57 - 1.59, solve in $\Re$ the given equations.

1.57. $(x-2)^2 + 5(x-2) - 6 = 0$

1.58. $3\left(\dfrac{1}{x+1}\right)^2 + 4\left(\dfrac{1}{x+1}\right) + 1 = 0$

1.59. $\left(\dfrac{2}{x-1}\right)^2 + \dfrac{2}{x-1} - 6 = 0$

In Problems 1.60 - 1.63, solve in $\Re$ the given systems of inequalities.

1.60. $\begin{cases} 5x - 2 < 0 \\ 2x - 3 \le 1 \end{cases}$

1.61. $\begin{cases} 2(x-2) \ge x - 3 \\ x \le 2 \\ x > 0 \end{cases}$

1.62. $\begin{cases} x^2 - 4 > 0 \\ x^2 - 3x \le -2 \end{cases}$

1.63. $\begin{cases} (x-1)(x-2) \leq 0 \\ 4x+5 \geq -3 \\ -x < 0 \end{cases}$

In Problems 1.64 - 1.70, solve in $\Re$ when possible .

1.64. $|x+5| > 0$

1.65. $|x+5| < 0$

1.66. $x^2+4 > 0$

1.67. $x^2+4 < 0$

1.68. $-x^2-4 < 0$

1.69. $-x^2-4 > 0$

1.70. $|x+4| > -2$

1.71. A factory produces electronic machines, the total cost, in thousands of dollars, of x tens of machines is given by $C(x) = 0.25x^2 - x + 3$ with $x \in [0, 10]$.

a. Calculate, in dollars, the cost of production of 100 machines then deduce the average cost of producing one machine.

b. Each machine is sold for 225 dollars, show that the profit, for selling all the production x, is given by : $P(x) = -0.25x^2 + 3.25x - 3$. Determine the number of machines produced for the factory to experience profit.

1.72. A factory produces electronic devices. The fixed cost is $ 1100 and the variable cost is $ 30 per device.

a. Determine the total cost function C .

b. Each unit is sold for $ 85, calculate the revenue and the profit obtained from producing and selling x devices.

c. Determine the required level of production in order to realize profit.

1.73. A Television manufacturer finds out that the weekly total cost function, in dollars, for producing x televisions is defined by :

$$C(x) = 20x^2 + 300x + 9000.$$

a. Knowing that each television is sold for $ 1200, find the revenue yielded by selling x television per week.

b. Find the weekly profit of the manufacturer.

c. What is the level of production for which the profit is null?

d. For what values of x, will the manufacturer attain profit? Experience loss?

1.74. A factory needs to minimize its cost for a certain product. The total cost is formed as follows:

A fixed cost of $ 2000.

A variable cost of $ 200 per unit.

A cost of storage that is inversely proportional to the quantity.

We know that for a quantity of 5 units, the cost of storage is $ 1000.

Show that the total cost function is given by $C(x) = \frac{1}{5}x + 2 + \frac{5}{x}$ where x is the quantity in units , C(x) is expressed in thousands of dollars.

1.75. Taan deposited $ 5000 in a bank paying a simple interest of 12 % per year. Let FV(t) be the sum acquired after t years.

a. Calculate FV (3) .

b. Express FV(t) in terms of t.

c. After how many years, will the capital exceed $ 9500 for the first time?

1.76. Reine borrowed an amount of money for 10 months, at an annual interest rate of 8 % . She paid $ 110 as simple interest for this amount. Find the amount of money that Reine borrowed.

1.77. A capital of $ 40 000 000 is deposited in a bank at an annual interest rate of 6 % for a period of six years. Determine the future value and the interests obtained by this capital in the following cases:

a. The interest is compounded monthly.

b. The interest is compounded quarterly.

1.78. Hanadi places an amount of $ 30 000 in a bank for 9 years at a simple interest paying an annual interest rate of 4 % .Rasha places an amount of $ 30 000 in a bank for 9 years at 4 % annual interest compound monthly. Which person gets more interest?

1.79. Reine borrows a sum of $ 20 000 000 from a bank and proposes to return the sum in monthly payments over a period of five years at an annual interest rate of 6 % compounded monthly.

a. Calculate the amount of each payment.

b. What will be the total interest amount paid by Reine.

1.80. Rania buys a car by paying $ 5000 cash and $ 200 as monthly payments for two years. The annual interest rate is 6 % compounded monthly.

a. What is the car 's price?

b. Determine the interest paid by Rania.

1.81. Walid wants to invest a sum of $ 10 000 000 during 10 years, he has to choose one of the following two offers :

Offer I

Put the sum in a bank paying a compound interest at an annual interest rate of 8 % compounded semiannually.

Offer II

Put the sum in another bank paying a compound interest at an annual interest rate of 8 % compounded monthly. Which of the two offers is more advantageous for Walid?

1.82. A series of seven annual successive payments, at an annual interest rate of 7 %, is determined as follows : 2 annuities of $ 400 each , 3 annuities of $ 500 each, and 2 annuities of $ 900 each .

a. Find the future value of this series of payments.

b. Find the present value of this series of payments.

Chapter 2 **Sequences**

Introduction

In economics, demography, and other domains of real life we encounter situations that can be modeled by lists or functions. Some of these functions are defined for positive integers such as the cost functions C (x) where x is the number of units , and the decay and growth function N (t) where t is the time in years. In what follows, we will focus our study in the function u (n), where n assumes integral positive and consecutive values. Also, we will focus our study in the function $u_{n+1} = u_n + k$ where k is a real number.

Why?

It is known that mathematics is the study of patterns that is of any kind of regularity in form or idea. Radio waves, molecular structures, sequences of numbers, orbits of celestial bodies, and the shape of a bee's cell. All have patterns that can be described and analyzed mathematically. Moreover, the method of the physical sciences, such as physics, chemistry, and biology, is that of induction or experimentation. Scientific induction is a method of reasoning from the particular to the general. A physicist, for example, may be interested in finding out the effect of tension upon a bar. To do this, he experiments with large numbers of metal bars and observes their behavior under stress. He draws conclusions based on what he sees. On the other hand, a mathematician does not accept a generalization based only on observation. He must develop hypotheses based on his observations, but he further requires proof based upon a logical scheme of deduction. From properties that characterize his system, he sets out to prove that his conclusions are true or have some probability value . Once he has done this , he has confidence that his deduction is always true when the conditions fit his properties.

Objectives

❖ After studying the material in this chapter, you should be able to:

- Use the proof by mathematical induction
- Identify numerical sequence
- Identify arithmetic and geometric Sequences
- Find the sum of the first n terms of an arithmetic and geometric sequences
- Study the limit of a general term u_n and the limit of the sum of the first n terms s_n in a geometric sequence when n approaches infinity
- Identify sequences defined recursively
- Study the sense of variations of a numerical sequence
- Identify bounded sequences

Section 2.1

Proof by Mathematical Induction

To prove a property of dimension n by mathematical induction, the following steps should be satisfied orderly:

1- Prove that the property is true for n = 0 or 1.

2- Assume it is true either for n = k -1 or n = k

3- Prove it is true for either n = k or n = k + 1 (depending on step 2).

Example 2.1-1

Use mathematical induction method to prove that :

$1 + 2 + 3 + \dots + n = \dfrac{n(n+1)}{2}$ for any natural integer n .

Proof

For n = 1, we have $1 + 2 + \dots + n = 1$ and $\dfrac{n(n+1)}{2} = \dfrac{1(1+1)}{2} = 1$. So the equality is true for n = 1.

Assume it is true for n = k – 1, that is $1 + 2 + 3 + \dots + (k-1) = \dfrac{(k-1)(k-1+1)}{2}$

$$= \frac{k(k-1)}{2}$$

Prove it is true for n = k; that is we need to prove that

$1 + 2 + 3 + ... + k = \frac{k(k+1)}{2}$.

We have

$$\begin{aligned} 1 + 2 + 3 + ... + k &= 1 + 2 + 3 + ... + (k-1) + k \\ &= \frac{k(k-1)}{2} + k \\ &= \frac{k(k-1)+2k}{2} \\ &= \frac{k[(k-1)+2]}{2} \\ &= \frac{k(k+1)}{2} \text{ claim .} \end{aligned}$$

Therefore , $1 + 2 + 3 + ... + n = \frac{n(n+1)}{2}$ for any natural integer n.

Example 2.1-2

Consider the sequence $\begin{cases} u_0 = e \\ and \\ u_{n+1} = 2u_n + 1 \end{cases}$ where $n \in N$ and e any real number.

Prove that for every n, $u_n = 2^n e + 2^n - 1$.

Proof

We will prove it by mathematical induction.

For n = 0 we have, $u_0 = e$ and $2^0 e + 2^0 - 1 = e$. Then it is true for n = 0.

Assume it is true for n = k; that is $u_k = 2^k e + 2^k - 1$

Prove it is true for n = k+1; that is we need to prove that

$u_{k+1} = 2^{k+1} e + 2^{k+1} - 1$.

We have

$$\begin{aligned} u_{k+1} &= 2u_k + 1 \\ &= 2(2^k e + 2^k - 1) + 1 \\ &= 2.2^k e + 2.2^k - 2 + 1 \\ &= 2^{k+1} e + 2^{k+1} - 1. \end{aligned}$$

Therefore , for every n , $u_n = 2^n e + 2^n - 1$.

Section 2.2

Numerical Sequence

Definition 2.2-1

A numerical sequence is a mapping u from a set N or a part of N into the set $\Re$. That is, $u : N \rightarrow \Re$ defined by : for every n in N, $u(n)$ is in $\Re$.

Example 2.2-1

$u_n = \frac{1}{n}$ from N^* into $\Re$.

$u_n = \sqrt{n-3}$ from $N - \{0,1,2\}$ into $\Re$.

Remark 2.2-1

Determination of sum terms in a sequence

Let (u_n) be a numerical sequence.

If the sequence is defined explicitly, then its terms are obtained by substituting n by its values in the general term u_n.

Example 2.2-2

Consider the sequence (u_n) defined by its general term $u_n = 2n - 1$.

To compute u_5 we let n = 5 and then get $u_5 = 2\times5 - 1 = 9$.

To compute u_0 we let n = 0 and then get $u_0 = 2 \times 0 - 1 = -1$.

Section 2.3

Arithmetic Sequence

Definition 2.3 - 1

A sequence $u_1, u_2, \ldots, u_n$ is said to be an arithmetic sequence if $u_n = u_{n-1} + d$ where d is said to be the common difference of the sequence ($d = u_n - u_{n-1}$).

Example 2.3-1

1, 2, 3,....., n is an arithmetic sequence of first term 1, of last term n, and with common difference $d = 2 - 1 = 3 - 2 = \dots = 1$.

Example 2.3-2

10, 8, 6, 4, 2, 0, -2, -4, -6 is an arithmetic sequence of first term 10, of last term -6, and with common difference $d = 8 - 10 = 0 - 2 = -2 - 0 = -4 - (-2) = -2$.

Property 2.3-1

If $u_1, u_2, \dots, u_n$ form an arithmetic sequence with common difference d, then $u_n = u_1 + (n-1)d$.

Proof

Proof by mathematical induction .

For $n = 1$, $u_n = u_1$ and $u_1 + (n-1)d = u_1 + (1-1)d = u_1$.

Then it is true for $n = 1$.

Assume it is true for $n = k$, that is $u_k = u_1 + (k-1)d$.

Prove it is true for $n = k + 1$, that is we need to prove that $u_{k+1} = u_1 + k\,d$.

$u_1, u_2, \dots, u_n$ form an arithmetic sequence with common difference d, then

$$\begin{aligned} u_{k+1} &= u_k + d \\ &= u_1 + (k-1)d + d \\ &= u_1 + k\,d - d + d \\ &= u_1 + k\,d. \end{aligned}$$

Hence, $u_n = u_1 + (n-1)d$.

Example 2.3-3

The sequence 2, 4, 6, 8, 10 is an arithmetic sequence of first term $u_1 = 2$, of last term $u_5 = 10$ and with common difference $d = 2$. By applying property 2.3-1 we get $u_5 = 2 + (5-1)(2) = 10$.

Remark 2.3-1

In an arithmetic sequence, the sum of the terms u_i and u_j equidistant from the extremes u_1 and u_n, is equal to the sum of these extremes. That is,
$u_i + u_j = u_1 + u_n$.
If the number of terms is odd, then there exists a mid term element in the middle of the sequence u_k, equidistant from u_1 and u_n, such that $u_k + u_k = u_1 + u_n$ and then $u_k = \frac{u_1 + u_n}{2}$. Hence, we can say that a necessary and sufficient condition for three numbers a, b and c, taken in this order, to form an arithmetic sequence is that $2b = a + c$ or $b = \frac{a+c}{2}$.

Example 2.3-4

In the arithmetic sequence 2, 5, 8, 11, 14, 17, 20, 23, 26 we have :

$$20 + 5 = 17 + 8 = 14 + 11 = 23 + 2.$$

$$5 = \frac{2+8}{2}$$

$$14 = \frac{2+26}{2}$$

Property 2.3-2

If $u_1, u_2, \ldots, u_n$ form an arithmetic sequence with common difference d, then the sum of the first n terms is given by : $s_n = \frac{n(u_1 + u_n)}{2}$.

Proof

$$\begin{cases} s_n = u_1 + u_2 + u_3 + \ldots\ldots + u_n \\ s_n = u_n + u_{n-1} + u_{n-2} + \ldots\ldots + u_1 \end{cases}$$, By adding them we get ;

$$2s_n = (u_1 + u_n) + (u_2 + u_{n-1}) + (u_3 + u_{n-2}) + \ldots\ldots + (u_n + u_1),$$

Regarding Remark 2.3-1, we get : that each of the above terms is equal to $u_1 + u_n$.

Then, $2s_n = (u_1 + u_n) + (u_1 + u_n) + (u_1 + u_n) + \ldots\ldots + (u_1 + u_n)$, n times,
Therefore, $2s_n = n(u_1 + u_n)$ implies that $s_n = \frac{n(u_1 + u_n)}{2}$.

Example 2.3-5

1, 2, 3,….., n is an arithmetic sequence of first term 1, of last term n, and with common difference d = 1.Then

$$s_n = 1 + 2 + 3 + \dots + n = \frac{n(n+1)}{2}.$$

Remark 2.3-2

If $u_0, u_2, \dots, u_n$ form an arithmetic sequence with common difference d, then :

1) $u_n = u_0 + n\,d$

2) $s_n = \dfrac{(n+1)\,(u_0 + u_n)}{2}$.

Section 2.4

Geometric Sequence

Definition 2.4 - 1

A sequence $u_1, u_2, \dots, u_n$ is said to be an geometric sequence if $u_n = u_{n-1} \cdot q$, where q is said to be the common ratio of the sequence ($q = u_n / u_{n-1}$).

Example 2.4 -1

1, 2, 4, 8, 16, 32, 64 is a geometric sequence of first term 1, of last term 64, and with common ratio $q = 2 / 1 = 4 / 2 = \dots = 64 / 32 = 2$.

Example 2.4-2

24 , 12 , 6 , 3 , 3 / 2 , 3 / 4 , 3 / 8 is a geometric sequence of first term 24, of last term 3 / 8 , with common ratio $q = 12 / 24 = 6 / 12 = \dots = (3/8) / (3/4) = 1/2$.

Property 2.4-1

If $u_1, u_2, \dots, u_n$ form a geometric sequence with common ratio q, then $u_n = u_1 \cdot q^{n-1}$.

Proof

Proof by mathematical induction .

For $n = 1$, $u_n = u_1$ and $u_1 . q^{n-1} = u_1 . q^{1-1} = u_1$

Then it is true for $n = 1$.

Assume it is true for $n = k$, that is $u_k = u_1 . q^{k-1}$

Prove it is true for $n = k + 1$, that is we need to prove that $u_{k+1} = u_1 . q^k$

$u_1, u_2, \ldots, u_n$ form a geometric sequence with common ratio q, then

$$\begin{aligned} u_{k+1} &= u_k . q \\ &= u_1 . q^{k-1} . q \\ &= u_1 . q^{k+1-1} \\ &= u_1 . q^k. \end{aligned}$$

Hence , $u_n = u_1 . q^{n-1}$ for every natural integer n.

Example 2.4-3

The sequence 2, 4, 8, 16,32, 64 form a geometric sequence of first term $u_1 = 2$, of last term $u_6 = 64$ and with common ratio $q = 2$. By applying property 2.4- 1 we get $u_6 = 2 . 2^{6-1} = 64$.

Remark 2.4 - 1

In a geometric sequence, the product of the terms u_i and u_j equidistant from the extremes u_1 and u_n, is equal to the product of these extremes. That is,

$u_i . u_j = u_1 . u_n$.

If the number of terms is odd , then there exists a mid term element in the middle of the sequence u_k, equidistant from u_1 and u_n , such that $u_k . u_k = u_1 . u_n$ and then $u_k^2 = u_1 . u_n$. Hence, we can say that a necessary and sufficient condition for three numbers a, b and c, taken in this order, to form a geometric sequence is that $b^2 = a . c$.

Example 2.4 - 4

In the geometric sequence 2, 4, 8, 16, 32, 64, 128 we have :

$$2 \times 128 = 4 \times 64 = 8 \times 32$$
$$16^2 = 2 \times 128$$
$$4^2 = 2 \times 8$$

Property 2.4 - 2

If $u_1, u_2, \ldots, u_n$ form a geometric sequence with common ratio q, then The sum of the first n terms is given by : $s_n = \dfrac{u_1(1-q^n)}{1-q}$.

Proof

$$\begin{cases} s_n = u_1 + u_2 + u_3 + \ldots\ldots + u_n = u_1 + u_1 q + u_1 q^2 + \ldots\ldots + u_1 q^{n-1} \\ q s_n = q u_1 + q^2 u_1 + q^3 u_1 + \ldots\ldots + q^n u_1 \end{cases},$$

By subtraction we get,

$(1-q)s_n = u_1 - q^n u_1$, then $(1-q)s_n = u_1(1-q^n)$

Hence ,

$$s_n = \frac{u_1(1-q^n)}{1-q} .$$

Example 2.4 – 5

1, -3, 9, -27, 81 form a geometric sequence of first term 1, of last term 81, and with common ratio q = - 3.Then

$$s_n = 1 - 3 + 9 - 27 + 81 = \frac{1[1-(-3)^5]}{1-(-3)} = 61.$$

Remark 2.4 – 2

If the absolute value of q is less than 1 (i.e. $|q|<1$), then $\lim_{n\to\infty} s_n = \dfrac{u_1}{1-q}$.

Example 2.4 – 6

Calculate the $\frac{1}{2} + \frac{1}{4} + \frac{1}{8} + \frac{1}{16} + \ldots\ldots$

Solution

$$\frac{1}{2} + \frac{1}{4} + \frac{1}{8} + \frac{1}{16} + \ldots\ldots = \frac{1}{2} + \frac{1}{2^2} + \frac{1}{2^3} + \frac{1}{2^4} + \ldots\ldots$$

$$= \frac{1}{2} + (\frac{1}{2})(\frac{1}{2}) + (\frac{1}{2^2})(\frac{1}{2}) + (\frac{1}{2^3})(\frac{1}{2}) + \ldots\ldots$$

$$= s_n$$

where s_n is the sum of an unlimited number of terms of a geometric sequence of first term $u_1 = \frac{1}{2}$ and with common ratio $|q| = \frac{1}{2} < 1$,

then its sum $= \lim_{n\to\infty} s_n = \frac{u_1}{1-q} = \frac{\frac{1}{2}}{1-\frac{1}{2}} = 1.$

Remark 2.4 - 3

If $u_0, u_2, \ldots, u_n$ form a geometric sequence with common ratio q, then :

1) $u_n = u_0 \cdot q^n$

2) $s_n = \frac{u_0(1-q^{n+1})}{1-q}$.

Section 2.5

Recurrent Sequence

Definition 2.5- 1

A recursive is a sequence that can be determined by stating its first term (initial condition) and a recurrence relation (recursion formula) that links two consecutive terms of the form $u_n = \varphi(u_{n-1})$, where φ is a function from $\mathfrak{R}$ into $\mathfrak{R}$ .In another word, it is defined by giving a recursion formula that expresses the later term of the sequence as a function of its preceding terms.

Example 2.5 – 1

The sequence (u_n) defined by : $\begin{cases} u_1 = 0 \\ and \\ u_n = \sqrt{u_{n-1} + 3} \end{cases}$

is a simple recursive sequence.

The second term is $u_2 = \sqrt{u_1 + 3} = \sqrt{0 + 3} = \sqrt{3}$.

The third term is $u_3 = \sqrt{u_2 + 3} = \sqrt{\sqrt{3}+3}$.

All terms of the sequence can be determined step by step.

Remark 2.5 – 1

a) Stating the first term and the recurrence relation does not always define a sequence.

b) When there exists a sequence satisfying $\begin{cases} u_1 = a \\ and \\ u_n = \varphi(u_{n-1}) \end{cases}$, then this sequence is unique.

Example 2.5 – 2

Consider the relation $\begin{cases} u_1 = 3/2 \\ and \\ u_n = \dfrac{1}{u_{n-1} - 1} \end{cases}$

The second term is $u_2 = 2$.

The third term is $u_3 = 1$.

But the fourth term u_4 and the following terms are not defined .So this relation does not define a sequence.

Remark 2.5 – 2

Some sequences are defined recursively by given values of the first few terms and a recurrence relation among the adjacent terms.

Example 2.5 – 3

Consider the sequence $\begin{cases} u_0 = 0 \ , u_1 = 2 \\ and \\ u_{n+1} = 2u_n + u_{n-1} + 1 \end{cases}$

The third term is $u_2 = 2u_1 + u_0 + 1 = 2(2) + 0 + 1 = 5$.

The forth term is $u_3 = 2u_2 + u_1 + 1 = 2(5) + 2 + 1 = 13$.

All terms of the sequence are determined step by step in the same manner.

Section 2.6

Monotone Sequence

Definition 2.6- 1

A sequence (u_n) starting at n_0 is increasing if for all positive integer $n \geq n_0$, we have $u_{n-1} \leq u_n$.

Example 2.6 – 1

The sequence (u_n) defined by its general term $u_n = \dfrac{-1}{n+2}$ for all integer $n \geq 0$ is increasing .We can prove this by using the proof of mathematical induction or by using the sense of variation method that we are going to see in Definition 2.6 – 5.

Definition 2.6- 2

A sequence (u_n) starting at n_0 is decreasing if for all positive integer $n \geq n_0$, we have $u_{n-1} \geq u_n$.

Example 2.6 – 2

The sequence (u_n) defined by its general term $u_n = \dfrac{1}{n+2}$ for all integer $n \geq 0$ is decreasing. We can prove this by using the proof of mathematical induction or by using the sense of variation method that we are going to see in Definition 2.6 – 5.

Remark 2.6 – 1

1) A sequence (u_n) starting at n_0 is strictly increasing if for all positive integer $n \geq n_0$, we have $u_{n-1} < u_n$.

2) A sequence (u_n) starting at n_0 is strictly decreasing if for all positive integer $n \geq n_0$, we have $u_{n-1} > u_n$.

Example 2.6 – 3

1) The sequence (u_n) defined by its general term $u_n = n + 2$ for all integer $n \geq 0$ is strictly increasing. We can prove this by using the proof of mathematical induction or by using the sense of variation method that we are going to see in Definition 2.6 – 5.

2) The sequence (u_n) defined by its general term $u_n = -2n + 4$ for all integer $n \geq 0$ is strictly decreasing .We can prove this by using the proof of mathematical induction or by using the sense of variation method that we are going to see in Definition 2.6 – 5.

Definition 2.6- 3

A sequence (u_n) is monotone if it is either increasing or decreasing.

Remark 2.6 – 2

1) A sequence (u_n) is monotone if it is either increasing or decreasing.

2) An alternative or oscillating sequence (u_n) with general term $u_n = (-3)^n$ is not monotone.

Definition 2.6- 4

A sequence (u_n) starting at n_0 is constant if for all positive integer $n \geq n_0$, we have $u_n = u_{n-1}$. It is a sequence with equal terms.

Example 2.6 – 4

The sequence (u_n) defined by its general term $u_n = 4$ for all integer $n \geq 0$ is a constant sequence.

Definition 2.6- 5

To study the sense of variation of a sequence (u_n), we compare two consecutive terms of the sequence by computing $u_n - u_{n-1}$.

1) If $u_n - u_{n-1} \geq 0$, then the sequence is increasing.
2) If $u_n - u_{n-1} > 0$, then the sequence is strictly increasing.
3) If $u_n - u_{n-1} = 0$, then the sequence is constant.
4) If $u_n - u_{n-1} \leq 0$, then the sequence is decreasing.
5) If $u_n - u_{n-1} < 0$, then the sequence is strictly decreasing.

Remark 2.6 – 3

If all terms of the sequence are positive, to study the sense of variation of a sequence (u_n), we may compare two consecutive terms of the sequence by calculating $\frac{u_n}{u_{n-1}}$ such that :

1) If $\frac{u_n}{u_{n-1}} \geq 1$, then the sequence is increasing.

2) If $\frac{u_n}{u_{n-1}} > 1$, then the sequence is strictly increasing.

3) If $\frac{u_n}{u_{n-1}} \leq 1$, then the sequence is decreasing.

4) If $\frac{u_n}{u_{n-1}} < 1$, then the sequence is strictly decreasing.

5) If $\frac{u_n}{u_{n-1}} = 1$, then the sequence is constant.

Example 2.6 – 5

Prove that the sequence (u_n) defined by its general term $u_n = \frac{1}{n+2}$ for all integers $n \geq 0$ is decreasing.

Proof

$$u_n - u_{n-1} = \frac{1}{n+2} - \frac{1}{n+1}$$

$$= \frac{(n+1)-(n+2)}{(n+2)(n+1)}$$

$$= \frac{-1}{(n+1)(n)} \leq 0 \text{ for all integer } n \geq 0.$$

Then $u_n \leq u_{n-1}$ and he sequence (u_n) is decreasing.

Example 2.6 – 6

Prove that the sequence (u_n) defined by its general term $u_n = n+2$ for all integers $n \geq 0$ is strictly increasing.

Proof

All terms of the sequence (u_n) are positive . We may calculate $\frac{u_n}{u_{n-1}}$.

$\frac{u_n}{u_{n-1}} = \frac{n+2}{n+1}$ by classical division we get,

$\frac{n+2}{n+1} = 1 + \frac{1}{n+1} > 1$ for all integer $n \geq 0$. So $\frac{u_n}{u_{n-1}} > 1$ and the sequence (u_n) is strictly increasing.

Remark 2.6 – 4

If the sequence (u_n) defined by its general term u_n as a function of n for all integer $n \geq 0$, then to study the sense of variation of the sequence (u_n), we may calculate the derivative of u_n such that :

1) If $u'_n < 0$, then (u_n) is a decreasing sequence.

2) If $u'_n > 0$, then (u_n) is an increasing sequence .

1) If $u'_n = 0$, then (u_n) is constant sequence .

Example 2.6 – 7

Prove that the sequence (u_n) defined by its general term $u_n = \frac{-1}{n+2}$ for all integer $n \geq 0$ is increasing.

Proof

$u'_n = \frac{1}{(n+2)^2} > 0$, then the sequence (u_n) is increasing.

Remark 2.6 – 5

1) An arithmetic sequence with common difference d is strictly increasing if $d > 0$.

2) An arithmetic sequence with common difference d is strictly decreasing if $d < 0$.

Section 2.7

Bounded Sequence

Definition 2.7- 1

A sequence (u_n) is bounded above if there exists a real number k independent of n such that $u_n \leq k$ for all n.

Example 2.7 – 1

The sequence defined by $u_n = \frac{1}{n}$ is bounded above by 1 for all $n \geq 1$.

We have $\frac{1}{n} \leq 1$. Then $u_n \leq 1$.

Definition 2.7- 2

A sequence (u_n) is bounded below if there exists a real number L independent of n such that $u_n \geq L$ for all n.

Example 2.7 – 2

The sequence defined by $u_n = \frac{1}{n}$ is bounded below by 0 for all $n \geq 1$.

We have $\frac{1}{n} > 0$. Then $u_n > 0$.

Definition 2.7- 3

A sequence (u_n) is bounded if it is bounded above and bounded below. That is, there exist two real numbers L and k independent of n such that $L \leq u_n \leq k$ for all $n \geq 0$.

Example 2.7 – 3

The sequence defined by $u_n = \frac{1}{n}$ is proved bounded below by 0 and bounded above by 1 for all $n \geq 1$. That is $0 < u_n \leq 1$. Then it is bounded.

We have $\frac{1}{n} > 0$. Then $u_n > 0$.

Summary

- ☒ A sequence $u_1, u_2, \ldots, u_n$ is said to be an arithmetic sequence if $u_n = u_{n-1} + d$ where d is said to be the common difference of the sequence.
- ☒ If $u_1, u_2, \ldots, u_n$ form an arithmetic sequence with common difference d, then $u_n = u_1 + (n-1)\,d$.
- ☒ If $u_1, u_2, \ldots, u_n$ form an arithmetic sequence with common difference d, then the sum of the first n terms is given by : $s_n = \dfrac{n(u_1 + u_n)}{2}$.
- ☒ A sequence $u_1, u_2, \ldots, u_n$ is said to be an geometric sequence if $u_n = u_{n-1} \cdot q$ where q is said to be the common ratio of the sequence.
- ☒ If $u_1, u_2, \ldots, u_n$ form a geometric sequence with common ratio q, then $u_n = u_1 \cdot q^{n-1}$.
- ☒ If $u_1, u_2, \ldots, u_n$ form a geometric sequence with common ratio q, then the sum of the first n terms is given by : $s_n = \dfrac{u_1(1-q^n)}{1-q}$.
- ☒ A sequence (u_n) starting at n_0 is increasing if for all positive integer $n \geq n_0$, we have $u_{n-1} \leq u_n$.
- ☒ A sequence (u_n) starting at n_0 is decreasing if for all positive integer $n \geq n_0$, we have $u_{n-1} \geq u_n$.
- ☒ A sequence (u_n) is monotone if it is either increasing or decreasing.
- ☒ A sequence (u_n) starting at n_0 is constant if for all positive integer $n \geq n_0$ we have $u_n = u_{n-1}$.
- ☒ A sequence (u_n) is bounded above if there exists a real number k independent of n such that $u_n \leq k$ for all n.
- ☒ A sequence (u_n) is bounded below if there exists a real number L independent of n such that $u_n \geq L$ for all n.
- ☒ A sequence (u_n) is bounded if it is bounded above and bounded below.

EXERCISES

In Problems 2.1 - 2.4 , use the mathematical induction method to prove the given relations.

2.1. $1^2+2^2+3^2+......+n^2 = \dfrac{n(n+1)(2n+1)}{6}$

2.2. $\dfrac{1}{1\times2}+\dfrac{1}{2\times3}+\dfrac{1}{3\times4}+......+\dfrac{1}{n(n+1)} = \dfrac{n}{n+1}$

2.3. $1^3+2^3+3^3+......+n^3 = \dfrac{1}{4}n^2(n+1)^2$

2.4. $1+3+5+....+(2n-1) = n^2$

In Problems 2.5 - 2.6 , for each sequence, defined by its general term u_n, Calculate u_{n+1}, u_{n-2} and u_{2n} .

2.5. $u_n = 2^n - n$

2.6. $u_n = 2n^2 - n + 2$

2.7. For which values of n the sequence $u_n = \sqrt{n-3}$ is defined ? Deduce the the first four terms of this sequence.

2.8. Consider the sequence (u_n) defined by $\begin{cases} u_0 = 1 \\ u_{n+1} = \dfrac{1}{2}u_n - 1 \end{cases}$, for all $n \in \mathbb{N}$.

Show, by using the mathematical induction method, that for all $n \in \mathbb{N}$, we have $u_n \geq -2$.

2.9. Consider the sequence (u_n) defined by $\begin{cases} u_0 = 2 \\ u_{n+1} = 2u_n - 3 \end{cases}$, for all $n \in \mathbb{N}$.

Show, by using the mathematical induction method , that for all $n \in \mathbb{N}$, we have $u_n = 3 - 2^n$.

2.10. Let (u_n) be the sequence defined by $\begin{cases} u_0 = 2 \\ u_{n+1} = \dfrac{u_n - 5}{u_n - 1} \end{cases}$, for all $n \in \mathbb{N}$.

a) Calculate u_1, u_2 and u_3.

b) Write u_{n+2} in terms of u_{n+1}, then verify the following equality

$u_{n+2} = u_n$.

2.11. Consider the sequence (u_n) defined by $\begin{cases} u_0 = 2 \\ u_{n+1} = 2u_n + 1 \end{cases}$, for all $n \in \mathbb{N}$.

a) Calculate u_1, u_2 and u_3.

b) Show, by using the mathematical induction method, that for all $n \in \mathbb{N}$, (u_n) is strictly increasing.

2.12. Consider the sequence (u_n) defined by $\begin{cases} u_0 = 7 \\ u_{n+1} = \sqrt{2 + u_n} \end{cases}$, for all $n \in \mathbb{N}$.

a) Calculate u_1 and u_2.

b) Show that u_n is positive.

c) Show, by using the mathematical induction method, that for all $n \in \mathbb{N}$, (u_n) is strictly decreasing.

2.13. Determine the real number x so that $a = x$, $b = 2x + 1$ and $c = 2x + 3$ are three consecutive terms of an arithmetic sequence.

2.14. Determine the real number x so that $a = x + 4$, $b = -2x - 8$ and $c = 3x - 4$ are three consecutive terms of a geometric sequence.

2.15. Let (u_n) be a sequence defined on N by its general term $u_n = \dfrac{n}{5} - \dfrac{1}{3}$.

Show that (u_n) is an arithmetic sequence whose first term and its common difference are to be determined.

2.16. Let (u_n) be a sequence defined by its general term $u_n = 5^{-n}$.

Show that (u_n) is a geometric sequence whose first term and its common ratio are to be determined.

2.17. Determine an arithmetic sequence if its eleventh term is 20 and its twenty-first term is 50.

2.18. Determine the geometric sequences with first term $u_0 = 4$ and such that $u_1 + u_2 = 8$.

In Problems 2.19 - 2.22, designate by (u_n) an arithmetic sequence of common difference d and with s_n the sum of the first n terms.

2.19. If $d = -5$ and $u_3 = 1$, calculate u_0 and s_3.

2.20. If $u_1 = 17$, $u_n = 74$ and $s_{74} = 228$, calculate d and n.

2.21. If $d = 14$, $u_1 = 5$ and $s_n = 329$, calculatc n and u_7.

2.22. If $u_2 = 15$ and $u_4 = 35$, calculate d and u_0.

In Problems 2.23 - 2.26, designate by (u_n) a geometric sequence of common ratio q and with s_n the sum of the first n terms.

2.23. If $u_1 = 3$ and $u_6 = \frac{1}{8}$, calculate q and s_6.

2.24. If $q = 2$, $u_0 = 1$ and $s_n = 15$, calculate n.

2.25. If $q = \frac{1}{3}$ and $u_6 = 108$, calculate u_1 and s_6.

2.26. If $q = -3$, $u_1 = 14$ and $s_n = 68894$, calculate n and u_n.

2.27. a) Show that the sequence (u_n) defined on N by its general term $u_n = n^2$ is not arithmetic.

b) Consider the two sequences (v_n) and (w_n) defined on N by $v_0 = 1$,

$$v_{n+1} = \frac{v_n}{1+v_n} \text{ and } w_n = \frac{1}{v_n}.$$

i) I s the sequence (v_n) arithmetic?

ii) I s the sequence (w_n) arithmetic?

In Problems 2.28 - 2.30, calculate the given sums.

2.28. $2^5 + 2^6 + 2^7 + \dots + 2^{20}$

2.29. $1 + x^2 + x^4 + \dots + x^{2n}$ where $n \in N$.

2.30. $\frac{1}{2} - \frac{1}{4} + \frac{1}{8} - \frac{1}{16} + \dots + \frac{1}{2^9}$.

2.31. Consider the sequence (u_n) defined on N by $\begin{cases} u_0 = 2 \\ u_{n+1} = \frac{1}{2} u_n + 3 \end{cases}$, for all $n \in N$.

a) Calculate u_1 and u_2.

b) Show that the sequence (u_n) is neither arithmetic nor geometric.

c) Let $v_n = u_n - 6$ for every n.

i) Show that the sequence (v_n) is geometric whose first term and its common ratio are to be determined.

ii) Calculate v_n then u_n in terms of n.

iii) calculate the sum $s = v_0 + v_1 + \dots + v_n$ in terms of n, then deduce the sum $s' = u_0 + u_1 + \dots + u_n$ in terms of n.

2.32. Consider the sequence (u_n) defined on N by $\begin{cases} u_0 = 3 \\ u_n = \frac{1}{3} u_{n-1} - 4 \end{cases}$, for all $n \in N$.

a) Show that $u_n - u_{n-1}$ keeps a constant sign.

b) Let (v_n) be the sequence defined by its general term $v_n = u_n + 6$.

i) Show that (v_n) is a geometric sequence whose first term and its common ratio are to be determined.

ii) Calculate v_n in terms of n.

iii) Calculate u_n in terms of n.

In Problems 2.33 - 2.37, discuss the sense of variations of the given recursive sequences.

2.33. $u_{n+1} = u_n + 2$ and $u_0 = -1$.

2.34. $u_n+1 = 3u_n$ and $u_0 = 1$.

2.35. $u_{n+1} = u_n^2 - 3u_n + 4$ and $u_0 = 1$.

2.36. $u_{n+1} = u_n + 2$ and $u_0 = -1$.

2.37. $u_{n+1} = u_n + \frac{1}{n+1}$ and $u_0 = 1$.

In Problems 2.38 - 2.42, study the sense of variations of the sequence (u_n) defined by its general term u_n, $n \in N$.

2.38. $u_n = n^2 + 2n - 3$.

2.39. $u_n = \frac{2n+3}{n+1}$

2.40. $u_n = \sqrt{n+1} - \sqrt{n}$

2.41. $u_n = \frac{2^n}{n+2}$

2.42. $u_n = (-1)^n$

2.43. Suppose that (u_n) is a geometric sequence of first term $u_0 = -1$ and with common ratio $q = \frac{1}{5}$.

a) What is the limit of this sequence?

b) Express, in terms of n, the sum s_n of the first n terms of this sequence.

c) Find the limit of s_n as n approaches infinity.

In Problems 2.44 - 2.46, determine whether the given sequences defined by their general terms are bounded above or below the given value of a.

2.44. $u_n = n^2 - n + 2$, $a = 2$

2.45. $u_n = \dfrac{n-1}{2n}$, $a = 1/2$

2.46. $u_n = 2 + e^{-n}$, $a = 2$

2.47. Show that the sequence (u_n) defined by its general term $u_n = \dfrac{2n-1}{n+1}$ for all positive integer n is bounded below by -1 and bounded above by 2.

2.48. The annual salary of a technician is raised to \$ 90 000 for the year 2006. His employer decides to increase his salary by 2 % plus \$ 5 000 bonus on each year. If s_0 represents the salary of the technician for the year 2006 , and s_n his salary for the year $(2006 + n)$ for every positive integer n . For example s_2 is the salary of the technician for the year 2008.

a) Evaluate s_1 and s_2.

b) Express s_{n+1} in terms of s_n.

c) Let (u_n) defined by the sequence $u_n = s_n + 250\,000$ for every positive integer n.

   i) Evaluate u_0.

   ii) Show that (u_n) is a geometric sequence of common ratio 1.02.

   iii) Express u_n in terms of n.

d) i) Express s_n in terms of n.

   ii) Deduce the expected salary for the year 2005.

e) When does the salary of the technician will be double?

2.49. a) Let P be a function defined on $\Re$ by $P(x) = x^2 + 9x - 4140$.

i) Evaluate P (60).

ii) Solve $P(x) = 0$.

iii) Draw the table of variations of P and deduce the sign of $P(x)$.

b) To dig a well of water in the desert we will spend \$ 414 00.

The cost of digging for he first meter is \$ 1000, for the second meter is \$ 1200, for the third meter is \$ 1400 and so on by increasing \$ 200 per dug meter.

Let u_n be the cost in dollars of the nth dug meter ($n \in N^*$).

1) State the nature of the sequence (u_n) and express u_n in terms of n.
2) Let S_n be the total cost in dollars for a well of n meters for all n positive integer different than zero . Show that $S_n = 100n^2 + 900n$.
3) By using part (1) , indicate the maximum depth that we can reach.

2.50. A family consumes 85 % of its revenue on daily necessities. The annual revenue of this family increases 4 % per year, and this family decided to reduce its expenses 2 % per year.

a) The basic annual revenue $R_0 =$ \$ 6 000 000

i) Evaluate the basic annual expenses C_0.

ii) Evaluate the annual revenue R_1 and the annual expenses C_1 for the next year.

b) Show that (R_n) is a geometric sequence . Express R_n in terms of n.
c) Express the annual expenses C_n in terms of n after n years.
d) Study the sense of variation of the sequence (C_n). Calculate the limit of C_n when n approaches $+\infty$.

2.51. A vehicle of type E, has a value of \$ 15000 that depreciates at a rate of 2.4 % per month. A vehicle of type D, has a value of \$ 12000 that depreciates at rate of 1.8 % per month. Denote by a_n the value of the vehicle of type E at the end of the nth month, so $a_0 = 15000$ and denote by b_n the value of the vehicle of type D at the end of the nth month, so $b_0 = 12000$.

a. 1) Calculate a_1 and a_2.

2) Express a_n in terms of n.

3) Show that $b_n = 12000(0.982)^n$.

b. At the end of which month will the value of the vehicle of type E will lose half its original price?

c. At the end of which month will the value of the vehicle of type D become more than that of E?

2.52. A customer deposits a sum of \$ 2000 in the year 2001 in a bank at an interest rate of 3.5 % compounded annually. He deposits on the 31st of December each year a supplementary amount of \$ 700 on his account. Designate by A_n the amount expressed in dollars on the first of January of the year 2000 + n. In this way $A_0 = \$ 2000$.

a. 1) Calculate A_1.

2) Find a relation between A_n and A_{n+1}.

b. Suppose that $B_n = A_n + 20000$, for all n .

1) Show that the sequence (B_n) is geometric.

2) Express B_n in terms of n and deduce that :

$A_n = 22000 (1.035)^n - 20000$.

3) Calculate the capital available on the first of January 2008.

2.53. Part I

Consider the sequence (a_n) defined by : $a_0 = 900$ and $a_{n+1} = 0.6\, a_n + 200$.

a. Calculate a_1 and a_2 .

b. Consider the sequence (b_n) defined by $b_n = a_n \; 500$.

1) Show that (b_n) is a geometric sequence whose first term and common ratio are to be determined.

2) Calculate b_n in terms of n and deduce a_n in terms of n.

3) Calculate $\lim\limits_{n \to \infty} a_n$.

Part II

In a country, two companies D and E share the telecommunications in the market place. The clients subscribe on the first of January, either to D or to E. An annual contract is made in which they are free to choose either D or E. in the year 2006, company D possesses 90 % of the market and company E only 10 %.We estimate that each year, 20 % of the clients of company D change to E and similarly 20 % of the clients of E change to D. Suppose that there are

1000 clients in the year 2006. Accordingly, 900 clients are those of D and 100 are the clients of E .We need to study the development of clients in the coming years.

a. 1) Verify that the clients of company D counts up to 740 in the year 2007 then calculate the number of clients in company D in 2008.

2) Denote by c_n the number of clients in company D in the year 2006 + n. Show that $c_{n+1} = 0.6\,c_n + 200$.

b. Using the result of Part I, What can you conclude with respect to the development of the telecommunication market in this country.

2.54. A person has in his bank account \$ 20000. This account undergoes no interest. At the beginning of each year , this person withdraws 10 % of his capital and at the end of each month he deposits a sum of \$ 1000 in his account.

a. Denote by C_0 the capital of this person on 31 December 2005 and C_n the capital at the end of n months.

1) Verify that $C_1 = 19000$ and that $C_2 = 18100$.

2) Show that $C_{n+1} = 0.9\,C_n + 1000$.

b. Consider the sequence (a_n) defined by $a_n = C_n - 10000$. Show that the sequence (a_n) is a geometric sequence whose first term and ratio are to be determined.

1) Express a_n in terms of n and deduce the expression of C_n in terms of n.

2) Calculate the value of C_{12} then the total sum that has been withdrawn during the year 2006.

2.55. The annual rent of a house raises up to \$ 4200. The resident wants to stay in the house for seven complete years. The owner proposes two contracts.

a. **First contract**

The resident accepts yearly an increase of 5 % on the rent fee. Let r_1 be the initial rent the first year and r_n the rent fee in the n th year.

1) Show that the sequence (r_n) is a geometric sequence and express r_n in terms of n.

2) Calculate the rent fee in the seventh year.

3) Calculate the total amount paid at the end of the seven years of residency.

b. **Second contract**

The resident accepts, each year, an annual inclusive rate of $ 340.

1) If u_1 is the initial rent fee, express the rent fee u_n in terms of n.
2) Calculate the rent fee in the seventh year.
3) Calculate the total amount paid at the end of the seven years of residency.

c. Which contract is more advantageous?

2.56. In the first of January 2000, a wage –earner deposits $ 10000 in his account paying an annual interest of 5 %.On the first of January of the following years, the interest is added to his capital. The wage-earner decides to withdraw from his account $ 200 on the first of January of the following years. Denote by w_0 the wage-earner 's capital on the first of January 2000 and by w_n his account on the first of January 2000 + n. That is, $w_0 = 10000$.

a. Calculate w_1, w_2 and w_3.

b. Show that $w_{n+1} = 1.05\, w_n - 200$.

c. Let $u_n = w_n - 4000$.

1) Prove that u_n is a geometric sequence whose common ratio and first term are to be determined.
2) Express u_n then w_n in terms of n.
3) What will the wage-earner capital be on the first of January 2010?

d. After how many years will the initial amount double?

2.57. A sport club proposes two categories for subscription.

Category I

An annual fee of $ 500 has to be paid with an additional fee of $ 10000 the first year.

Category II

An initial annual fee of that follows the subscription of $ 1000 that increases 10 % per year.

From the year that follows the subscription and to motivate customers, a reduction of $ 50 is implemented on the annual fees. If A_n is the amount , expressed in dollars, of the annual fees of the nth year then

$A_1 = 1000$.

a. Determine the sum s_n paid by a member during n years and using category I.

b. Let (E_n) be the sequence defined by $E_n = A_n + a$ for all $n \geq 1$ and where a is a real number.

1) Express A_{n+1} in terms of A_n.
2) Determine the real number a for the sequence (E_n) to be geometric of common ratio 1.1 then precise the first term of this sequence.

c. Suppose a = - 500.

1) Expressed E_n then A_n in terms of n.
2) Let t_n be the sum paid by a member to the club during n years following category II. Show that $t_n = 5000\,[\,(1.1)^n - 1\,] + 500\,n$.
3) What is the minimum number of years of subscription of a member for category I to be more advantageous than that of category II?

2.58. We expect that a new truck will lose 20 % of its value the first year, then 15 % the second year and 10 % each of the subsequent years.

a. A new truck costs \$ 120000. Determine the truck's value, to the nearest dollar, at the end of :

1) the 1 st year .
2) the 1 st two years.
3) the 1 st four years.

b. A new truck costs P_0 dollars. Let P_n be the truck's value, in dollars, at the end of n years.

1) Express P_n in terms of P_0 and of n when $n \geq 3$.
2) At the end of 4 years, the truck's value is \$ 75000. What was its initial price to the nearest dollar ?
3) What is the smallest integer n such that $0.68 \times 0.9^{n-2} \leq 0.5$?
4) A truck was bought in the year 2000. Deduce from question b. 3) the year after which its value will, for the first time, exceeds half its original price. Justify your answer.

Chapter 3 Functions

Introduction

By the end of the sixteenth century, the development of astronomy, navigation, international trade,……, led to long calculations (multiplications and divisions of large numbers).It is John Napier (1550-1617), a Scottish mathematician, who, from kinematics considerations, constructed a table of correspondence that transforms the multiplication into addition: This is the table of the logarithmic function. Logarithmic functions as well as other functions play a fundamental role in the most varied domains: Physics, Mechanics, Economy, Finance, Demography, Geography, Geology, etc…..

Why?

Learning mathematics depends on the mastery of concepts and skills. Students may grow in this mastery by performing the exercises of different functions. The importance of functions increases when we apply it in other scientific fields such as physics (study the variations of the velocity function in terms of the motion x), in chemistry (study the variations of the steam function in terms of the temperature T), in Finance (study the variations of the demand functions in terms of the marketing units of product q) etc….. So as we see functions play a major role in our daily life.

Objectives

❖ After studying the material in this chapter, you should be able to:

- Identify all remarkable functions (polynomial, rational, irrational, Logarithm, exponential, cosine, sine, tangent,…..)
- Find the domain of definition and the range of certain functions .
- Draw some remarkable functions.
- Find the combination and the composite of two functions .
- Identify one-to-one and onto functions .

Section 3.1

Remarkable Functions

The most well known functions in mathematics are the following functions:

Polynomial, rational, irrational, logarithm, exponential, sine, cosine, and tangent.

Polynomial Functions

Definition 3.1-1

Any function that can be written in the form of
$P(x) = a_n x^n + a_{n-1} x^{n-1} + \dots + a_1 x + a_0$ is a polynomial function with one variable x. $a_0, a_1, \dots, a_n$ are all real numbers not all of them zeros .They are called coefficients . " n " is a natural integer.

Remark 3.1-1

1) If $a_n \neq 0$, then the degree of the polynomial P(x) is n.

2) If $a_n \neq 0$, then $M(x) = a_n x^n$ is a monomial with variable x.

3) If $a_n \neq 0$ and $a_{n-1} \neq 0$, then $B(x) = a_n x^n + a_{n-1} x^{n-1}$ is a binomial with variable x.

4) If $a_n \neq 0$, $a_{n-1} \neq 0$ and $a_{n-2} \neq 0$, then
$T(x) = a_n x^n + a_{n-1} x^{n-1} + a_{n-2} x^{n-2}$ is a trinomial with variable x.

Example 3.1-1

- $f(x) = x^3 + 2x^2 + 4x - 3$ is a polynomial with degree 3.
- $M(x) = 23.5\, x^5$ is a monomial with degree 5.
- $B(x) = \frac{3}{2} x^2 - 7$ is a binomial with degree 2.
- $T(x) = -10x^2 + \sqrt{5}\, x^9 + 4x$ is a trinomial with degree 9.

Domain of definition of a polynomial function

All polynomial functions are defined for every $x \in \Re$. Then the domain of definition of a polynomial function y = f (x) is $\Re$. It is denoted by $D_f = \Re$.

Property 3.1-1

If a is a root of P(x) = 0 (i.e. p(a) = 0), then (x – a) is a factor of the polynomial function P (x) . In another word, P(x) is divisible by (x – a).

Remark 3.1-1

1) To find a factor for a polynomial function P(x), check which of the divisors of the constant term in P(x) allows P(x) to be zero.

2) To factorize a polynomial function P(x), find at least one factor of P(x) and then divide P(x) by this factor using classical division method.

3) If the division of a polynomial function A(x) by B(x) gives as quotient Q(x) and a remainder R(x), then A(x) = B(x) × Q(x) + R(x).

Example 3.1-1

Consider the polynomial function $P(x) = x^3 - 2x^2 + x - 2$.

Factorize P(x).

Solution

The divisors of the constant term " – 2 " in P(x) are $\{ \pm 1, \pm 2 \}$.

$P(-1) = (-1)^3 - 2(-1)^2 + (-1) - 2$

$= -6 \neq 0$, then (x + 1) is not a factor of P(x).

$P(1) = (1)^3 - 2(1)^2 + (1) - 2$

$= -2 \neq 0$, then (x - 1) is not a factor of P(x).

$P(-2) = (-2)^3 - 2(-2)^2 + (-2) - 2$

$= -20 \neq 0$, then (x + 2) is not a factor of P(x).

$P(2) = (2)^3 - 2(2)^2 + (2) - 2 = 0$, then (x - 2) is a factor of P(x).

Classical division

$x^3 - 2x^2 + x - 2$ $\ominus x^3 \underset{-}{\oplus} 2x^2$	$x - 2$
$x - 2$ $\ominus\ x \underset{-}{\oplus} 2$	$x^2 + 1$
0	

Then, $x^3 - 2x^2 + x - 2 = (x - 2)(x^2 + 1) + 0$ which implies that

$$P(x) = (x - 2)(x^2 + 1).$$

Polynomial function identical to zero

Property 3.1-2

Polynomial function P (x) is said to be identical to zero ($P(x) \equiv 0$) if and only if all the coefficients of P (x) are equal to zero.
That is, $P(x) = a_n x^n + a_{n-1} x^{n-1} + \ldots + a_1 x + a_0 \equiv 0$ if and only if $a_0, a_1, \ldots, a_n$ are all equal to zero.

Example 3.1-2

Consider the polynomial function $P(x) = (m - 1)x^4 + -2a x^3 + n x$.

Determine the real numbers m, a and n so that $P(x) \equiv 0$.

Solution

$$P(x) \equiv 0 \Leftrightarrow \begin{cases} m - 1 = 0 \\ -2a = 0 \\ n = 0 \end{cases}$$

$$\Leftrightarrow \begin{cases} m = 1 \\ a = 0 \\ n = 0 \end{cases}$$

Identical Polynomials

Property 3.1-2

Two polynomial P (x) and G (x) are said to be identical if and only if their corresponding coefficients are equal. That is,

$$a_n x^n + a_{n-1} x^{n-1} + \ldots + a_1 x + a_0 \equiv b_n x^n + b_{n-1} x^{n-1} + \ldots + b_1 x + b_0$$

if and only if $a_0 = b_0$, $a_1 = b_1$, ….., $a_n = b_n$.

Proof

$$a_n x^n + a_{n-1} x^{n-1} + \ldots + a_1 x + a_0 \equiv b_n x^n + b_{n-1} x^{n-1} + \ldots + b_1 x + b_0 \Leftrightarrow$$

$$(a_n - b_n) x^n + (a_{n-1} - b_{n-1}) x^{n-1} + \ldots + (a_1 - b_1) x + (a_0 - b_0) \equiv 0$$

$$\Leftrightarrow \begin{cases} a_n - b_n = 0 \\ a_{n-1} - b_{n-1} = 0 \\ \ldots\ldots\ldots\ldots \\ a_1 - b_1 = 0 \\ a_0 - b_0 = 0 \end{cases} \Leftrightarrow \begin{cases} a_n = b_n \\ a_{n-1} = b_{n-1} \\ \ldots\ldots\ldots\ldots \\ a_1 = b_1 \\ a_0 = b_0 \end{cases}.$$

Example 3.1-3

Consider the two polynomials $P(x) = (3m - 3)x^3 + 5ax^2 + x - b + 4$, and $G(x) = ux^4 + 5x^3 + kx - 2$.

Determine the real numbers m, a, b, u, and k so that $P(x) \equiv G(x)$.

Solution

$$P(x) \equiv G(x) \Leftrightarrow \begin{cases} 3m - 3 = 5 \\ 5a = 0 \\ k = 1 \\ -b + 4 = -2 \\ u = 0 \end{cases} \Leftrightarrow \begin{cases} m = \dfrac{8}{3} \\ a = 0 \\ k = 1 \\ b = 6 \\ u = 0 \end{cases}$$

Constant Functions

Definition 3.1-2

Any function that can be written in the form of $P(x) = k$, where k is a real number, is a constant function.

Example 3.1- 4

The following functions are constant functions :

- $P(x) = 4$
- $G(x) = -\frac{4}{7}$
- $H(x) = \sqrt{5}$

Remark 3.1-2

Note that the constant function $P(x) = k$ is parallel to the x-axis in a normal system of axes.

Remark 3.1-3

Any polynomial function with one variable x and of degree two is called Parabola.

The parabola $P(x) = a x^2 + b x + c$ has the point $S\left(\frac{-b}{2a}, \frac{-b^2+4ac}{4a}\right)$ as vertex.

Example 3.1-5

- $P(x) = 5x^2 - 3x + 4$ is a polynomial with vertex $\left(\frac{3}{10}, \frac{71}{20}\right)$
- $G(x) = x^2 - 2x$ is a polynomial with vertex $(1, -1)$
- $H(x) = -x^2 + \frac{7}{5}$ is a polynomial with vertex $\left(0, \frac{7}{5}\right)$
- $L(x) = 6x^2$ is a polynomial with vertex $(0, 0)$

Rational Functions

Definition 3.1-3

Any function f (x) that can be written in the form of $f(x) = \frac{P(x)}{Q(x)}$, where P(x) is a polynomial function with one variable x and Q(x) is also a polynomial function with one variable x such that Q(x) is different from zero, then f (x) called rational function with one variable x.

Example 3.1-6

- $f(x)=\frac{x+3}{x}$ *with* $x \neq 0$
- $g(x) = \frac{7x^{3}+3x-1}{3x+4}$ *with* $x \neq \frac{-4}{3}$

Domain of definition of a rational function

All rational functions are defined if the expression in the denominator is different from zero. Then the domain of definition of a rational function $f(x) = \frac{p(x)}{q(x)}$ is $\Re$ without the values of x that make q (x) = 0. It is denoted by D_f = $\Re - \{ x : q(x) = 0 \}$.

Example 3.1-7

Find the domain of definition of the following rational functions :

a. $f(x) = \frac{x^{3}+x}{5x}$

b. $g(x) = \frac{x}{x-3}$

c. $h(x) = \frac{x \quad -x+1}{x^{2}-1}$

Solution

a. f is defined if $5x \neq 0$ implies that $x \neq 0$. Then $D_f = \Re - \{ 0 \}$ or $\Re^{*}$.

b. g is defined if $x - 3 \neq 0$ implies that $x \neq 3$. Then $D_g = \Re - \{3\}$.

c. h is defined if $x^2 - 1 \neq 0$ implies that $x \neq \pm 1$. Then $D_h = \Re - \{1, -1\}$.

Remark 3.1-4

- Any rational function $f(x) = \dfrac{ax+b}{cx+d}$,such that a, b, and d are real numbers, $c \neq 0$ and $x \neq \dfrac{-d}{c}$, then this rational function is called hyperbola.

- $x = \dfrac{-d}{c}$ is a vertical asymptote and $y = \dfrac{a}{c}$ is a horizontal asymptote.

Example 3.1-8

- $P(x) = \dfrac{2x-3}{5x+2}$ is a hyperbola function with vertical asymptote $x = \dfrac{-2}{5}$, and horizontal asymptote $y = \dfrac{2}{5}$.

- $G(x) = \dfrac{7x}{2x-1}$ is a hyperbola function with vertical asymptote $x = \dfrac{1}{2}$, and horizontal asymptote $y = \dfrac{7}{2}$.

- $F(x) = \dfrac{-7}{-3x+4}$ is a hyperbola function with vertical asymptote $x = \dfrac{4}{3}$, and horizontal asymptote $y = 0$ (the x – axis).

- $L(x) = \dfrac{x-9}{5x}$ is a hyperbola function with vertical asymptote $x = 0$ (the y – axis), and horizontal asymptote $y = \dfrac{1}{5}$.

- $J(x) = \dfrac{8}{x}$ is a hyperbola function with vertical asymptote the y – axis, and horizontal asymptote the x – axis.

Economical Applications To Rational Functions

In general, when the price p of an item increases, the quantity demanded q will decrease, and vice – versa. The so-called total cost function for producing a quantity q of a certain product usually depends on :

- The variable costs which vary with the produced quantity q.
  That is, each extra unit produced will require additional expenses of labor, materials, etc.
- The fixed costs (rent, electricity, insurance, ….) are independent of the quantity produced q.
  That is , Total Cost = Fixed Costs + Variables Costs.
  The average cost $\overline{C}(q)$ is the total cost C (q) divided by the quantity

produced q. That is, $\overline{C}(q) = \dfrac{C(q)}{q}$. Thus, if C is a polynomial function , then its average $\overline{C}$ is a typical example of a rational function.

Example 3.1- 9

If the total cost C, in dollars, for producing a quantity q of a certain product is defined by C (q) = $q^2 + 50q + 1600$, then the average cost per unit is $\overline{C}(q) = \dfrac{q^2+50q+1600}{q}$,

which is a rational function in q. Figure 3.1-1 is the representative curve of the function on] 0 , 90 [.

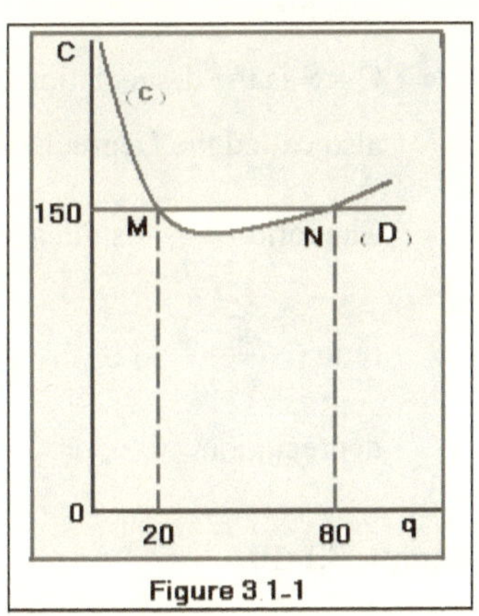

Figure 3 1-1

Straight Line Depreciation

Definition 3.1-4

The straight line depreciation is a function for calculating the value of an asset after it has been in service for a certain period of time under the assumption that its depreciation is the same from year to year.

Total Depreciation

Definition 3.1- 5

The total depreciation of the asset after t years in service can be modulated by a linear function $TD(t) = \frac{C-S}{n} \times t \; for \; t \in [0, n]$. C is the original cost of an asset, n is the useful lifetime of the asset, and S is its value at the end of its useful lifetime. S is called salvage or scrap value.

Definition 3.1- 6

The depreciated value of the asset after t years can be modulated by a linear function $DV(t) = C - \frac{C-S}{n} \times t$ for $t \in [0, n]$. That is, $DV(t) = C - TD(t)$.

Remark 3.1-5

- The cost C means the total cost.
- Useful Life time of an asset is the number of years the asset is expected to be useable.
- C – S is the depreciation over n years of the useful lifetime of the asset. It may also called the Depreciable value.
- The ratio $\frac{C-S}{n}$ is the annual depreciation of the asset. The negative of this ratio ($-\frac{C-S}{n}$) is the slope of the Depreciated Value. It is called the rate of depreciation.

Example 3.1-10

A small business office buys a computer for \$ 2000 that has a useful lifetime of 4 years and a salvage value of \$ 400 at the end of 4 years.

a. Find the depreciation over 4 years.
b. Find the annual depreciation of the computer.
c. Find the total depreciation of the computer after t years.
d. Find the depreciated value of the asset after t years.
e. Sketch the graph of DV. What do you conclude?

Solution

a. C = 2000, S = 400 and n = 4

The depreciation over 4 years is C – S = 2000 – 400 = \$ 1600

b. The annual depreciation of the computer is $\frac{C-S}{n} = \frac{1600}{4}$ = \$ 400

c. The total depreciation of the computer after t years is

$$TD(t) = \frac{C-S}{n} \times t = 400\,t \qquad \text{for } t \in [0, 4].$$

d. The depreciated value of the asset after t years is DV (t) = C – TD (t) which implies that DV (t) = 2000 – 400 t for $t \in [0, 4]$.

e. Graph of DV. We have DV (t) = 2000 – 400 t for $t \in [0, 4]$.

Figure 3.1 – 2 describe the graph of DV (t).

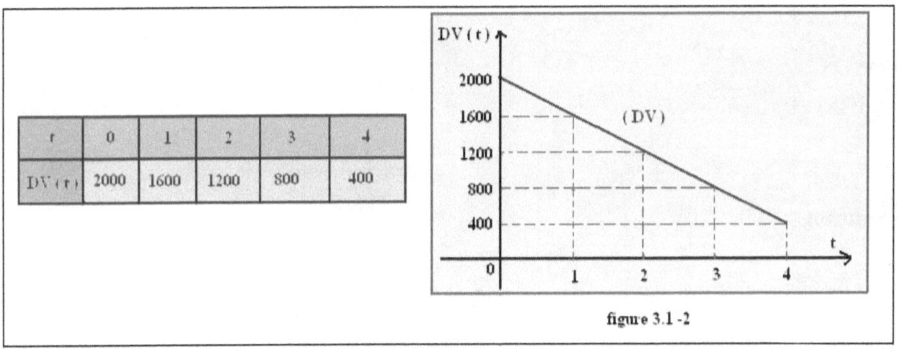

t	0	1	2	3	4
DV (t)	2000	1600	1200	800	400

figure 3.1 -2

Deduction

Figure 3.1 -2 shows that the same amount of money is lost every year.

Simple Irrational Functions

Definition 3.1- 7

Any function that can be written in the form of $f(x) = \sqrt{P(x)}$, where P(x) is a positive polynomial function with one variable x (i.e. $P(x) \geq 0$), is called simple irrational function with one variable x.

Example 3.1-11

- $F(x) = \sqrt{2x}$ is a simple irrational function if $x \geq 0$.
- $P(x) = \sqrt{3x - 3}$ is a simple irrational function if $x \geq 1$.

Domain of definition of a simple irrational function

All simple irrational functions are defined if the expression in the radicand (the expression under the radical) is greater or equal to zero. Then the domain of definition of the irrational function $f(x) = \sqrt{ax + b}$ is the set of values of x that make $a x + b \geq 0$. It is denoted by $D_f = [\frac{-b}{a}, +\infty)$.

Example 3.1-12

Find the domain of definition of the following irrational functions:

a. $f(x) = \sqrt{x}$

b. $g(x) = \sqrt{x + 5}$

c. $h(x) = \sqrt{-x + 3}$

d. $L(x) = \sqrt{x^2 - 2}$

Solution

a. f is defined if $x \geq 0$. then the domain of definition of f is $D_f = [0, +\infty)$.

b. g is defined if $x + 5 \geq 0$ which implies that $x \geq -5$, then the domain of definition of g is $D_g = [-5, +\infty)$.

c. h is defined if $-x + 3 \geq 0$ which implies that $x \leq 3$, then the domain of definition of h is $D_h = (-\infty, 3]$.

d. L is defined if $x^2 - 2 \geq 0$ which implies that $x \leq -\sqrt{2}$ or $x \geq \sqrt{2}$, then the domain of definition of L is $D_L = (-\infty, -\sqrt{2}] \cup [\sqrt{2}, +\infty)$.

Remark 3.1-6

A simple irrational function $f(x) = \sqrt{ax + b}$ with $a \neq 0$ admits a limit point or end point with abscissa $x = \frac{-b}{a}$.

Example 3.1-13

In Example 3.1 – 11, the function F (x) admits the point (0 , 0) as a limit point, and P (x) admits the point (1 , 0) as a limit point.

Logarithm Function

Definition 3.1- 8

The primitive of a function g defined by $g(x) = \frac{1}{x}$ on $(0, +\infty)$ is a function defined on the same interval. This primitive is called natural logarithm or Napierian logarithm and is denoted by " ln ".

According to this definition the image of a real number x in $(0, +\infty)$ is ln (x) or simply ln x .

Remark 3.1-7

The number " e "

The function ln is a bijection from $(0, +\infty)$ onto $\Re$. Then there exists a unique real number $e \in (0, +\infty)$ such that $\ln e = 1$. The approximate value of e is 2.718 to the nearest 10^{-3}. This number e was first introduced by the Swiss mathematician Leonard Euler (1707 – 1783) in 1736 . $e \approx 2.7182818284$.......

Logarithm Function at base a

Definition 3.1- 9

The logarithmic function at base a, denoted by " $\log_a$ ", is a function defined on $(0, +\infty)$ by $\log_a x = \frac{\ln x}{\ln a}$, where the base a is a strictly positive real number different from 1. That is, $a \in \Re^{+*} - \{1\}$.

Domain of definition of Logarithmic Function

All logarithm functions are defined if the expression in the logarithmic function is strictly greater than zero. Then the domain of definition of the logarithm function $f(x) = \log_a u$, where u is a function of x and $a \in \Re^{+*} - \{1\}$, is the set of values of x that make $u(x) > 0$. If $u(x) = mx + n$ where m is a strictly positive real number, then the domain of definition of f is denoted by

$D_f = (\frac{-n}{m}, +\infty)$.

Example 3.1-14

Find the domain of definition of the following logarithmic functions:

a. $f(x) = \ln x$

b. $g(x) = \ln(x + 7)$

c. $h(x) = \log(-x - 6)$

d. $L(x) = \log_3(-x^2 + 4)$

Solution

a. f is defined if $x > 0$, then the domain of definition of f is $D_f = (0, +\infty)$.

b. g is defined if $x + 7 > 0$ which implies that $x > -7$, then the domain of definition of g is $D_g = (-7, +\infty)$.

c. h is defined if $-x - 6 > 0$ which implies that $x < -6$, then the domain of definition of h is $D_h = (-\infty, -6)$.

d. L is defined if $-x^2 + 4 > 0$ which implies that $-2 < x < 2$, then the domain of definition of L is $D_L = (-2, 2)$.

Property 3.1-3

- $\log x$ is defined if and only if $x > 0$
- $\log_a x$ is defined if and only if $x > 0$ where a is a positive number different from 1
- If $0 < x < 1$, then $\ln x < 0$
- If $x > 1$, then $\ln x > 0$
- $\ln x = \log_e x$
- $\log x = \log_{10} x$
- $\log_a (x y) = \log_a x + \log_a y$
- $\log_a (x / y) = \log_a x - \log_a y$
- $\log_a x^n = n \log_a x$
- $\log_a (1 / x) = - \log_a x$
- $\log_a x = \ln x / \ln a$
- $\log_a a = 1$
- Same procedure for ln and log.

Example 3.1-15

1) $\log (x - 3)$ is defined if $x - 3 > 0$, then $x > 3$

2) $\ln e = 1$

3) $\ln 1 = 0$

4) $\log 1 = 0$

5) $\log 10 = 1$

6) $\text{Log}_5 1 = 0$

Exponential Function

Definition 3.1- 10

The inverse function of the natural logarithmic function ln is called the exponential function with base e. It is denoted by " exp ". The image of a real variable x by exp is $\exp(x) = e^x$. It is read as " exponential x ". For all real x, e^x is defined and it is strictly positive.

Why " e^x " is the inverse of " ln x " ?

For all strictly positive real number x, we have : $y = \ln x$ is equivalent to $1 \times y = \ln x$ which is equivalent to $\ln e \times y = \ln x$ which is equivalent to $\ln e^y = \ln x$ which is equivalent to $e^y = x$. Hence, the inverse of $y = \ln x$ is $y = e^x$.

Remark 3.1-8

R	$\xrightarrow{\exp}$ / $\xleftarrow{\ln}$	$(0, +\infty)$
$y = \exp(x)$	equivalent to	$x = \ln y$
$x \in R$		$y \in (0, +\infty)$

Exponential Functions at base a

Definition 3.1- 11

For " a " strictly positive real number different from 1, the function $\log_a$ is a bijection from $(0, +\infty)$ onto $\mathfrak{R}$. Then it admits an inverse function, denoted $\exp_a$, from $\mathfrak{R}$ on $(0, +\infty)$ defined by $\exp_a(x) = a^x$. For all real number x, $a^x > 0$.

Example 3.1-16

a. 2^x is an exponential function at base 2.

b. $\left(\frac{2}{3}\right)^x$ is an exponential function at base $\frac{2}{3}$.

Domain of definition of Exponential Functions

All exponential functions are defined for every $x \in \mathfrak{R}$. Then the domain of definition of an exponential function $f(x) = a^u$, or $f(x) = e^u$, where u is a polynomial function of x and $a \in \mathfrak{R}^{+*} - \{1\}$, is $\mathfrak{R}$. It is denoted by $D_f = \mathfrak{R}$

Example 3.1-17

Find the domain of definition of the following exponential functions:

a. $f(x) = e^{-2x}$

b. $g(x) = 5^{8x^3+2x^2}$

Solution

a. f is defined for every x in $\Re$. Its domain of definition is $\Re$.

b. g is defined for every x in $\Re$. Its domain of definition is $\Re$.

Property 3.1-4

$a \in \Re^{+*} - \{1\}$ and $x > 0$.

- $y = \log_a x$ if and only if $a^y = x$
- $10^{\log x} = x$
- $e^{\ln x} = x$
- $a^{\log_a x} = x$
- $e^0 = 1$

Example 3.1-18

- $3 = \log_2 8$ if and only if $2^3 = 8$
- $10^{\log 4} = 4$
- $e^{\ln 3} = 3$
- $5^{\log_5 7} = 7$

Section 3.2

Polar Coordinates

Definition 3.2 -1

If $\left(O,\vec{i},\vec{j}\right)$ is a direct orthonormal system, M (x , y) is any point, θ is the angle ($\vec{i}$, $\overrightarrow{OM}$), and r is the length OM, then :

- θ is called the polar angle of M.
- r is called the radius vector.
- The axis $\left(O,\vec{i}\right)$ is called the polar axis.
- R and θ are called the polar coordinates of M .

In figure 3.2 – 1, the unit vector $\vec{u} = \frac{1}{r}\overrightarrow{OM}$ is such that $\vec{u} = \cos\theta\,\vec{i} + \sin\theta\,\vec{j}$

$\overrightarrow{OM} = r\,\vec{u} = r\cos\theta\,\vec{i} + r\sin\theta\,\vec{j}$

But $\overrightarrow{OM} = x\,\vec{i} + y\,\vec{j}$, so by identification we get : x = $r\cos\theta$ and y = $r\sin\theta$.

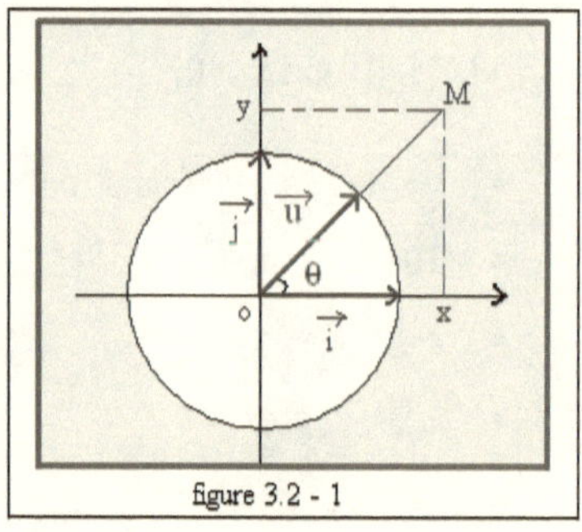

figure 3.2 - 1

Remark 3.2-1

The calculation of the polar coordinates of a point M (x , y) can be done without difficulty . r = $\left\|\overrightarrow{OM}\right\| = \sqrt{x^2+y^2}$, and the angle θ can be obtained from the relations $\cos\theta = \frac{x}{r}$ and $\sin\theta = \frac{y}{r}$ by simultaneously solving trigonometric equations of this type , which will be done later .

Trigonometric Formulas

Property 3.2 – 1

$\cos^2 x + \sin^2 x = 1$

Proof

In Remark 3.2-1, we have $\cos\theta = \dfrac{x}{r}$ and $\sin\theta = \dfrac{y}{r}$.

Then $\cos^2\theta + \sin^2\theta = (\dfrac{x}{r})^2 + (\dfrac{y}{r})^2$

$= \dfrac{x^2+y^2}{r^2}$ (but $r = \sqrt{x^2+y^2} \Rightarrow r^2 = x^2+y^2$)

$= 1$. Claim

Property 3.2 – 2

$\cos(a-b) = \cos a \cos b + \sin a \sin b$

Proof

Consider the trigonometric circle (C) with center O and origin A (figure 3.2 – 2). Let $(O, \vec{i}, \vec{j})$ be a direct orthonormal system. M (x , y) and N (x' , y') are two points of (C) such that $(\vec{i}, \overrightarrow{OM}) = a$ and $(\vec{i}, \overrightarrow{ON}) = b$, where a and b are two real numbers.

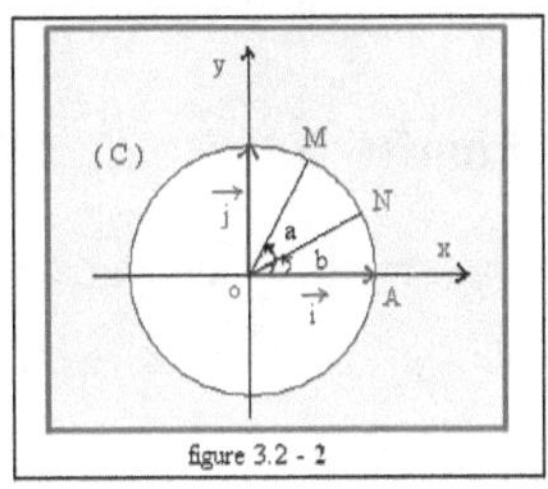

figure 3.2 - 2

$\overrightarrow{OM} = x\vec{i} + y\vec{j}$

$= r\cos a\,\vec{i} + r\sin a\,\vec{j}$, but $r = \|\overrightarrow{OA}\| = 1$ because (C) is a trigonometric circle. Then , $\overrightarrow{OM} = \cos a\,\vec{i} + \sin a\,\vec{j}$.

Hence, $\overrightarrow{OM}$ (cos a , sin a).

Moreover, $\overrightarrow{OM} \cdot \overrightarrow{ON} = OM . ON \operatorname{Cos}(\overrightarrow{OM}, \overrightarrow{ON})$.

$= 1 \;.\; 1 \;.\operatorname{Cos}(\overrightarrow{OM}, \overrightarrow{ON})$

$= \operatorname{Cos}(\overrightarrow{OM}, \overrightarrow{ON})$

$(\overrightarrow{OM}, \overrightarrow{ON}) = (\overrightarrow{OA}, \overrightarrow{OM}) - (\overrightarrow{OA}, \overrightarrow{ON}) + 2k\pi$ where $k \in Z$.

$= a - b + 2k\pi$, then

$$\overrightarrow{OM} \cdot \overrightarrow{ON} = \cos(a - b + 2k\pi)$$
$$= \cos(a - b)$$

On the other hand, $\overrightarrow{OM} \cdot \overrightarrow{ON} = xx' + yy'$

$$= (r\cos a \cdot r\cos b) + (r\sin a \cdot r\sin b)$$
$$= \cos a \cdot \cos b + \sin a \cdot \sin b, \text{ because } r = 1$$

Thus, by identification, $\cos(a - b) = \cos a \cdot \cos b + \sin a \cdot \sin b$

Property 3.2 – 3

$\cos(a + b) = \cos a \cos b - \sin a \sin b$

Proof

$$\cos(a + b) = \cos[a - (-b)]$$
$$= \cos a \cos(-b) + \sin a \sin(-b)$$
$$= \cos a \cos b - \sin a \sin b \text{ . Claim .}$$

Property 3.2 – 4

$\sin(a + b) = \sin a \cos b + \cos a \sin b$

$\sin(a - b) = \sin a \cos b - \cos a \sin b$

Same proof as above

Corollary 3.2 – 1

$$\text{Sin } a \cos b = \frac{1}{2}[\sin(a + b) + \sin(a - b)]$$

$$\cos a \sin b = \frac{1}{2}[\sin(a + b) - \sin(a - b)]$$

$$\cos a \cos b = \frac{1}{2}[\cos(a + b) + \cos(a - b)]$$

$$\sin a \sin b = \frac{1}{2}[\cos(a - b) - \cos(a + b)$$

Property 3.2 – 5

$\text{Cos}(2a) = \cos^2 a - \sin^2 a = 2\cos^2 a = 1 - 2\sin^2 a$

$\text{Sin}(2a) = 2\sin a \cos a$

Proof

$$\begin{aligned}
\cos(2a) &= \cos(a + a) \\
&= \cos a \cos a - \sin a \sin a \\
&= \cos^2 a - \sin^2 a \\
&= \cos^2 a - (1 - \cos^2 a) \\
&= 2\cos^2 a - 1 \\
&= (1 - \sin^2 x) - \sin^2 x \\
&= 1 - 2\sin^2 x \quad . \text{ Claim } .
\end{aligned}$$

Similarly to prove $\text{Sin}(2a) = 2\sin a \cos a$

Corollary 3.2 – 2

$\text{Cos } a = \cos^2 \frac{a}{2} - \sin^2 \frac{a}{2} = 2\cos^2 \frac{a}{2} - 1 = 1 - 2\sin^2 \frac{a}{2}.$

$\text{Sin } a = 2\sin \frac{a}{2} \cos \frac{a}{2}.$

$\text{Cos}^2 a = \frac{1 + \cos 2a}{2}$ and $\text{Sin}^2 a = \frac{1 - \cos 2a}{2}$

$\text{Cos}^2 \frac{a}{2} = \frac{1 + \cos a}{2}$ and $\text{Sin}^2 \frac{a}{2} = \frac{1 - \cos a}{2}$

Property 3.2 – 6

$$\tan(a + b) = \frac{\tan a + \tan b}{1 - \tan a \tan b}$$

$$\tan(a - b) = \frac{\tan a - \tan b}{1 + \tan a \tan b}$$

$$\tan(2a) = \frac{2\tan a}{1-\tan^2 a}$$

Proof

$$\tan(a+b) = \frac{\sin(a+b)}{\cos(a+b)} = \frac{\sin a\cos b + \cos a\sin b}{\cos a\cos b - \sin a\sin b}$$

Divide the numerator and the denominator by cos a cos b, we get :

$$\tan(a+b) = \frac{\tan a + \tan b}{1-\tan a\tan b}.$$

Similarly to prove the other properties.

Corollary 3.2 – 3

$$\tan a = \frac{2\tan\frac{a}{2}}{1-\tan^2\frac{a}{2}}$$

Remark 3.2-2

If the tangent exists and $t = \tan\frac{a}{2}$, then

$$\cos a = \frac{1-t^2}{1+t^2}$$

$$\sin a = \frac{2t}{1+t^2}$$

$$\tan a = \frac{2t}{1-t^2}$$

Example 3.2-1

1) Express in terms of cos x or sin x :

a. $\cos(x+\frac{2\pi}{3}) + \cos(x+\frac{4\pi}{3})$

b. $\sin(x+\frac{\pi}{6}) + \sin(x-\frac{\pi}{6})$

2) a is a real number such that $\cos a = \frac{-3}{5}$. Suppose that $\sin a > 0$.

Evaluate cos 2a, sin 2a, tan a, and tan 2a.

Solution

1) a. By applying property 3.2 – 3, we get :

$$\cos(x+\frac{2\pi}{3})+\cos(x+\frac{4\pi}{3}) = \cos x\cos\frac{2\pi}{3} - \sin x\ \sin\frac{2\pi}{3} + \cos x\ \cos\frac{4\pi}{3} - \sin x \sin\frac{4\pi}{3}$$

$$= \cos x\cos(\pi-\frac{\pi}{3}) - \sin x\ \sin(\pi-\frac{\pi}{3}) + \cos x\cos(\pi+\frac{\pi}{3}) - \sin x\ \sin(\pi+\frac{\pi}{3})$$

$$= -\cos x\cos(\frac{\pi}{3}) - \sin x\ \sin(\frac{\pi}{3}) + -\cos x\cos(\frac{\pi}{3}) + \sin x\ \sin(\frac{\pi}{3})$$

$$= -2\cos x\cos(\frac{\pi}{3})$$

$$= -2(\frac{1}{2})\cos x$$

$$= \cos x$$

b. By applying property 3.2 – 4, we get :

$$\sin(x+\frac{\pi}{6}) + \sin(x-\frac{\pi}{6}) = \sin x\cos\frac{\pi}{6} + \cos x\sin\frac{\pi}{6} + \sin x\cos\frac{\pi}{6} - \cos x\sin\frac{\pi}{6}$$

$$\sin(x+\frac{\pi}{6}) + \sin(x-\frac{\pi}{6}) = 2\ \sin x\cos\frac{\pi}{6}$$

$$= 2(\frac{\sqrt{3}}{2})\ \sin x$$

$$= \sqrt{3}\ \sin x.$$

2) $\cos a < 0$ and $\sin a > 0$, then a is found in the second quadrant, that is

$$a \in \left[\frac{\pi}{2}, \pi\right] + 2k\pi \text{ where } k \in Z.$$

- By applying property 3.2 – 5, we get :

$$\cos 2a = 2\cos^2 a - 1 = 2\left(\frac{-3}{5}\right)^2 - 1 = \frac{-7}{25}$$

- By applying property 3.2 – 1, we get :

$$\sin^2 2a = 1 - \cos^2 2a = 1 - \left(\frac{-7}{25}\right)^2 = \frac{576}{625}$$

$a \in \left[\frac{\pi}{2}, \pi\right] + 2k\pi$ where $k \in Z$ implies that $2a \in [\pi, 2\pi] + 4k\pi$

which means that $\sin 2a < 0$.

Hence, $\sin 2a = -\sqrt{\frac{576}{625}}$ implies that $\sin 2a = -\frac{24}{25}$.

- $\tan a = \frac{\sin a}{\cos a} = \frac{\sqrt{1-\cos^2 a}}{\cos a} = \frac{\sqrt{1-\left(\frac{-3}{5}\right)^2}}{\frac{-3}{5}} = \frac{-4}{3}$.

- By applying property 3.2 – 6, we get :

$$\tan 2a = \frac{2\tan a}{1-\tan^2 a} = \frac{2\times\frac{-4}{3}}{1-\left(\frac{-4}{3}\right)^2} = \frac{24}{7}.$$

Domain of definition of Trigonometric Functions

- A function f (x) = sin x is defined for every $x \in \Re$. Then its domain of definition is $D_f = \Re$. It is periodic with period 2π, and it is an odd function; that is sin (- x) = - sin x. Therefore, its graph admits the origin O (0 , 0) as center of symmetry.
- A function f (x) = cos x is defined for every $x \in \Re$. Then its domain of definition is $D_f = \Re$. It is periodic with period 2π, and it is an even

function; that is cos (- x) = cos x. Therefore, its graph admits the y – axis as axis of symmetry.

- A function f (x) = tan x is defined for every $x \neq \frac{\pi}{2} + k\pi$ where $k \in Z$. Then its domain of definition is $D_f = \Re - \left\{ \frac{\pi}{2} + k\pi \right\}$. It is periodic with period π, and it is an odd function ; that is tan (- x) = - tan x. Therefore, its graph admits the origin O (0 , 0) as center of symmetry.

Example 3.2 - 2

Find the domain of definition of the following exponential functions:

a. $f(x) = \sin(3x + \frac{\pi}{4})$

b. $g(x) = \tan 4x$

Solution

a. f is defined for $(3x + \frac{\pi}{4}) \in \Re$. Then $D_f = \Re$.

b. g is defined for $4x \in \Re - \left\{ \frac{\pi}{2} + k\pi \right\}, k \in Z$. Then $D_f = \Re - \left\{ \frac{\pi}{8} + \frac{k\pi}{4} \right\}$.

Section 3.3

Range and Sketch of Functions

Definition 3.3 - 1

Consider the function f defined from a set A into a set B. The range of f is the set of all elements y in B or a subset of B such that for every x in A, y = f (x). That is, it is the set of images of the elements in A. It is denoted by Ran (f).

Example 3.3 – 1

Consider the following graphs (figure 3.3 – 1) of certain functions. Find the domain of definition and the range of each function.

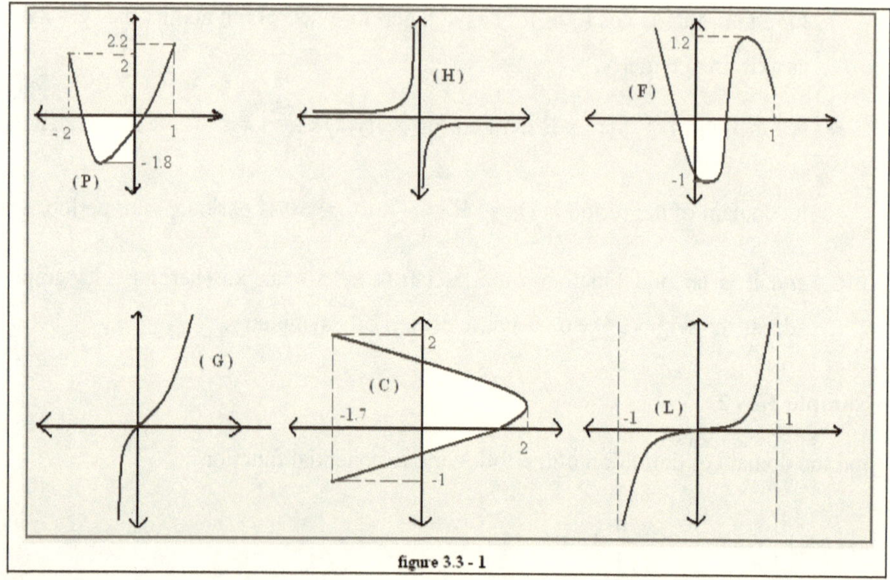

figure 3.3 - 1

Solution

- $D_P = [-2, 1]$ and $Ran(P) = [-1.8, 2.2]$
- $D_H = \Re^*$ and $Ran(H) = \Re^*$
- $D_F = (-\infty, 1]$ and $Ran(F) = [-1, +\infty)$
- $D_G = \Re$ and $Ran(G) = \Re$
- $D_C = [-1.7, 2]$ and $Ran(C) = [-1, 2]$
- $D_L = (-1, 1)$ and $Ran(L) = (-\infty, +\infty)$

Some Rules to Sketch a Function

To sketch a function f, it is preferable to follow the following rules:

- Find the domain of definition D_f of the function.
- Choose some values x from D_f, including the lower limit and the upper limit of D_f if possible, and find their images y to get the coordinates (x , y) of the points that describe the graph of f.
- Locate these points in an orthonormal system and join them.

Example 3.3 – 2

Sketch the parabola function $f(x) = x^2 - 3x + 2$.

Solution

This function is defined for every $x \in \Re$. Then $D_f = \Re$.

This parabola has a vertex $S\left(\dfrac{-b}{2a}, \dfrac{-b^2+4ac}{4a}\right)$, that is

$S\left(\dfrac{3}{2}, \dfrac{-1}{4}\right)$. Figure 3.3 – 2 describes the graph of (f).

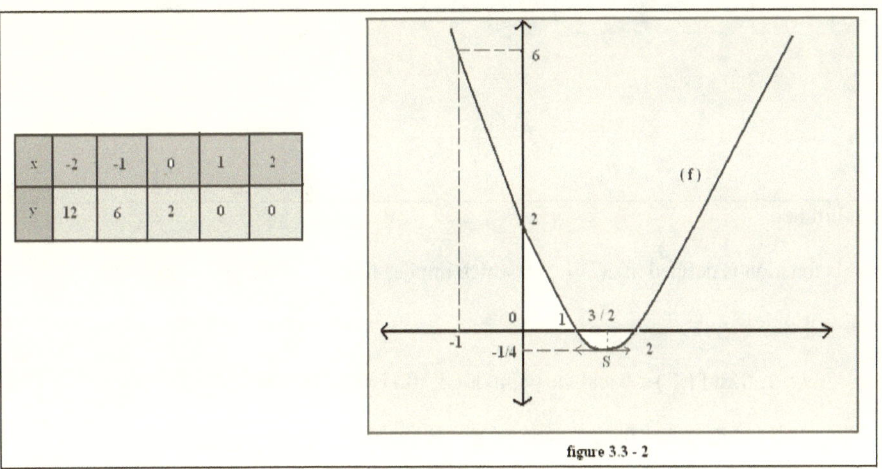

x	-2	-1	0	1	2
y	12	6	2	0	0

figure 3.3 - 2

Example 3.3 – 3

Sketch the Hyperbola function $f(x) = \dfrac{2x+6}{x-1}$.

Solution

This function is defined if $x - 1 \neq 0$ which implies that $x \neq 1$. Then

$D_f = \Re - \{1\}$.

This hyperbola has two asymptotes :

A horizontal asymptote (h) : $y = \dfrac{a}{c} = \dfrac{2}{1} = 2$.

A vertical asymptote (v) : $x = \dfrac{-d}{c} = \dfrac{-(-1)}{1} = 1$.

Figure 3.3 – 3 describes the graph of (f).

Example 3.3 – 4

Sketch the simple irrational function $f(x) = \sqrt{x-1}$.

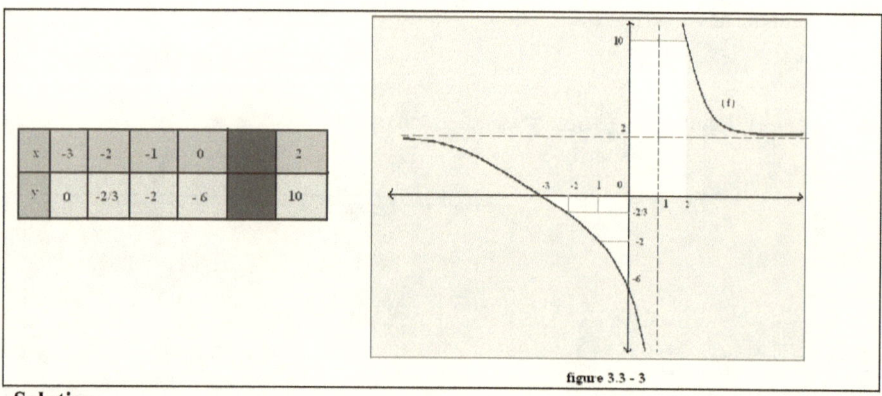

x	-3	-2	-1	0		2
y	0	-2/3	-2	-6		10

figure 3.3 - 3

Solution

This function is defined if x - 1 ≥ 0 which implies that x ≥ 1. Then

D_f = [1 , + ∞).

If x = 1, then f (1) = 0 and the point E (1 , 0) is a limit point.

Figure 3.3 – 4 describes the graph of (f).

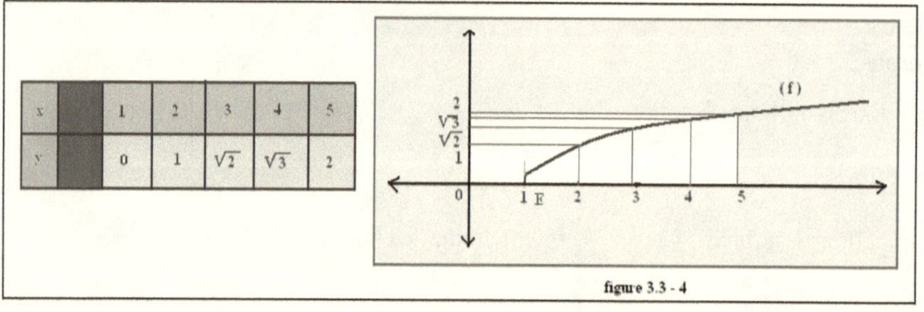

x		1	2	3	4	5
y		0	1	$\sqrt{2}$	$\sqrt{3}$	2

figure 3.3 - 4

Example 3.3 – 5

Sketch the logarithm function $f(x) = \ln x$.

Solution

This function is defined if $x > 0$. Then $D_f = (0, +\infty)$.

As x tends to 0^+, $f(x)$ tends to $-\infty$. Figure 3.3 – 5 describes the graph of (f).

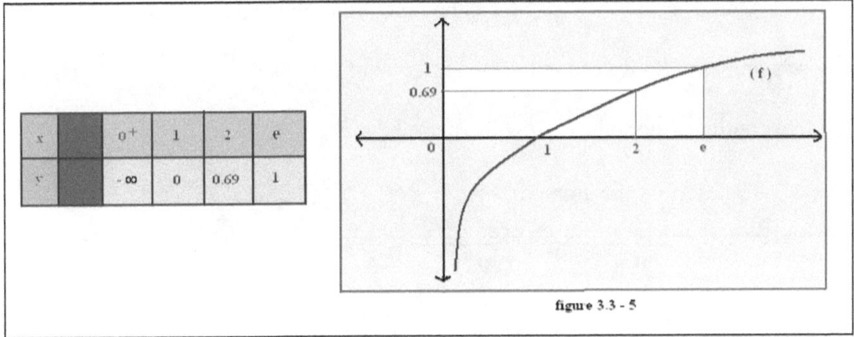

x		0^+	1	2	e
y		$-\infty$	0	0.69	1

figure 3.3 - 5

Example 3.3 – 6

Sketch the exponential function $f(x) = e^x$.

Solution

This function is defined for every $x \in \Re$. Then $D_f = \Re$.

Figure 3.3 – 6 describes the graph of (f).

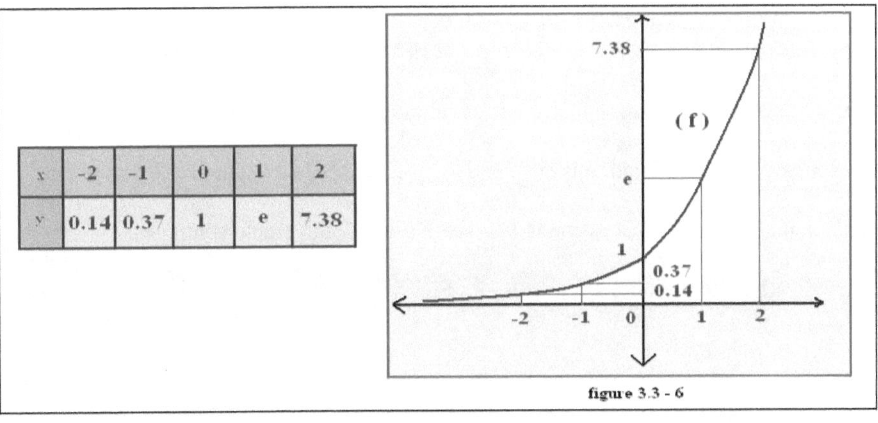

x	-2	-1	0	1	2
y	0.14	0.37	1	e	7.38

figure 3.3 - 6

Example 3.3 – 7

Sketch the trigonometric function $f(x) = \sin x$.

Solution

This function is defined for every $x \in \Re$. Then $D_f = \Re$.

It is periodic with period 2π, then we can reduce the interval of discussion to $[0, 2\pi]$. Moreover, $f(-x) = -f(x)$, the function is odd and the origin O is center of symmetry .So we can reduce the interval of discussion to $[0, \pi]$. Finally, the graph continues periodically with period $2k\pi$, $k \in Z$.

Figure 3.3 – 7 describes the graph of (f).

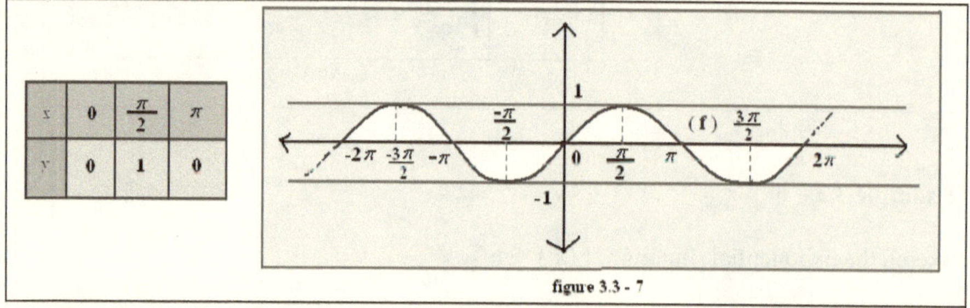

x	0	$\frac{\pi}{2}$	π
y	0	1	0

figure 3.3 - 7

Example 3.3 – 8

Sketch the trigonometric function $f(x) = \cos x$.

Solution

This function is defined for every $x \in \Re$. Then $D_f = \Re$. It is periodic with period 2π, then we can reduce the interval of discussion to $[0, 2\pi]$. Moreover, $f(-x) = f(x)$, the function is even and the y-axis is axis of symmetry. So we can reduce the interval of discussion to $[0, \pi]$. Finally, the graph continues periodically with period $2k\pi$, $k \in Z$. Figure 3.3 – 8 describes the graph of (f).

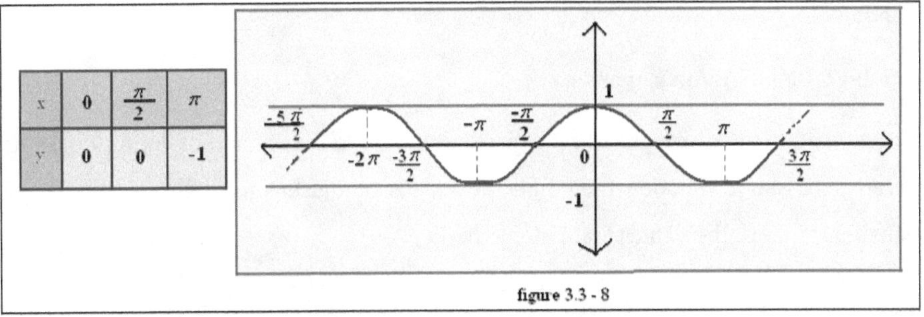

x	0	$\frac{\pi}{2}$	π
y	0	0	-1

figure 3.3 - 8

Example 3.3 – 9

Sketch the trigonometric function $f(x) = \tan x$.

Solution

This function is defined for every $x \in \Re - \left\{\frac{\pi}{2} + k\pi\right\}$, $k \in Z$. then $D_f = \Re - \left\{\frac{\pi}{2} + k\pi\right\}$. It is periodic with period π, then we can reduce the interval of discussion to $[0, \frac{\pi}{2}) \cup (\frac{\pi}{2}, \pi]$. Moreover, $f(-x) = -f(x)$, the function is odd and the origin O is center of symmetry. So we can reduce the interval of discussion to $[0, \frac{\pi}{2})$. Finally, the graph continues periodically with period $k\pi, k \in Z$. This function admits asymptotes that are equal to $\frac{\pi}{2} + k\pi, k \in Z$.

Figure 3.3 – 9 describes the graph of (f).

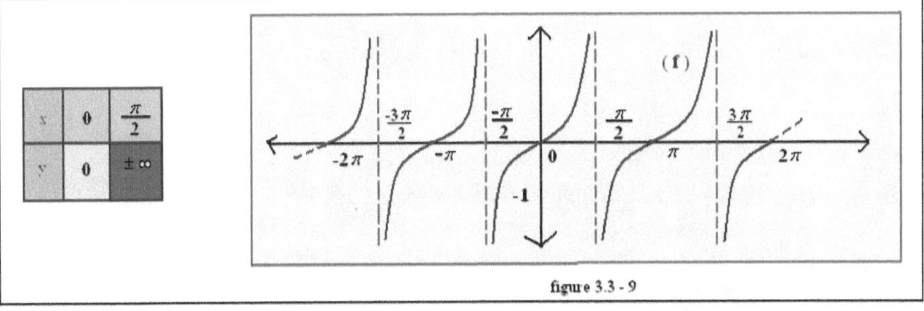

x	0	$\frac{\pi}{2}$
y	0	$\pm\infty$

figure 3.3 - 9

Section 3.4

Solving Logarithmic Equations

To solve a logarithmic equation, it is preferable to find the domain of definition of the given logarithmic function then find the values of the unknown that fit the given equation. That is, the values that belong to D_f.

Example 3.4-1

Solve the following equations :

a) $\ln(x+4) = 0$

b) $\ln(2x) + \ln 4 = 1$

c) $2\log(5x) = 2$

d) $\log_2 x + \log_2(x-1) = 2$

Solution

a. This equation is defined if $x+4>0$, implies that $x>-4$. Then,

$D_f = (-4, +\infty)$.

$\ln(x+4) = 0 \Rightarrow x+4 = e^0 \Rightarrow x = -3 \in D_f$ acceptable.

b. This equation is defined if $2x>0$, implies that $x>0$. Then,

$D_f = (0, +\infty)$.

$\ln(2x) + \ln 4 = 1 \Rightarrow \ln 8x = 1 \Rightarrow 8x = e^1 \Rightarrow x = \frac{e}{8} \approx 0.34 \in D_f$

acceptable.

c. This equation is defined if $5x>0$, implies that $x>0$. Then,

$D_f = (0, +\infty)$.

$2\log(5x) = 2 \Rightarrow \log(5x)^2 = 2 \Rightarrow 25x^2 = 10^2 \Rightarrow x^2 = \frac{100}{25} \Rightarrow x = \pm\sqrt{4}$

$$\Rightarrow \begin{cases} x = 2 \in D_f & acceptable \\ or & \\ x = -2 \notin D_f & rejected \end{cases}$$

d. This equation is defined if $x > 0$ and $x > 1$, implies that $x > 1$. Then,

$D_f = (1, +\infty)$.

$$\log_2 x + \log_2 (x-1) = 2 \Rightarrow \log_2 x(x-1) = 2 \Rightarrow x(x-1) = 2^2$$

$\Rightarrow x^2 - x - 4 = 0$ .Using calculator, the roots are $\begin{cases} x \approx 2.56 \in D_f & acceptable \\ or & \\ x \approx -1.56 \notin D_f & rejected \end{cases}$

Solving Exponential Equations

Example 3.4-2

Solve the following equations :

a) $e^x = 1$

b) $e^{x-3} + e = 4$

c) $2^{x+4} - 2^{3x-1} = 0$

d) $3^{x^2-3x+3} = 27$

Solution

a) $e^x = 1 \Rightarrow e^x = e^1 \Rightarrow x = 1.$

b) $e^{x-3} + e = 4 \Rightarrow e^{x-3} = 4 - e \Rightarrow x - 3 = \ln(4-e) \Rightarrow x = 3 + \ln(4-e).$

c) $2^{x+4} - 2^{3x-1} = 0 \Rightarrow 2^{x+4} = 2^{3x-1} \Rightarrow x + 4 = 3x - 1 \Rightarrow x = \dfrac{5}{2}.$

d) $3^{x^2-3x+3} = 27 \Rightarrow 3^{x^2-3x+3} = 3^3 \Rightarrow x^2 - 3x + 3 = 3 \Rightarrow x^2 - 3x = 0$

$\Rightarrow x(x-3) = 0 \Rightarrow x = 0$ or $x = 3.$

Solving Trigonometric Equations

Example 3.4-2

Solve the following equations :

a) $\cos 2x = \cos x$

b) $\sin(3x + \frac{\pi}{3}) - \sin(x - \frac{\pi}{6}) = 0$

c) $\cos^2 x - \sin^2 x = -1$

d) $\tan 4x = \sqrt{3}$

Solution

a) $\cos 2x = \cos x \Rightarrow 2x = \pm x + 2k\pi$, *where* $k \in Z$.

$$\Rightarrow \begin{cases} 2x = x + 2k\pi \\ or \\ 2x = -x + 2k\pi \end{cases} \Rightarrow \begin{cases} x = 2k\pi \\ or \\ 3x = 2k\pi \end{cases} \Rightarrow \begin{cases} x = 2k\pi \\ or \\ x = \dfrac{2k\pi}{3} \end{cases}.$$

b) $\sin(3x + \frac{\pi}{3}) - \sin(x - \frac{\pi}{6}) = 0 \Rightarrow \sin(3x + \frac{\pi}{3}) = \sin(x - \frac{\pi}{6})$

$$\Rightarrow \begin{cases} 3x + \dfrac{\pi}{3} = x - \dfrac{\pi}{6} + 2k\pi \\ or \\ 3x + \dfrac{\pi}{3} = \pi - (x - \dfrac{\pi}{6}) + 2k\pi \end{cases} where\, k \in Z \Rightarrow \begin{cases} 2x = -\dfrac{\pi}{3} - \dfrac{\pi}{6} + 2k\pi \\ or \\ 4x = -\dfrac{\pi}{3} + \pi + \dfrac{\pi}{6} + 2k\pi \end{cases}$$

$$\Rightarrow \begin{cases} 2x = -\dfrac{\pi}{2} + 2k\pi \\ or \\ 4x = \dfrac{5\pi}{6} + 2k\pi \end{cases} \Rightarrow \begin{cases} x = -\dfrac{\pi}{4} + k\pi \\ or \\ x = \dfrac{5\pi}{24} + \dfrac{k\pi}{2} \end{cases}.$$

c) $\cos^2 x - \sin^2 x = -1 \Rightarrow \cos(2x) = \cos\pi \Rightarrow 2x = \pm\pi + 2k\pi$, *where* $k \in Z$

$$\Rightarrow \begin{cases} 2x = \pi + 2k\pi \\ or \\ 2x = -\pi + 2k\pi \end{cases} \Rightarrow \begin{cases} x = \dfrac{\pi}{2} + k\pi \\ or \\ x = \dfrac{-\pi}{2} + k\pi \end{cases}.$$

d) $\tan 4x = \sqrt{3} \Rightarrow \tan 4x = \tan\frac{\pi}{3} \Rightarrow 4x = \frac{\pi}{3} + k\pi$, *where* $k \in Z$

$$\Rightarrow x = \frac{\pi}{12} + \frac{k\pi}{4}.$$

Section 3.5

Combination and Inverse Functions

Definition 3.5 – 1

Let f and g be two functions defined on the two intervals I and J, respectively. The combination of the functions f and g is a function h defined by a mean of addition, subtraction, multiplication, or division between f and g. That is,

$h(x) = (f+g)(x) = f(x) + g(x)$,

$h(x) = (f-g)(x) = f(x) - g(x)$,

$h(x) = (f.g)(x) = f(x) . g(x)$, or

$h(x) = (f/g)(x) = f(x) / g(x)$, provided that $g(x) \neq 0$.

Example 3.5 - 1

Let f and g be two functions defined on $\Re$ by :

$f(x) = x^2$ and $g(x) = x + 4$. Find :

a. $(f+g)(x)$

b. $(f-g)(x)$

c. $(f.g)(x)$

d. $(f/g)(x)$

e. $(f.g)(-1)$

Solution

a. $(f+g)(x) = f(x)+g(x) = x^2 + x + 4$

b. $(f-g)(x) = f(x)-g(x) = x^2 - x - 4$

c. $(f.g)(x) = f(x).g(x) = x^2(x+4) = x^3 + 4x^2$

d. $(f/g)(x) = f(x)/g(x) = \dfrac{x^2}{x+4}$

e. $(f.g)(-1) = (-1)^3 + 4(-1)^2 = 3$.

Composite Functions

Consider two functions f and g that are defined on the two respective intervals I and J by : If for all x in I, y = f (x) is in J, then we can apply g on y, hence :

$$x \xrightarrow{f} f(x) = y \xrightarrow{g} g(y) = z$$

$$x \xrightarrow{\quad\quad h \quad\quad} h(x) = z$$

As we see here the function x permits to pass directly from x to z. h is called the composite function of f and g defined on I by $h(x) = g(y) = g[f(x)]$.

Definition 3.5 – 2

Let f and g be two functions defined on the two intervals I and J respectively. The composite function of f and g is a function $h = g \circ f$ defined on I by : $g \circ f(x) = g[f(x)]$.

Remark 3.5 - 1

The function " $\circ$ " is not commutative. That is, $g \circ f \neq f \circ g$.

Example 3.5 - 2

Let f and g be two functions defined on $\Re$ by $f(x) = 2x - 1$ and $g(x) = x^2 + 1$.

Find $g \circ f(x)$ *and* $f \circ g(x)$ then compare their obtained results.

Solution

- $g \circ f(x) = g[f(x)] = g(2x-1) = (2x-1)^2+1 = 4x^2 - 4x + 2.$

- $f \circ g(x) = f[g(x)] = f(x^2+1) = 2(x^2+1)-1 = 2x^2+1.$

Hence, $g \circ f(x) \neq f \circ g(x)$.

Inverse Functions

Any function f has an inverse function f^{-1} if and only if f is a one – to – one function.

Definition 3.5 – 3

A function f is said to be a one – to – one function if and only if for all x_1 and x_2 in D_f, if $x_1 \neq x_2$, then $f(x_1) \neq f(x_2)$ or

if $f(x_1) = f(x_2)$, then $x_1 = x_2$.

Example 3.5 – 3

Which of the following functions is one – to – one function? Justify your answer.

a. $f(x) = (2x+4)^2$, for $x \leq 2$.

b. $g(x) = (2x+4)^2$.

c. $h(x) = |x-3|$

Solution

a. Let x_1 and x_2 be two different numbers that are less than or equal to 2.

$$x_1 \neq x_2 \Rightarrow 2x_1 \neq 2x_2$$
$$\Rightarrow 2x_1+4 \neq 2x_2+4$$
$$\Rightarrow (2x_1+4)^2 \neq (2x_2+4)^2$$
$$\Rightarrow f(x_1) \neq f(x_2)$$

Hence, f is a one – to – one function.

b. The function g is not a one – to – one function.

Proof by counter example,

Let $x_1 = \frac{-5}{2}$ and $x_2 = -\frac{3}{2}$. Then $x_1 \neq x_2$; however,

$g(x_1) = g(\frac{-5}{2}) = (2 \times \frac{-5}{2} + 4)^2 = 1$, and

$g(x_2) = g(-\frac{3}{2}) = (2 \times \frac{-3}{2} + 4)^2 = 1$, which means that $g(x_1) = g(x_2)$

and this contradict definition 3.5 - 3.

c. The function h is not a one – to – one function.

Proof by counter example ,

Let $x_1 = 4$ and $x_2 = 2$. Then $x_1 \neq x_2$; however,

$h(x_1) = h(4) = |4 - 3| = 1$, and

$h(x_2) = h(2) = |2 - 3| = 1$, which means that $g(x_1) = g(x_2)$

and this contradict definition 3.5 - 3.

Definition 3.5 – 4

Let f be a one – to – one function from a set A onto a set B. The function that assigns to each y in B a real number x in A such that $y = f(x)$, is the inverse function of f and it is denoted by f^{-1}.That is $y = f(x)$ if and only if $x = f^{-1}(y)$.

$$f : A \longrightarrow B, \quad x \longrightarrow y = f(x) \quad \Longleftrightarrow \quad f^{-1} : B \longrightarrow A, \quad y \longrightarrow x = f^{-1}(y)$$

That is,

$$x \xrightarrow{f} y = f(x) \xrightarrow{f^{-1}} f^{-1}(y) = x$$

$$x \xrightarrow{f^{-1} \circ f} x$$

Remark 3.5 – 2

- The function f^{-1} is also a one – to – one function from B onto A.
- f^{-1} is not $\frac{1}{f}$.
- In an orthonormal system, the graphs of f and its inverse f^{-1} are symmetric with respect to the first bisector of equation : y = x.
- For all x in A, we have $(f^{-1} \circ f)(x) = f^{-1}(f(x)) = f^{-1}(y) = x$.
- For all y in B, we have $(f \circ f^{-1})(y) = f(f^{-1}(y)) = f(x) = y$.
- $f^{-1} \circ f$ is the identity function on A and $f \circ f^{-1}$ is the identity function on B.
- If f is defined on an interval I, then its inverse f^{-1} is defined on f(I).

Example 3.5 – 4

Consider the one – to –one function f defined on $[0, +\infty)$ by $y = f(x) = x^2$. Find the inverse function f^{-1} of f. Sketch in an orthonormal system the graph of f and f^{-1}.

Solution

Since f is a one – to –one function defined on $[0, +\infty)$, then it admits an inverse f^{-1} symmetric of f with respect to the first bisector. To find it, we have to find x in terms of y from the given function. That is,

$x^2 = y$ implies that $x = \pm\sqrt{y}$. But $x \in [0, +\infty)$, which means that x is positive, then $x = \sqrt{y}$. Hence, the inverse function f^{-1} of f is defined by $f^{-1}(x) = \sqrt{x}$.

$f(0) = 0^2 = 0$ and as x tends to $+\infty$, x^2 tends to $+\infty$. So the domain of definition of f^{-1} is $[0, +\infty)$.

Figure 3.5 – 1 describes the graph of (f) and (f^{-1}). Note that the graph of f is drawn first, and then the graph of f^{-1} is deduced by symmetry of f with respect to the first bisector.

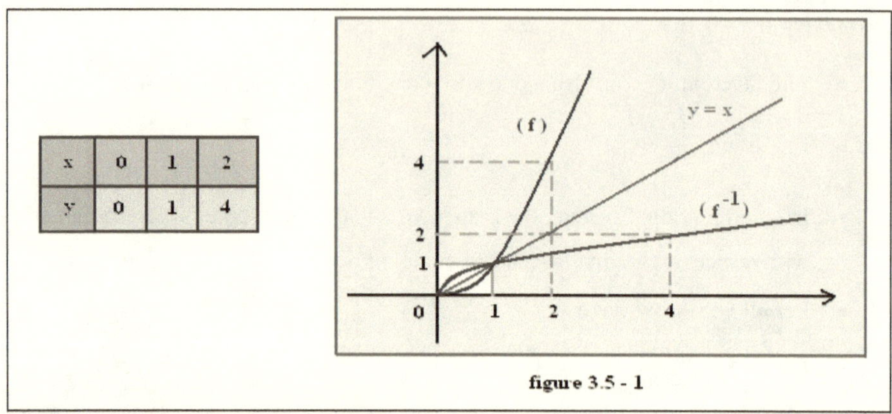

x	0	1	2
y	0	1	4

figure 3.5 - 1

Inverse Trigonometric Functions

The trigonometric functions sine, cosine and tangent are one –to –one functions, then they admit inverse functions.

Property 3.5 – 1

The function f defined over the interval $\left[\frac{-\pi}{2}, \frac{\pi}{2}\right]$ by $f(x) = \sin x$ admits an inverse function f^{-1} defined over the interval $[-1, 1]$ by $f^{-1}(x) = \sin^{-1}(x)$.

$$\left[\frac{-\pi}{2}, \frac{\pi}{2}\right] \underset{\sin^{-1}}{\overset{\sin}{\rightleftarrows}} [-1, 1]$$

The $\sin^{-1}$ function is symmetric of the sine function with respect to the first bisector. Figure 3.5 – 2 describes the graph of sine and $\sin^{-1}$.

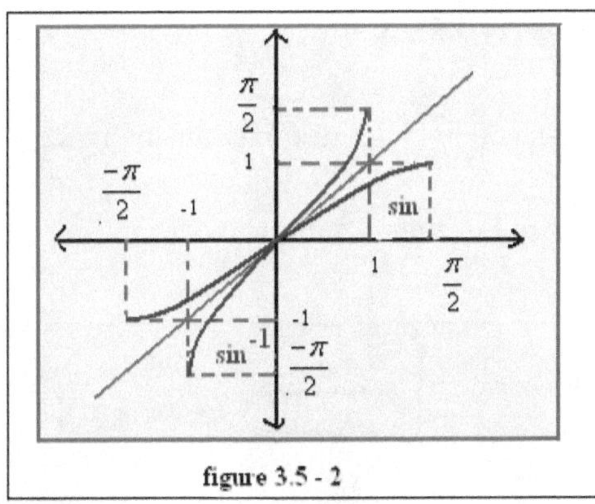

figure 3.5 - 2

Property 3.5 – 2

The function f defined over the interval $[0\,,\,\pi]$ by $f(x) = \cos x$ admits an inverse function f^{-1} defined over the interval $[-1\,,1]$ by $f^{-1}(x) = \cos^{-1}(x)$.

$$[0\,,\,\pi] \underset{\cos^{-1}}{\overset{\cos}{\rightleftarrows}} [-1\,,1]$$

The $\cos^{-1}$ function is symmetric of the cosine function with respect to the first bisector. Figure 3.5 – 3 describes the graph of cosine and $\cos^{-1}$.

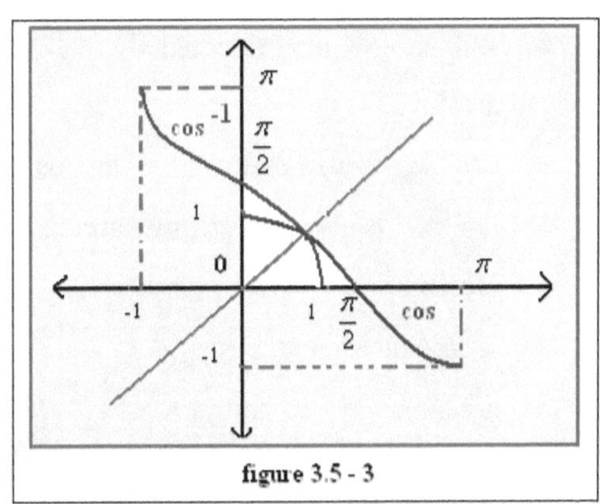

figure 3.5 - 3

Property 3.5 – 3

The function f defined over the interval $\left(\dfrac{-\pi}{2}, \dfrac{\pi}{2}\right)$ by $f(x) = \tan x$ admits an inverse function f^{-1} defined over $(-\infty, +\infty)$ by $f^{-1}(x) = \tan^{-1}(x)$.

$$\left(\frac{-\pi}{2}, \frac{\pi}{2}\right) \underset{\tan^{-1}}{\overset{\tan}{\rightleftarrows}} (-\infty, +\infty)$$

The $\tan^{-1}$ function is symmetric of the cosine function with respect to the first bisector. Figure 3.5 – 4 describes the graph of tan and $\tan^{-1}$

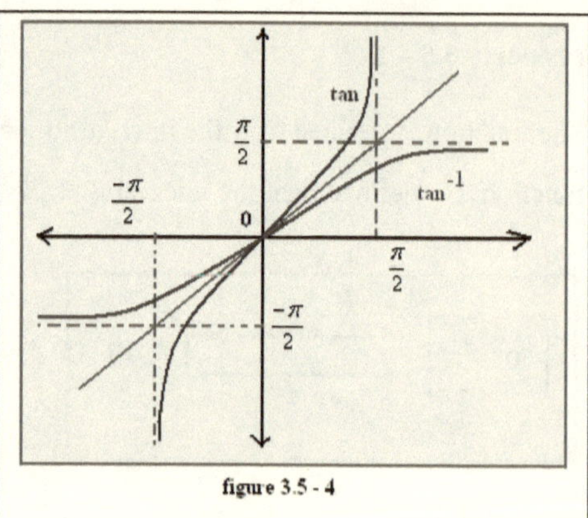

figure 3.5 - 4

Remark 3.5 – 3

- $\sin^{-1} x$ is also denoted by arcsin x.
- $\cos^{-1} x$ is also denoted by arccos x.
- $\tan^{-1} x$ is also denoted by arctan x.
- $\arcsin x = y \Leftrightarrow \sin y = x$
- $\arccos x = y \Leftrightarrow \cos y = x$
- $\arctan x = y \Leftrightarrow \tan y = x$

Example 3.5 – 5

- $\cos \frac{\pi}{3} = \frac{1}{2} \Leftrightarrow \cos^{-1} \frac{1}{2} = \frac{\pi}{3}$ or $\arccos \frac{1}{2} = \frac{\pi}{3}$.
- $\sin \frac{\pi}{3} = \frac{\sqrt{3}}{2} \Leftrightarrow \sin^{-1} \frac{\sqrt{3}}{2} = \frac{\pi}{3}$ or $\arcsin \frac{\sqrt{3}}{2} = \frac{\pi}{3}$.
- $\tan \frac{\pi}{4} = 1 \Leftrightarrow \tan^{-1} \frac{\pi}{4} = 1$ or $\arctan \frac{\pi}{4} = 1$.

Summary

☒ If a is a root of $P(x) = 0$, then $(x - a)$ is a factor of the polynomial function $P(x)$

☒ $P(x) = a_n x^n + a_{n-1} x^{n-1} + \ldots + a_1 x + a_0 \equiv 0$ if and only if $a_0, a_1, \ldots, a_n$ are all equal to zero.

☒ $a_n x^n + a_{n-1} x^{n-1} + \ldots + a_1 x + a_0 \equiv b_n x^n + b_{n-1} x^{n-1} + \ldots + b_1 x + b_0$

☒ if and only if $a_0 = b_0, a_1 = b_1, \ldots, a_n = b_n$

☒ $\log x$ is defined if and only if $x > 0$

☒ $\log_a x$ is defined if and only if $x > 0$ where a is a positive number different from 1

☒ If $0 < x < 1$, then $\ln x < 0$

☒ If $x > 1$, then $\ln x > 0$

☒ $\ln x = \log_e x$

☒ $\log x = \log_{10} x$

☒ $\log_a (xy) = \log_a x + \log_a y$

☒ $\log_a (x / y) = \log_a x - \log_a y$

☒ $\log_a x^n = n \log_a x$

☒ $\log_a (1 / x) = - \log_a x$

☒ $\log_a x = \ln x / \ln a$

☒ $y = \log_a x$ if and only if $a^y = x$

☒ $10^{\log x} = x$

☒ $e^{\ln x} = x$

☒ $a^{\log_a x} = x$

☒ $e^0 = 1$

- $\log_a a = 1$
- $\cos^2 x + \sin^2 x = 1$
- $\cos(a - b) = \cos a \cos b + \sin a \sin b$
- $\sin(a + b) = \sin a \cos b + \cos a \sin b$
- $\sin(a - b) = \sin a \cos b - \cos a \sin b$
- $\text{Sin } a \cos b = \frac{1}{2}[\sin(a + b) + \sin(a - b)]$
- $\cos a \sin b = \frac{1}{2}[\sin(a + b) - \sin(a - b)]$
- $\cos a \cos b = \frac{1}{2}[\cos(a + b) + \cos(a - b)]$
- $\sin a \sin b = \frac{1}{2}[\cos(a - b) - \cos(a + b)]$
- $\text{Cos}(2a) = \cos^2 a - \sin^2 a = 2\cos^2 a = 1 - 2\sin^2 a$
- $\text{Sin}(2a) = 2 \sin a \cos a$
- $\text{Cos } a = \cos^2 \frac{a}{2} - \sin^2 \frac{a}{2} = 2\cos^2 \frac{a}{2} - 1 = 1 - 2\sin^2 \frac{a}{2}$
- $\text{Sin } a = 2 \sin \frac{a}{2} \cos \frac{a}{2}$
- $\text{Cos}^2 a = \frac{1 + \cos 2a}{2}$ and $\text{Sin}^2 a = \frac{1 - \cos 2a}{2}$
- $\text{Cos}^2 \frac{a}{2} = \frac{1 + \cos a}{2}$ and $\text{Sin}^2 \frac{a}{2} = \frac{1 - \cos a}{2}$
- $\tan(a + b) = \frac{\tan a + \tan b}{1 - \tan a \tan b}$
- $\tan(a - b) = \frac{\tan a - \tan b}{1 + \tan a \tan b}$
- $\tan(2a) = \frac{2 \tan a}{1 - \tan^2 a}$
- $\tan a = \frac{2 \tan \frac{a}{2}}{1 - \tan^2 \frac{a}{2}}$

- $h(x) = (f+g)(x) = f(x) + g(x)$
- $h(x) = (f-g)(x) = f(x) - g(x)$
- $h(x) = (f.g)(x) = f(x) . g(x)$
- $h(x) = (f/g)(x) = f(x) / g(x)$
- $g \circ f(x) = g[f(x)]$
- A function f is said to be a one – to – one function if and only if for all x_1 and x_2 in D_f, if $x_1 \neq x_2$, then $f(x_1) \neq f(x_2)$ or if $f(x_1) = f(x_2)$, then $x_1 = x_2$
- f^{-1} is the inverse of f, $y = f(x)$ if and only if $x = f^{-1}(y)$
- $\arcsin x = y \Leftrightarrow \sin y = x$
- $\arccos x = y \Leftrightarrow \cos y = x$
- $\arctan x = y \Leftrightarrow \tan y = x$
- $\text{arccot}\, x = y \Leftrightarrow \cot y = x$

EXERCISES

In Problems 3.1 - 3.35, find the domain of definition of the given functions .

3.1 . $f(x) = 5x^2 - 3x + 1$

3.2 . $f(x) = \dfrac{3x^5 - 5x^3 + x}{\sqrt{5}}$

3.3 . $f(x) = \dfrac{2x^4 - x^3 + 1}{x}$

3.4 . $f(x) = \dfrac{x^2 - x}{3x - 9}$

3.5 . $f(x) = \dfrac{x^8 + 6x^3 - 5}{(x-2)(x+3)}$

3.6 . $f(x) = \dfrac{1}{x^2 - 4}$

3.7 . $f(x) = \dfrac{-2x^3 + 1}{3x^2 + x - 4}$

3.8 . $f(x) = \dfrac{-x^5 + x + 4}{x^3 - 2x^2 + 1}$

3.9 . $f(x) = \dfrac{2x^4 + 1}{x^2 + 3}$

3.10 . $f(x) = \dfrac{x^2 + 1}{-x^4 - 2}$

3.11 . $f(x) = \sqrt{x+2}$

3.12 . $f(x) = \sqrt{-x(x-3)}$

3.13 . $f(x) = \sqrt{x^2 - 9}$

3.14 . $f(x) = \sqrt{x^2 + 7x + 6}$

3.15 . $f(x) = \dfrac{2}{\sqrt{x^2 - 1}}$

3.16 . $f(x) = \ln x + \ln(2x + 4)$

3.17 . $f(x) = x \log(-x) + \log(x^2 + 4)$

3.18 . $f(x) = \log_2 x(1 + x)$

3.19 . $f(x) = \log_3 (-4x^2 + 16)$

3.20 . $f(x) = \ln(2e^{2x} - 1)$

3.21 . $f(x) = \dfrac{e^x + 1}{e^x - 1}$

3.22 . $f(x) = e^{\ln(5x-2)}$

3.23 . $f(x) = 8^{\ln(-x-3)}$

3.24 . $f(x) = \sin(4x - 7) + \cos x$

3.25 . $f(x) = \dfrac{\sin 2x}{\cos(x-2)}$

3.26 . $f(x) = \dfrac{7x}{\tan(x-2)}$

3.27 . $f(x) = \dfrac{\sqrt{x+2}}{\sqrt{x-2}}$

3.28 . $f(x) = \dfrac{\sqrt{x} + 2}{\sqrt{x-2}}$

3.29 . $f(x) = \dfrac{3+x}{2+|x|}$

3.30 . $f(x) = \dfrac{x}{1-|x|}$

3.31 . $f(x) = \sqrt{4-x} + \sqrt{x+5}$

3.32 . $f(x) = \sqrt{\dfrac{x-2}{x+2}}$

3.33 . $f(x) = \dfrac{\ln x + 4}{e^{x} - 1}$

3.34 . $f(x) = \dfrac{\ln x^{2} + 4}{e^{x} + 1}$

3.35 . $f(x) = \dfrac{x - 7}{\sin(x - 2)}$

3.36 . Determine, if possible, a and b so that the following polynomials are identical to each others:

a. $P(x) = (a + 3)x^{3} + (3b - 1)x^{2} + 1.$
$q(x) = (b - 2)x^{3} + (a + 3)x^{2} + 1.$

b. $P(x) = 2x^{3} - 3x^{2} - 4x - 1.$
$q(x) = (ax + b)(x^{2} - 2x - 1).$

3.37 . Without calculating the division, show that P(x) is divisible by q(x) in each of the following cases:

a. $P(x) = x^{3} + 2x^{2} - 8x - 9$, $q(x) = x + 1.$

b. $P(x) = 6x^{2} + x - 2$, $q(x) = 2x - 1.$

3.38 . a. Show that 1 is a root of the polynomial $P(x) = x^{3} + x - 2.$

b. Determine the real numbers a, b, c such that $P(x) = (x - 1)(ax^{2} + bx + c)$

3.39 . a. Determine the real number a such that $x + 2$ is a factor of the polynomial

$P(x) = 2x^{3} + 12x^{2} + ax - 84.$

b. Find a polynomial q(x) so that $P(x) = (x + 2)\,q(x).$

c. Calculate q(3) and deduce a factorization of P(x) in three factors.

3.40 . Factorize $P(x) = x^{3} + 3x^{2} - 4x - 12$ and $L(x) = x^{4} - 4x^{3} - 5x^{2} - 36x - 36$

3.41 . Use the classical division method to divide P(x) by q(x) in each of the following cases :

a. $P(x) = 2x^{3} + x + 3$, $q(x) = x + 1.$

b. $P(x) = x^{3} + 27$, $q(x) = x + 3.$

3.42 . let $P(x) = x^n - 1$ and $f(x) = x^{n+1} - 1$.

a. Show that $P(x)$ and $f(x)$ are divisible by $x - 1$.

b. Show that $A(x) = (x^n - 1)(x^{n+1} - 1)$ is divisible by $B(x) = (x + 1)(x - 1)^2$.

3.43 . Determine the real numbers a, b and c such that for all $x \neq 1$, we have

$$\frac{-x^2 - x - 8}{(x-1)(x^2+4)} = \frac{ax+b}{x^2+4} + \frac{c}{x-1}$$

3.44 . Consider the function f defined by $f(x) = \dfrac{(x-1)^3}{x^2+x+1}$. Show that there exists two real numbers a and b such that $f(x) = x - 4 + \dfrac{ax+b}{x^2+x+1}$.

3.45 . Consider the function f defined by $f(x) = \dfrac{x^2-2}{x+1}$. Show that f(x) can be written in the form $f(x) = ax + b + \dfrac{c}{x+1}$ where a, b, c are real numbers to be determined.

3.46 . Consider the function f defined by $f(x) = \dfrac{x^4+x^2+4}{x(x^2+2)}$. Find the real numbers a, b and c such that for all real numbers $x \neq 0$, $f(x) = ax + \dfrac{b}{x} + \dfrac{cx}{2+x^2}$.

3.47 . Determine the real numbers a, b and c such that

$$f(x) = \frac{3x^2 - 4x + 2}{x-1} = ax + b + \frac{c}{x-1}.$$

3.48 . A zoologist estimates that for the coming 12 years, starting today the population of a species, will be given by $P(x) = 3600 \times \dfrac{2t+1}{t+3}$, where t designates the number of years and P(t) the number of individuals in the species.

a. What is the percent population?
b. What is the domain of definition of the function P.
c. Would this species expand or become extinct for the next 12 years?

3.49. If the total cost C, in dollars, for producing a certain product is given by

$$C(q) = \frac{98\times 106}{400-q^{2}},$$ where q is the quantity produced between 0 and 17.

a. Calculate the total cost of 10 units and the average cost of one unit $\overline{C}(q)$.

b. Study the variations of C on [0 , 17]. Sketch its representative curve in an orthogonal system (1 cm for one unit produced as abscissa and 1 cm for $ 10^{5} as ordinate).

3.50. Suppose that a company has a group of 14 identical sport cars. If each of these cars is depreciating at a rate of $ 1000 per year, what is the rate of depreciation of the whole group?

3.51. Suppose that a new office equipment is purchased for $ 1000, has a scrap value of $ 200 after 2 years.

a. Find the linear function that models the total depreciation D(t).
b. Find the linear function that models the total depreciated value V(t).
c. What would be the value of the equipment after the first year and the second year ?
d. Sketch V for $t \in [0, 2]$.

3.52. Suppose that new living room furniture is purchased for $ 2000, has a useful lifetime of six years, and has no scrap value. After how many years will it have a value of $ 1200 ?

3.53. A manufacturer scientific calculators has fixed costs of $ 300 per week and variable costs of $ 5 per calculator.

a. Find the total cost function of producing x calculators per week.
b. Sketch this function.

3.54. A small business office buys a computer for \$ 4000 that has a useful lifetime of 2 years and a salvage value of \$ 200 at the end of 2 years.

a. Find the depreciation over 2 years.

b. Find the annual depreciation of the computer.

c. Find the total depreciation of the computer after t years.

d. Find the depreciated value of the asset after t years.

e. Sketch the graph of DV .What do you conclude ?

3.55. f is a function defined over $\Re$ by $f(x) = x^2 + 2x - 3$.

a. Determine the range of f for all $x \in [-4, 2]$.

b. Show that for all real numbers x , $f(x) = (x+1)^2 - 4$.

c. Let a and b be two real numbers such that $a < b$. Show that:

i) If a and b belong to $[-1, +\infty[$, then $f(a) < f(b)$.

ii) If a and b belong to $]-\infty, -1]$, then $f(a) > f(b)$.

d. Deduce the sense of variation of f.

3.56. Sketch the graph of $f(x) = \frac{1}{x+1}$. Deduce the graphical representation of

$g(x) = -1 + \frac{1}{x+1}$.

3.57. Sketch the graph of $f(x) = |x|$. Deduce the graphical representation of

$g(x) = |x-2|$.

3.58. Sketch the graph of $f(x) = \sqrt{x}$. Deduce the graphical representation of

$g(x) = \sqrt{x+2}$.

3.59. Sketch the graph of $f(x) = x^2 - 6x + 5$ after determining its vertex.

3.60. Sketch the graph of $f(x) = x^2 + 2x + 1$. Deduce the graphical representation of $g(x) = (x+1)^2 - 1$.

3.61. Find the domain of definition of $f(x) = \ln(-x)$, then sketch the graphical representation of f on its domain of definition.

3.62. Sketch the graph of $f(x) = 2^x$. Deduce the graphical representation of $g(x) = 2^x - 1$.

3.63. Sketch the graph of $f(x) = e^{2x}$. Deduce the graphical representation of $g(x) = e^{2x} + 1$.

3.64. Sketch the graph of $f(x) = \left(\frac{1}{2}\right)^x$. Deduce the graphical representation of

$g(x) = \left(\frac{1}{2}\right)^x + 1$.

3.65. Sketch the graph of the function $f(x) = \cot x$. Deduce the graph of its inverse function $g(x) = \cot^{-1} x$.

3.66. Sketch the graph of the following trigonometric functions :

a. $g(x) = \sec x$.

b. $h(x) = \csc x$.

3.67. In the adjoining figure, A B C D is a square with center O such that A D = 1.

This square is referred to the system $(A, \vec{i}, \vec{j})$ such that $\vec{i} = \overrightarrow{AB}$ and $\vec{j} = \overrightarrow{BC}$

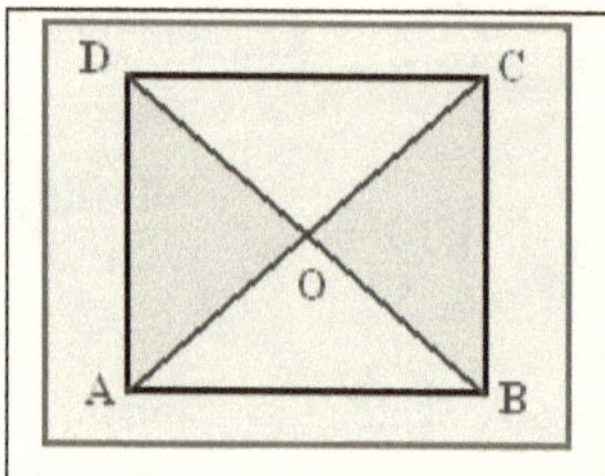

a. Give the coordinates of O.

b. Give the polar coordinates of O.

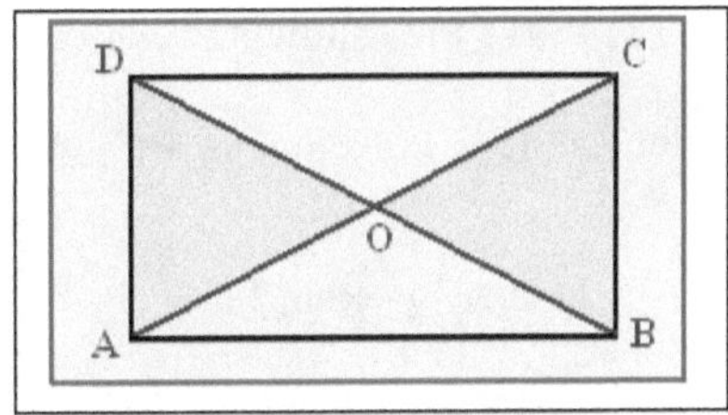

3.68. A B C D is a rectangle with center O, with side $AB = \sqrt{3}$ and $AD = 1$.

a. Determine a measure of each of the following directed angles:

$\left(\overrightarrow{AC},\overrightarrow{AD}\right)$, $\left(\overrightarrow{BA},\overrightarrow{BD}\right)$, $\left(\overrightarrow{OA},\overrightarrow{OB}\right)$, and $\left(\overrightarrow{OC},\overrightarrow{AD}\right)$.

b. Let I be a point such that $\overrightarrow{AI} = \frac{1}{\sqrt{3}}\overrightarrow{AB}$.

i) Determine a direct orthonormal system from the figure.

ii) What are the polar coordinates of the points O, B, C, and D?

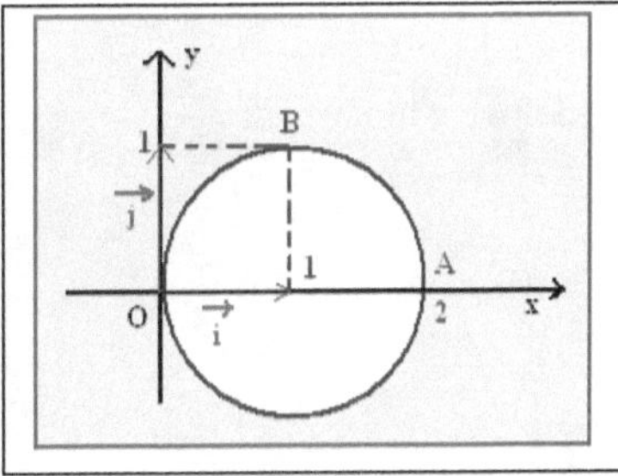

3.69. The plane is referred to a direct orthonormal system $\left(O,\vec{i},\vec{j}\right)$. Consider the point A (2 , 0) and the circle (C) with diameter [O A].

a. Determine the polar coordinates of the point B (1 , 1) of (C).

b. Let M be the point of (C) with polar angle θ. Calculate O M in terms of θ and find the polar coordinates of B again.

3.70. Simplify the following expressions:

a. $\cos(x+y)\cos y + \sin(x+y)\sin y$

b. $\sin(y+x)\cos(y-x) + \cos(y+x)\sin(y-x)$

c. $\dfrac{1}{\tan x} - \dfrac{2}{\tan 2x}$

3.71. Prove the following relations for any x:

a . $(\cos x + \sin x)^2 + (\cos x - \sin x)^2 = 2.$

b. $\cos^4 x + \sin^4 x = 1 - \frac{1}{2}\sin^2(2x).$

3.72. Factorize:

a. $1 - \cos 4x + 2\sin 2x$

b. $\cos x - \cos^2 \frac{x}{2}$

In Problems 3.73 - 3.108, Solve the given equations

3.73. $\cos x = \cos \frac{\pi}{4}$

3.74. $\sin x = \sin \frac{2\pi}{3}$

3.75. $\tan x = \tan \frac{\pi}{3}$

3.76. $\cos x = -\frac{1}{2}$

3.77. $\sin x = -\frac{\sqrt{3}}{2}$

3.78. $\tan x = -1$

3.79. $(\cos^2 x - \frac{1}{2})(\sin^2 x - \frac{1}{2}) = 0$

3.80. $\sin 3x = \cos x$

3.81. $\sqrt{3}\tan(5x + \frac{\pi}{6}) = 1$

3.82. $\tan^2 x - 3 = 0$

3.83. $\sin 2x - \tan x = 0$

3.84. $\cos x - \sqrt{3}\sin x = 1$

3.85. $\cos^4 x - \sin^4 x = 0$

3.86. $\cos^3 x + \cos^3 x = 0$

3.87. $2\cos^2 x - 3\cos x + 1 = 0$

3.88. $\sqrt{3}\sin^2 x + 2\sqrt{3}\sin x + \sqrt{3} = 0$

3.89. $2\tan^2 x - 3\tan x + 1 = 0$

3.90. $(x+1)\ln(x+1) = 0$

3.91. $\log(2x+1) = -2$

3.92. $\log_3(x-1) = \log_3(x+1)$

3.93. $\ln(x+2) + \ln(x-1) = 2\ln x$

3.94. $\log|x| + \log|x-1| = 0$

3.95. $\log(\cos x) = 0$, $x \in (0, \pi)$.

3.96. $\log_x 10 + 2\log_{10x} 10 + 3\log_{100x} 10 = 0$

3.97. $\ln(x^2 - 2x - 3) = \ln(x+7)$

3.98. $\ln\sqrt{3-x} + \ln\sqrt{x+1} = \ln\sqrt{10-6x}$

3.99. $(e^x - 1)(e^x + 2) = 0$

3.100. $e^x - 4e^{-x} = -3$

3.101. $\dfrac{e^x - e^{2x} - 14e^x + 24}{e^x - 1} = 0$

3.102. $e^{2x} + e^x - 2 = 0$

3.103. $(2^x - 1)(2^x + 2) = 0$

3.104. $2^{2x-1} + 3^x + 4^{x+\frac{1}{2}} - 9^{\frac{x}{2}+1} = 0$

3.105. $\text{Arccos}(3x-1) + \text{Arcsin}\, x = \dfrac{\pi}{2}$

3.107. $2 \text{ Arctan } x = \text{Arctan } \frac{2}{x}$

3.108. $\text{Arcsin } \frac{1}{2} - \text{Arccos } \frac{1}{3} = \text{Arcsin } x$

In Problems 3.109 - 3.112, express f [g(x)] in terms of x.

3.109. $f(x) = - x + 4$ and $g(x) = 2x - 1$

3.110. $f(x) = - x^2 + 1$ and $g(x) = x - 1$

3.111. $f(x) = 2x$ and $g(x) = 3x + 3$

3.112. $f(x) = \frac{2+x}{x}$ and $g(x) = -x + 1$

In Problems 3.113 - 3.115, determine f ∘ g and g ∘ *f* in terms of x.

3.113. $f(x) = x + 1$ and $g(x) = x^2$

3.114. $f(x) = - x + 3$ and $g(x) = \sqrt{x}$

3. 115. $f(x) = 2x$ and $g(x) = \frac{1}{x}$

3. 116. Let $f(x) = x + 2$ and $g(x) = x^2 - 1$ Find :

a. $f[g(t)]$

b. $f[g(5)]$

c. $g[f(x)]$

d. $g[f(-2)]$

e. $f \circ g(x)$

f. $g \circ f(x)$

j. $f \circ g(\frac{1}{2})$

h. $g \circ f(\sqrt{2})$

In Problems 3.117 - 3.119, determine the range of the given intervals by the function f, then Determine the inverse function f^{-1}:

3.117. $f(x) = x^2 + 2x + 1$; $]-5, -1[$

3.118. $f(x) = \dfrac{3x}{x+1}$; $[0, 4]$

3.119. $f(x) = \dfrac{-x+1}{x}$; $]0, +\infty[$

In Problems 3.120 - 3.124, justify if the given functions are one to one functions.

3. 120. $f(x) = x^2$

3. 121. $f(x) = x^2$ for $x > 0$.

3. 122. $f(x) = (3x - 9)^2$ for $x < 3$.

3. 123. $f(x) = |2x - 1|$

3. 124. $f(x) = \ln(x - 4)$ for $x > 4$.

In Problems 3.125 - 3.130, determine the inverse function f^{-1} when it exists.

3. 125. $f(x) = \sqrt{1+x}$, $x \in [-1, +\infty)$

3. 126. $f(x) = (2x - 4)^2$, $x \in \Re$

3. 127. $f(x) = (2x - 4)^2$, $x \geq 2$

3. 128. $f(x) = \dfrac{x+2}{x-1}$, $x \in \Re - \{1\}$

3. 129. $f(x) = \sin x$, $x \in (0, \frac{\pi}{2})$

3. 130. An enterprise decides to produce a large series of a given article.

The product cost of each article is \$ 200 knowing that the initial fixed cost of production is \$ 1 500 000 .

a. What is the production cost of n articles ?
b. The demand of this article in the market is a function of its unit sale price. The studying of the market shows that for the unit sale price p,

the number of the demanded articles is 2 100 000 – 6000 p, where p is a positive integer given in dollars belonging to the interval [200 , 350]. Show that the corresponding total profit, in dollars, is $-6 \times 1000\, p^2 + 33 \times 10\,0000\; p - 4215 \times 10\,0000$.

c. a. Study the variations of the numerical function f defined on [200 , 350] by $f(x) = -6 \times 10^3 x^2 + 33 \times 10^5\; x - 4215 \times 10\,0000$.

b. Determine the unit sale price that assures the maximum profit. Then calculate this profit and the corresponding total number of sold articles.

3.131. Jim buy a machine for his factory for \$ 15000. This machine has a scrap \$ 5000 and a useful lifetime of 10 years.

a. Find the linear function that represents the total depreciation TD(t) of this machine.

b. What will the value of this machine be after 3 years?

3.132. An enterprise observes that the sold quantity q of a certain article is related to its unit sale price p by the relation $q = 4500 - 5\,p$. The enterprise pays 30 % of the receipts as taxes and a fixed charge equal to \$ 1500. Let R be the enterprise receipt and B the realized profit.

a. Express R in terms of p , B in terms of R then p in terms of p.

b. Let f be a function defined on [0 , 500] by $f(p) = 4500\,p - 5\,p^2$ and let g be the function defined on $[\,0\,,\,10^6\,]$ by $g(R) = 0.7\,R - 1500$. Let h be the composite function of f and g . Show that $h(p) = B$.

c. Deduce the maximum profit.

3.133. An office equipment is bought for \$ 8000, shipping costs \$ 100.The useful lifetime of this equipment is 10 years and its scrap value is \$ 500. After how many years will this equipment have a value \$ 5060 ?

3.134. A machine costs \$ 12100, has a scrap value of \$ 2000 and has a useful lifetime of 5 years.

a. Determine the machine's total depreciation TD(t) in terms of the time t.

b. What will the value of this machine be after 2 years?

3.135. An electrical generator costs \$ 12000. Shipping costs \$ 100 and installation costs \$ 400. The useful lifetime of this generator is 10 years and its scrap value is \$ 4000.

a. Determine the total depreciation TD(t) of this generator in terms of time t.
b. What will the generator's value be after 7 years?

3.136. Two capitals having a sum of \$ 20 000 000 are deposit in a bank for 9 years. The first capital is placed at a simple interest of annual interest rate 9 %. The second capital is placed at an annual interest rate of 7 % compounded annually. Determine the two capitals knowing that they have the same future value.

3.137. A capital of \$ 80 000 000 is deposit in a bank paying an annual interest rate of 10 % compounded semiannually. After how many years will the capital double ?

3.138. Determine the settlement date of an amount of \$ 200 000 compounded annually aiming at paying two loans one of \$ 100 000 repaid in two years and another of \$ 90 000 repaid in 5 years. The two loans are at an annual interest rate of 10 %.

Chapter 4 Limits and Continuities

Introduction

A common sense definition for a limit is the intended height of a function. Questions are posed how I suppose to know what height of a function intends to reach ? How do you determine a function's height?

Why ?

If a function is defined on an interval of the form [a , ∞[, then it is interesting in knowing the mostly big values in this interval. There exist different types of functions whose values increase more and more depending on the increase of the values of their variables. On the other hand, certain functions are close to real values by the increase of the variable. So it is important in such situations to define a method of study called by: determination of the limit in the neighborhood of ∞ and at a point of abscissa x_0.

Objectives

- After studying the material in this chapter, you should be able to
  - calculate the limit of a function at a given point.
  - calculate the limit of a function at ∞ .
  - calculate the limit of polynomial functions.
  - calculate the limit of rational functions.
  - calculate the limit of irrational functions .
  - calculate the limit involving inequalities.
  - calculate the limit of logarithmic functions.
  - calculate the limit of exponential functions.
  - calculate the limit of trigonometric functions.
  - Identify continuous and discontinuous functions.

Section 4.1

Limit of a function at a given point of abscissa a :

Consider the function $f(x) = x^2$ presented by figure 4.1- 1.
Notice that different x values have different heights.
For x = 2, f(2) = 4. It means that the graph of f(x) has a height of 4 when x = 2. That is as x is being close to 2, f(x) is being close to 4 and we write as x tends to 2, f(x) tends to 4. (as $x \to 2$, $f(x) \to 4$) or we write

$$\lim_{x \to 2} f(x) = 4.$$

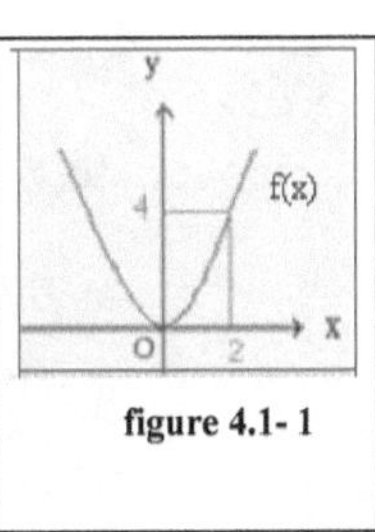

figure 4.1- 1

Remark 4.1-1

Functions that don't reach their intended heights

Consider the function $g(x) = \dfrac{x^2 - 6x + 8}{x - 2}$ presented by figure 4.1- 2.

The hole exists at the point (2 , -2) because $g(2) = \dfrac{0}{0}$.

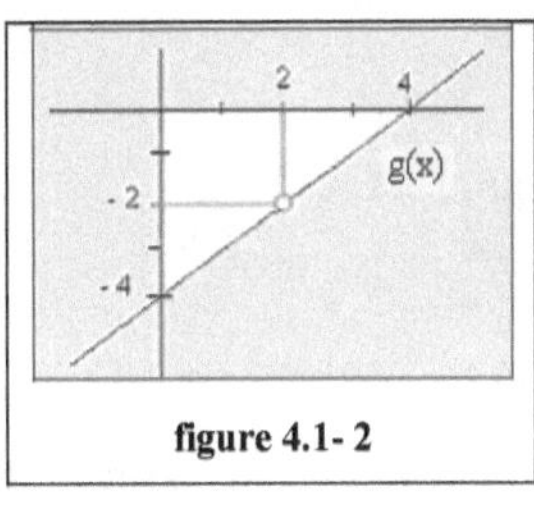

figure 4.1- 2

You can't divide by zero, and zero divided by zero ? How ugly is that !
However, a limit of the function is still exists at x = 2.

The graph intends to reach a height equals $|-2|$ when x = 2 and $\lim_{x \to 2} g(x) = -2.$

The question is raised now when does a limit exists ?

In the above figure, the two travelers will meet at the same place. However, one will reach from the left and the other will reach from the right. Mathematically, this is basically what limit is. A limit exists if when you travel from the left side or from the right side for different values of x, you will meet at the same point. That is:

If $\lim_{x \to a^+} f(x) = \lim_{x \to a^-} f(x) = L \in \Re$.

$\lim_{x \to a^+} f(x) = L$ is the Right- Hand Limit and $\lim_{x \to a^-} f(x) = L$ is the Left-Hand Limit .

Example 4.1-1

1) Look at figure 4.1- 3.Travel towards 2 from the right, the height of the function is 2

   i.e. $\lim_{x \to 2^+} f(x) = 2$; and travel towards 2 from the left, the height of the function is also 2

   i.e. $\lim_{x \to 2^-} f(x) = 2$. Moreover, hence, the limit

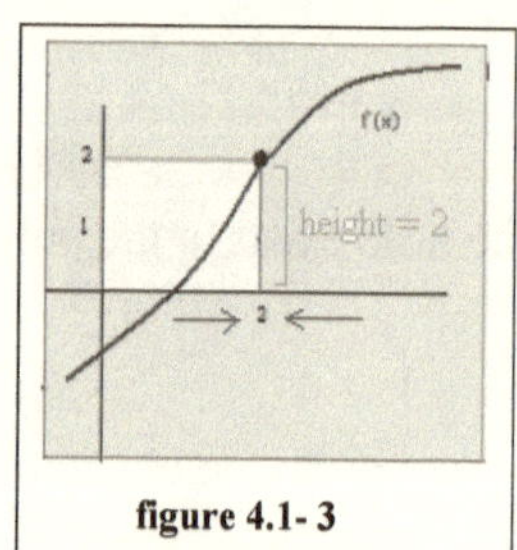

figure 4.1- 3

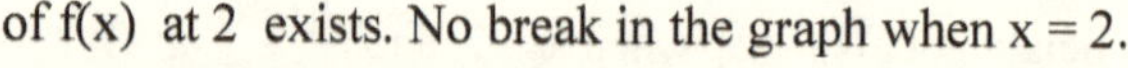
of f(x) at 2 exists. No break in the graph when x = 2.

Knowing that the limit of f (x) exists at x = 1, 3, 4. So if a graph doesn't break at a given x value, then a limit exists there. However, a limit can still exist even if your ultimate destination is a hole in the graph.

2) Look at figure 4.1- 4.Travel towards 2 from the right, the height of the function is 2 i.e. $\lim_{x \to 2^+} f(x) = 2$; and travel towards 2 from the left, the height of the function is 2 i.e. $\lim_{x \to 2^-} f(x) = 2$. Hence, the limit of f(x) at 2 exists even though there is a hole in the graph when x = 2. Notice that if the function travel towards 2 from the left-hand side or the right – hand side, it reaches a height of 2.

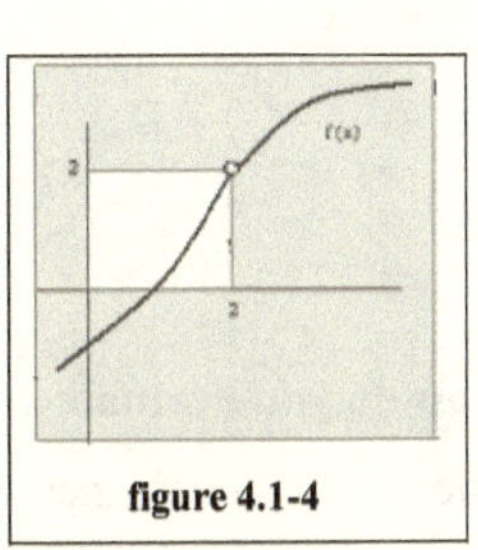

figure 4.1-4

3) Look at figure 4.1- 5.Travel towards 2 from the right, the height of the function is 2 (i.e. $\lim_{x \to 2^+} f(x) = 2$; however, travel towards 2 from the left, the height of the function is 1 (i.e. $\lim_{x \to 2^-} f(x) = 1$). Hence, the limit of f(x) at 2 doesn't exist.

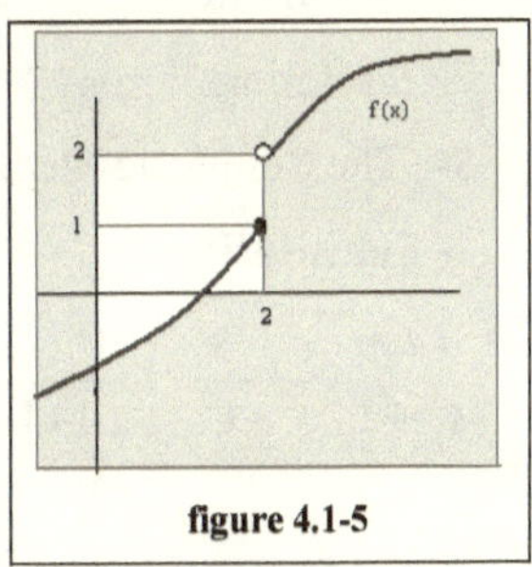

figure 4.1-5

In conclusion, a general limit exists on f(x) when x = c if the right-hand and the left-hand limits are both equal there.

Definition 4.1-1

We say that the function f tends to L as x tends to the left of a if:

$$\lim_{x \to a^-} f(x) = L.$$

Definition 4.1-2

We say that the function f tends to L as x tends to the right of a if:

$$\lim_{x \to a^+} f(x) = L.$$

Definition 4.1-3

If $\lim_{x \to a^+} f(x) = \lim_{x \to a^-} f(x) = L$, then L is the limit of the function f at a.

We write $\lim_{x \to a} f(x) = L$. The point (a , L) is a limit point for f(x).

How do you evaluate a limit ?

The word evaluate means you should give a numerical answer.

Techniques for evaluating limits

1- Substitution

2- Factoring

3- The Conjugate Method

1- Substitution

$\lim_{x \to 1} \frac{x^3 - 3}{x + 1} = \frac{1^3 - 3}{1 + 1} = \frac{-2}{2} = -1 \in \Re$. So the limit of the function

$f(x) = \frac{x^3 - 3}{x + 1}$ exists at x = 1 and it is equal to – 1.

However, substitution won't always work ! For example,

$\lim_{x \to 2} \frac{x^2 - 6x + 8}{x - 2} = \frac{2^2 - 6 \times 2 + 8}{2 - 2} = \frac{0}{0}$ Indeterminate form while graphically, as

shown in figure 4.1- 2 , $\lim_{x \to 2} \frac{x^2 - 6x + 8}{x - 2} = -2$. Then in this case, we have to use

factoring.

2- Factoring

$$\lim_{x \to 2} \frac{x^2 - 6x + 8}{x - 2} = \lim_{x \to 2} \frac{(x-4)(x-2)}{x - 2}$$

$$= \lim_{x \to 2} x - 4$$

$$= 2 - 4$$

$= -2 \in \Re$, the limit exists.

3- The Conjugate Method

What is conjugate ?

The conjugate of any irrational expression is the factor when multiply it by an expression, the radical sign disappeared.

1) the conjugate of $\sqrt{x}$ is $\sqrt{x}$ because $\sqrt{x} \cdot \sqrt{x} = x$.
2) the conjugate of $\sqrt{x} - 2$ is $\sqrt{x} + 2$ because $(\sqrt{x} - 2)(\sqrt{x} + 2) = x - 4$.
3) the conjugate of $\sqrt[3]{x} - 4$ is $[(\sqrt[3]{x})^2 + 4\sqrt[3]{x} + 4^2]$ because

$$(\sqrt[3]{x} - 4)[(\sqrt[3]{x})^2 + 4\sqrt[3]{x} + 4^2] = (\sqrt[3]{x})^3 - 4^3 = x - 64.$$

Example 4.1-2

$$\lim_{x \to 4} \frac{\sqrt{x} - 2}{x - 4} = \frac{\sqrt{4} - 2}{4 - 4} = \frac{0}{0} \text{ indeterminate form.}$$

$$= \lim_{x \to 4} \frac{\sqrt{x} - 2}{x - 4} \times \frac{\sqrt{x} + 2}{\sqrt{x} + 2}$$

$$= \lim_{x \to 4} \frac{x - 4}{(x - 4)(\sqrt{x} + 2)}$$

$$= \lim_{x \to 4} \frac{1}{\sqrt{x} + 2}$$

$$= \frac{1}{\sqrt{4} + 2}$$

$$= \frac{1}{4} \in \Re. \text{ The limit exists.}$$

Section 4.2

Limit of a Function at ∞:

" x gets infinity large " translates mathematically as $\lim_{x \to \infty}$ (limit at infinity).

Practice :

Given the following functions :

$f(x) = 2x$, $g(x) = x^2$, $h(x) = \sqrt{x}$ and $p(x) = \frac{1}{x}$.

1) Look at the following table :

x	1	10	10^3	10^{10}	10^{20}	10^{30}	10^{40}
f(x)	2	20	2000	$2.\ 10^{10}$	$2.\ 10^{20}$	$2.\ 10^{30}$	$2.\ 10^{40}$
g(x)	1	100	10^6	10^{20}	10^{40}	10^{60}	10^{80}
h(x)	1	$\sqrt{10}$	$10\sqrt{10}$	100000	10^{10}	10^{15}	10^{20}
p(x)	1	0.1	0.001	10^{-10}	10^{-20}	10^{-30}	10^{-40}

1) What do you recognize?
2) Can you pretend the maximum value of the functions f(x), g(x), and h(x) for infinitely many x?
3) Can you pretend the maximum value of the function p(x) for infinitely many x?

Definition 4.2-1

Let f be a function defined on $[a, +\infty)$, where a is a given real number. We say that f tends to $+\infty$ as x tends to $+\infty$ if for every positive real number k, we can have $f(x) > k$ for a big value of x. We write $\lim_{x \to +\infty} f(x) = +\infty$.

*Graphically, this means that the representative curve of f exceeds all straight lines of equation y = k, for every given value of $k > 0$.
(Figure 4.2- 1)

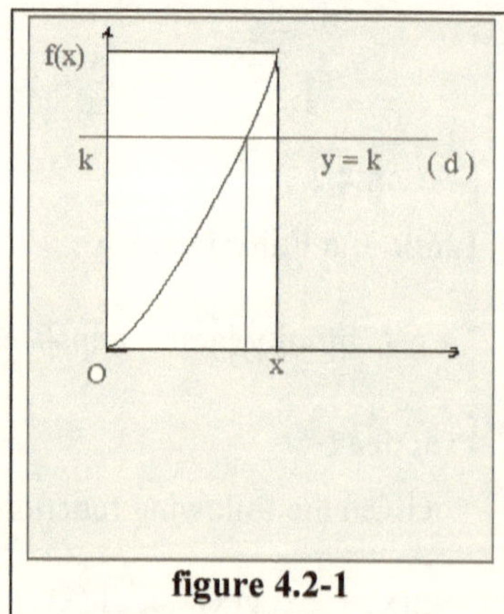

figure 4.2-1

Property 4.2-1

We have : $\lim\limits_{x \to +\infty} x^2 = +\infty$ and $\lim\limits_{x \to +\infty} \sqrt{x} = +\infty$

Definition 4.2-2

We say that f tends to $-\infty$ as x tends to $+\infty$ if the limit of $-f$ as x tends to $+\infty$ is equal to $+\infty$: $\lim\limits_{x \to +\infty} f(x) = -\infty$ if and only if $\lim\limits_{x \to +\infty} (-f(x)) = +\infty$.

Example 4.2-1

The representative curve of $f(x) = -x^2$ is the symmetric of $f(x) = x^2$ with respect the x-axis.

Property 4.2-2

We have $\lim\limits_{x \to +\infty} (-x^2) = -\infty$ and $\lim\limits_{x \to +\infty} \left(-\sqrt{x}\right) = -\infty$.

Definition 4.2-3

We say that f tends to 0 as x tends to $+\infty$ if f(x) comes very close to zero for a big value of x. We write: $\lim\limits_{x \to +\infty} f(x) = 0$.

Property 4.2-3

We have $\lim\limits_{x \to +\infty} (\frac{1}{x}) = 0$ and $\lim\limits_{x \to +\infty} \left(\frac{1}{\sqrt{x}}\right) = 0$.

Definition 4.2- 4

We say that f has a limit L (real number) at $+\infty$ if: $\lim\limits_{x \to +\infty} (f(x)-L) = 0$. We write $\lim\limits_{x \to +\infty} f(x) = L$.

Example 4.2-2

Let $f(x) = \frac{2x+1}{x}$. We have $\lim_{x \to +\infty} f(x) = 2$ because $\lim_{x \to +\infty} (f(x) - 2) = \lim_{x \to +\infty} \frac{1}{x} = 0$.

Remark 4.2-1

* If $\lim_{x \to +\infty} f(x) = L$, then this limit is unique.

* If f is monotone in the neighborhood of $+\infty$, and if $\lim_{x \to +\infty} f(x) = L$, then its representative curve (C) takes one of the two cases in figure 4.2-2 :

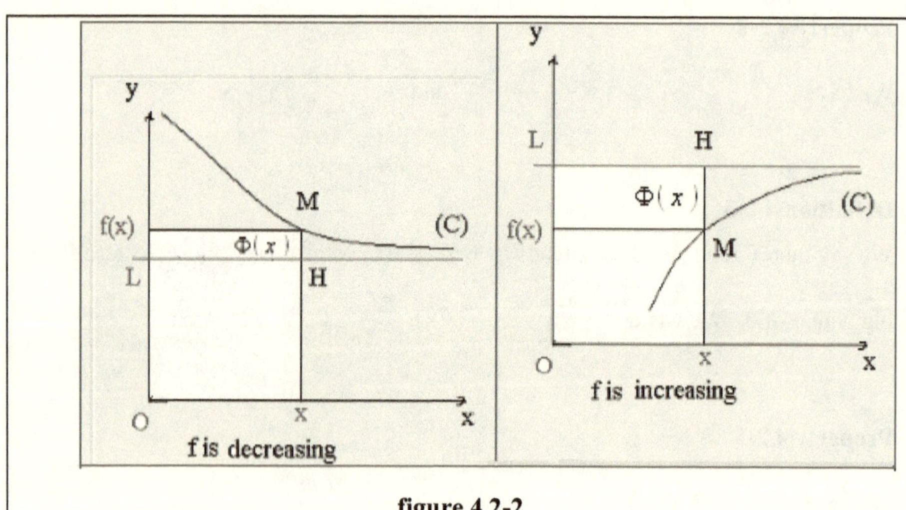

figure 4.2-2

$MH = |f(x) - L| = |\Phi(x)| \to 0$ as $x \to +\infty$.

Definition 4.2-5

If $\lim_{x \to \pm\infty} f(x) = L$ then the straight line (d) of equation $y = L$ is a horizontal asymptote.

Example 4.2-3

$\lim_{x \to +\infty} \frac{1}{x} = 0$ then the straight line y = 0 (x-axis) is a horizontal asymptote to the function $f(x) = \frac{1}{x}$.

* Let f be a function defined on $(-\infty, a)$, where a is a given real number.

Definition 4.2-6

$$\lim_{x \to -\infty} f(x) = \lim_{x \to +\infty} f(-x)$$

Example 4.2-4

If $f(x) = x^2$, then

$$\lim_{x \to -\infty} f(x) = \lim_{x \to +\infty} (-x)^2 = \lim_{x \to +\infty} x^2$$

$= +\infty$. (Figure 4.2-3)

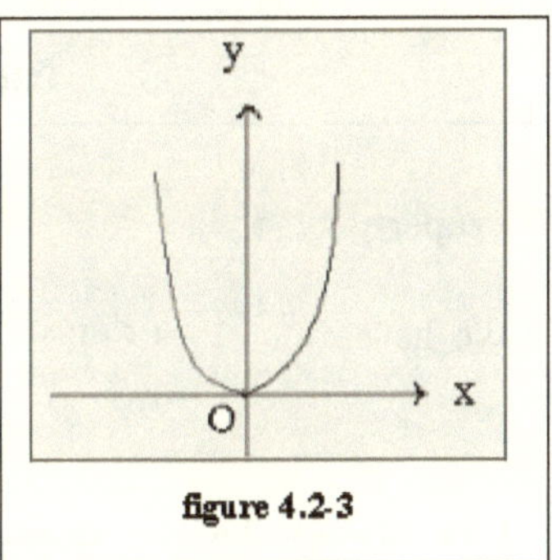

figure 4.2-3

Definition 4.2-7

We say that f tends to $\pm\infty$ as x tends to a if for every positive real number k, we can have $f(x) > k$ for x is on the neighborhood of a. We write :

$\lim_{x \to a} f(x) = \pm\infty$, and y = a is a vertical asymptote.

Example 4.2-5

In figure 4.2-4, the straight line (d) is a vertical asymptote.

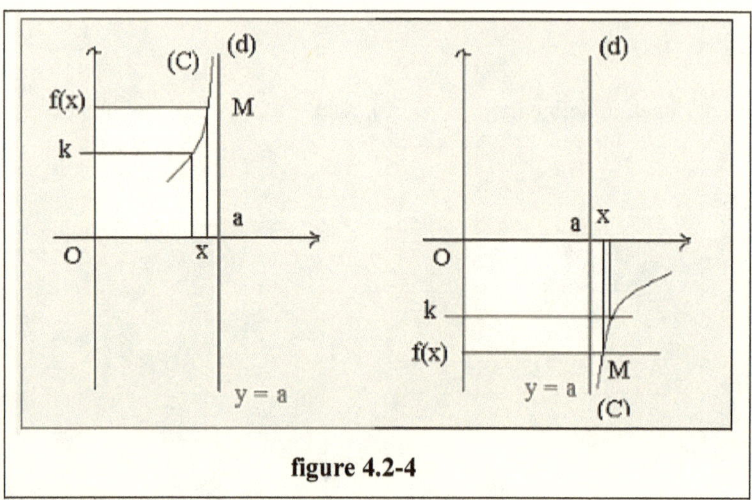

figure 4.2-4

Property 4.2-4

We have $\lim_{x \to 0^+} \left(\frac{1}{x}\right) = +\infty$ and $\lim_{x \to 0^+} \left(\frac{1}{\sqrt{x}}\right) = +\infty$.

Conventions : look at Appendix -A to see the behavior of indeterminate forms.

Section 4.3

Operations on the Limits

* $\lim_{x \to a} [\ f(x) + g(x)\ ] = \lim_{x \to a} f(x) + \lim_{x \to a} g(x).$

* $\lim_{x \to a} [\ f(x) \times g(x)\ ] = \lim_{x \to a} f(x) \times \lim_{x \to a} g(x).$

* $\lim_{x \to a} \frac{f(x)}{g(x)} = \frac{\lim_{x \to a} f(x)}{\lim_{x \to a} g(x)}.$

* $\lim_{x \to a} \sqrt{f(x)} = \sqrt{\lim_{x \to a} f(x)}$, where $f(x) \geq 0$.

* $\lim_{x \to a} [\, f(x) + c \,] = \lim_{x \to a} f(x) + c$ where c is constant.

* $\lim_{x \to a} [\, c\, f(x) \,] = c \lim_{x \to a} f(x)$.

* $\lim_{x \to a} \dfrac{c}{f(x)} = \dfrac{c}{\lim_{x \to a} f(x)}$.

Evaluating Limits at ∞

To evaluate a limit at infinity, compare the degrees in the numerator and denominator of the fraction. If the fraction of two polynomials has an indeterminate form ($\frac{\infty}{\infty}, \frac{0}{0}$), then the limit of f(x) is equal to the limit of highest power of both numerator and denominator polynomials.

Property 4.3-1

Consider the polynomials:

$p(x) = a_n x^n + a_{n-1} x^{n-1} + \dots + a_0$; $n \in N$ and $a_n \neq 0$.

$q(x) = b_m x^m + b_{m-1} x^{m-1} + \dots + b_0$; $m \in N$ and $b_m \neq 0$.

As x tends to $+\infty$ or $-\infty$, we have :

* $\lim_{x \to a} p(x) = \lim_{x \to a} (a_n x^n)$

* $\lim_{x \to a} q(x) = \lim_{x \to a} (b_m x^m)$

* $\lim_{x \to a} \dfrac{p(x)}{q(x)} = \lim_{x \to a} \dfrac{a_n x^n}{b_m x^m}$

Proof

$$\lim_{x \to a} p(x) = \lim_{x \to a} a_n x^n + a_{n-1} x^{n-1} + \dots + a_0$$

$$= \lim_{x \to a} (a_n x^n)\left(1 + \frac{a_{n-1}}{a_n}\frac{1}{x} + \dots + \frac{a_0}{a_n}\frac{1}{x_n}\right)$$

$$= \lim_{x \to a} (a_n x^n) \times \lim_{x \to a} \left(1 + \frac{a_{n-1}}{a_n} \frac{1}{x} + \dots + \frac{a_0}{a_n} \frac{1}{x_n} \right)$$

$$= \lim_{x \to a} (a_n x^n) \times 1$$

$$= \lim_{x \to a} (a_n x^n) .$$

Example 4.3-1

a. $\lim_{x \to -\infty} \frac{2x^2 - 3x + 1}{-x + 2} = \lim_{x \to -\infty} \frac{2x^2}{-x} = \lim_{x \to -\infty} (-2x) = +\infty .$

b. $\lim_{x \to +\infty} \frac{3x^2 - 2}{x^2 + x} = \lim_{x \to +\infty} \frac{3x^2}{x^2} = 3 .$

c. $\lim_{x \to -\infty} \frac{x - 1}{x^2 - 2x} = \lim_{x \to -\infty} \frac{x}{x^2} = \lim_{x \to -\infty} \frac{1}{x} = 0.$

d. $\lim_{x \to 0} \frac{1}{x^n} = +\infty$ if n is even .

e. $\lim_{x \to 0^+} \frac{1}{x^n} = +\infty$ and $\lim_{x \to 0^-} \frac{1}{x^n} = -\infty$ if n is odd .

f. $\lim_{x \to 2} \frac{x^2 - 4}{x - 2} = \lim_{x \to 2} \frac{(x-2)(x+2)}{x - 2} = \lim_{x \to 2} (x + 2) = 2 + 2 = 4.$

Warning

Saying that a limit equals infinity is equivalent to saying that the limit does not exist, because a limit must be a real number. So it is better to say that the limit increases infinitely rather than saying that the limit equals infinity.

Section 4.4

Limits and Inequalities

Let f, g, u be functions defined on the intervals of the form $[a, +\infty)$ or $(-\infty, a]$.

Property 4.4-1 (Sandwich Rule)

* If $f \le u \le g$ and if $\lim_{x \to a} f(x) = \lim_{x \to a} g(x) = L$, then

$\lim_{x \to a} u(x) = L$.

where L is a real number or one of $-\infty$ or $+\infty$.

Example 4.4-1

Show that $\lim_{x \to +\infty} \frac{\sin x}{x+1} = 0$.

Solution

For every $x > 0$, we have:

$$\frac{-1}{x+1} \le \frac{\sin x}{x+1} \le \frac{1}{x+1} \quad \text{and} \quad \lim_{x \to +\infty} \frac{-1}{x+1} \le \lim_{x \to +\infty} \frac{\sin x}{x+1} \le \lim_{x \to +\infty} \frac{1}{x+1}$$

then $0 \le \lim_{x \to +\infty} \frac{\sin x}{x+1} \le 0$ and then $\lim_{x \to +\infty} \frac{\sin x}{x+1} = 0$.

Property 4.4-2

f is a function defined over $\Re^*$.

1) $\lim_{x \to 0} \frac{\sin x}{x} = 1$.

2) $\lim_{x \to 0} \frac{x}{\sin x} = 1$.

3) $\lim_{x \to 0} \frac{\tan x}{x} = 1$.

4) $\lim_{x \to 0} \frac{x}{\tan x} = 1$.

Section 4.5

Continuity of Functions

Definition 4.5-1

Let f be a function defined on an interval I .

* we say that f is continuous at a point of abscissa a in I.

if $\lim_{x \to a} f(x) = f(a)$.

* We say that f is continuous on the interval I if f is continuous at every point of I.

Remark 4.5.1

1) Continuous functions are predictable:

* No breaks in the graph
* No holes
* No jumps

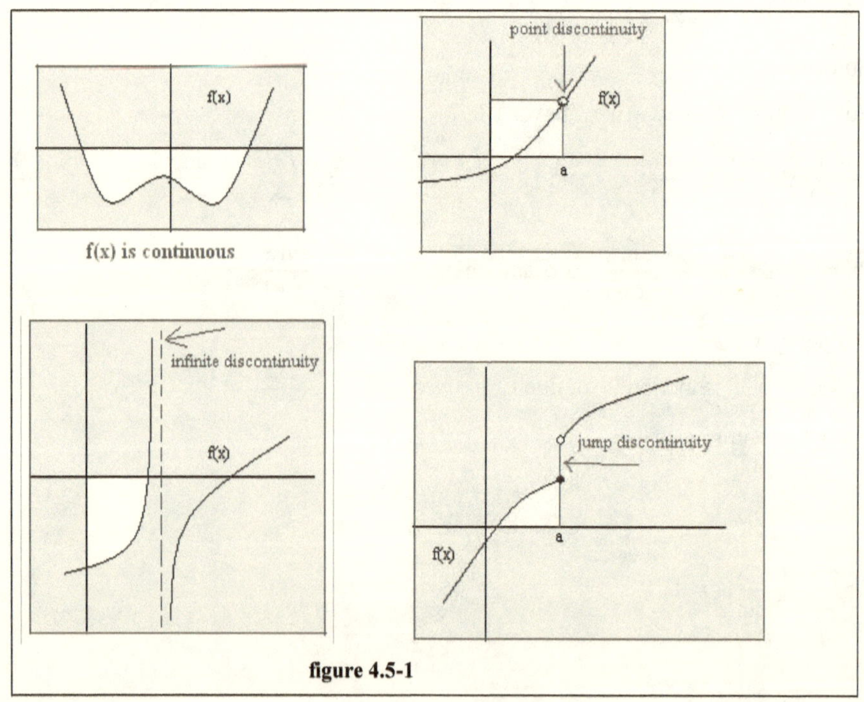

figure 4.5-1

Remark 4.5-2

If f is not continuous at a, then we say that f is discontinuous at this point. Graphically, the continuity of f means that the representative curve (C) is drawn by a continuous line (neither interruption nor jumping) figure 4.5-2 .

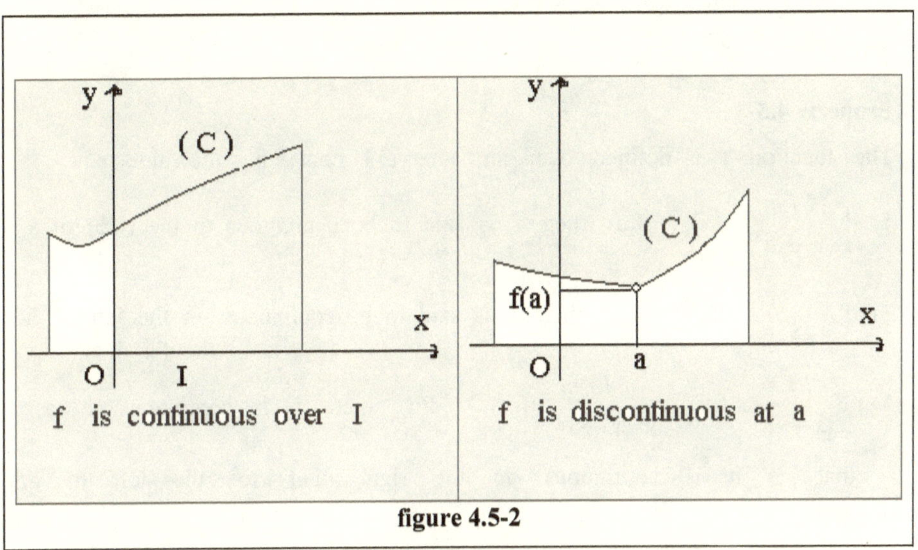

figure 4.5-2

Continuous functions can be drawn with a single, unbroken pencil stroke.

Predictable Continuous Functions

What is that mean ?

It means that :

1. No break in the graph, i.e. a limit must exist at every x-value or else the graph will break.
2. No holes in the graph, i.e. the function cannot have undefined points or vertical asymptotes.

Mathematically speaking , if f is continuous then for every x = c in the function

$\lim_{x \to c} f(x) = f(c)$. The function exists at the height indicated by the function.

If you can evaluate any limit on the function using only the substitution method, then the function is continuous.

Property 4.5-1

Every polynomial function is continuous over $\Re$.

Property 4.5-2

The following functions are continuous over their domain of definitions :

$f(x) = \frac{1}{x}$ and $f(x) = \sqrt{x}$.

Property 4.5-3

The function f is defined over an interval I containing the value a.

1) if $\lim_{x \to a^+} f(x) = f(a)$, then f is said to be continuous to the right of a.

2) if $\lim_{x \to a^-} f(x) = f(a)$, then f is said to be continuous to the left of a.

3)) if $\lim_{x \to a^+} f(x) = \lim_{x \to a^-} f(x) = f(a)$, then f is continuous at a,

that is it is continuous to the right and to the left of a.

Remark 4.5-3

If at least one of the above conditions is not satisfied, then f is said to be discontinuous at a.

Example 4.5-1

Study the continuity of $f(x) = |x|$ at the point of abscissa 0.

Solution

f(x) is defined for every $x \in \Re$. $f(x) = \begin{cases} +x & if\ x \geq 0 \\ -x & if\ x \leq 0 \end{cases}$. $f(0) = |0| = 0$.

$\lim_{x \to 0^+} f(x) = \lim_{x \to 0^+} x = 0 = f(0)$, then f is continuous to the right of 0.

$\lim_{x \to 0^-} f(x) = \lim_{x \to 0^-} -x = 0 = f(0)$, then f is continuous to the left of 0.

Since f is continuous to the right and to the left of 0, then f is continuous at 0.

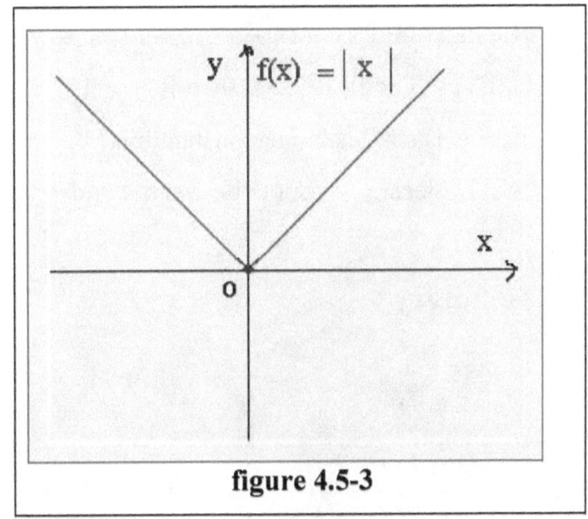

figure 4.5-3

Remark 4.5-4

Figure 4.5-3 describes the results of the above example.

Remark 4.5-5

If a function has as a limit ∞ or more than one limit, then we say that this function has no limit .

Property 4.5-4(The Intermediate Value Theorem)

Let f be a continuous function on [a , b]. For every h between f(a) and f(b), there exists at least one real root e for f(x) = 0 between a and b so that f(e) = h . (figure 4.5-4).

If h is between f(a) and f(b) , then a corresponding e, between a and b, exists so that f(e) = h.

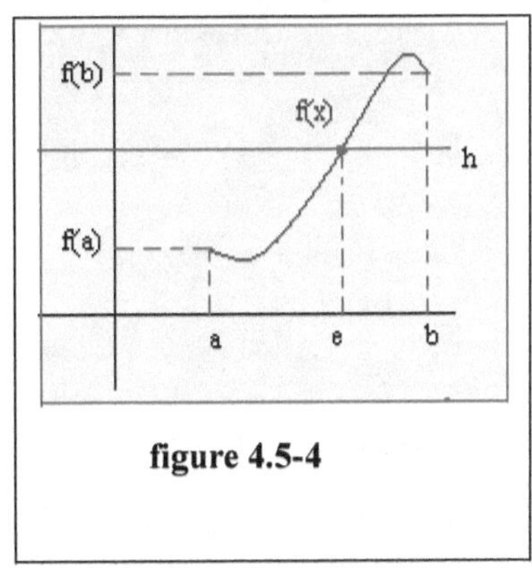

figure 4.5-4

Example 4.5-2

Prove that there exists at least one real root for the equation $2x^3-3x^2+5x-1 = 0$ between 0 and 1.

Proof

The function $f(x) = 2x^3-3x^2+ 5x-1$ is a polynomial function which is continuous on $\Re$ .f(1) = 3 and f(0) = -1, then $f(1) \neq f(0)$. Moreover, $-1 = f(0) < 0 < f(1) = 3$. Hence, there exists at least one constant $e \in (0, 1)$ so that f(e) = 0 which means that there exists at least one root e between 0 and 1.

Summary

- $\lim_{x \to a^+} f(x) = \lim_{x \to a^-} f(x) = L$, then L is the limit of the function f at a
- $\lim_{x \to a} f(x) = L$. The point (a , L) is a limit point for f(x)
- $\lim_{x \to +\infty} x^2 = +\infty$ and $\lim_{x \to +\infty} \sqrt{x} = +\infty$
- $\lim_{x \to +\infty} f(x) = -\infty$ if and only if $\lim_{x \to +\infty} (-f(x)) = +\infty$
- $\lim_{x \to +\infty} (-x^2) = -\infty$ and $\lim_{x \to +\infty} \left(-\sqrt{x}\right) = -\infty$
- $\lim_{x \to +\infty} (\frac{1}{x}) = 0$ and $\lim_{x \to +\infty} \left(\frac{1}{\sqrt{x}}\right) = 0$
- $\lim_{x \to +\infty} (f(x)-L) = 0$ if and only if $\lim_{x \to +\infty} f(x) = L$
- $\lim_{x \to \pm\infty} f(x) = L$, then the straight line (L) of equation y = L is a horizontal asymptote
- $\lim_{x \to a} f(x) = \pm\infty$, then the straight line of equation x = a is a vertical asymptote
- $\lim_{x \to 0^+} (\frac{1}{x}) = +\infty$ and $\lim_{x \to 0^+} \left(\frac{1}{\sqrt{x}}\right) = +\infty$
- Consider the polynomials :

  $p(x) = a_n x^n + a_{n-1} x^{n-1} + ... + a_0$; $n \in N$ and $a_n \neq 0$

  $q(x) = b_m x^m + b_{m-1} x^{m-1} + ... + b_0$; $m \in N$ and $b_m \neq 0$

  As x tends to $+\infty$ or $-\infty$, we have :

* $\lim_{x \to \pm\infty} p(x) = \lim_{x \to \pm\infty} (a_n x^n)$
* $\lim_{x \to \pm\infty} q(x) = \lim_{x \to \pm\infty} (b_m x^m)$
* $\lim_{x \to \pm\infty} \frac{p(x)}{q(x)} = \lim_{x \to \pm\infty} \frac{a_n x^n}{b_m x^m}$

☒ If $f \leq u \leq g$ and if $\lim_{x \to \pm\infty} f(x) = \lim_{x \to \pm\infty} g(x) = L$, then $\lim_{x \to \pm\infty} u(x) = L$

☒ $\lim_{x \to 0} \frac{\sin x}{x} = 1$

☒ $\lim_{x \to 0} \frac{x}{\sin x} = 1$

☒ $\lim_{x \to 0} \frac{\tan x}{x} = 1$

☒ $\lim_{x \to 0} \frac{x}{\tan x} = 1$

☒ f is continuous at a point of abscissa a if $\lim_{x \to a} f(x) = f(a)$

☒ f is continuous on the interval I if f is continuous at every point of I

☒ Every polynomial function is continuous over $\Re$

☒ if $\lim_{x \to a^+} f(x) = f(a)$, then f is said to be continuous to the right of a

☒ if $\lim_{x \to a^-} f(x) = f(a)$, then f is said to be continuous to the left of a

☒ if $\lim_{x \to a^+} f(x) = \lim_{x \to a^-} f(x) = f(a)$, then f is continuous at

☒ **The Intermediate Value Theorem :** Let f be a continuous function on [a , b]. For every h between f(a) and f(b), there exists at least one real root e for f(x) = 0 between a and b so that f(e) = h

EXERCISES

In problems 4.1 – 4.31, calculate the given limits when possible :

4.1. $\lim_{x \to +\infty} -x^4 - 2x^2 - x - 1$

4.2 . $\lim_{x \to -\infty} \frac{-x^5 + 2x}{x^5 - 2x + 1}$

4.3 . $\lim_{x \to 1} \frac{2x^3 - x^2 - 1}{x^2 + x - 2}$

4.4 . $\lim_{x \to -\infty} \left(\frac{x^4}{x^3 + 2} - x \right)$

4.5 . $\lim_{x \to +\infty} \left(\sqrt{x^2 + 1} - x \right)$

4.6. $\lim_{x \to 0} \frac{\sin 2x}{x}$

4.7 . $\lim_{x \to 0} \frac{\sqrt{4+x} - 2}{\sqrt{1+x} - 1}$

4.8 . $\lim_{x \to +\infty} \frac{x - 1}{\sqrt{x^2 + 4x - 5}}$

4.9 . $\lim_{x \to 0} \frac{\sin x}{\sqrt{x}}$

4.10 . $\lim\limits_{x \to +\infty} (\sqrt{3x^2 - 2x + 5} - 3x)$

4.11 . $\lim\limits_{x \to \pm\infty} \frac{|x - 4|}{x+2}$

4.12 . $\lim\limits_{x \to +\infty} \frac{3x - 2\tan 3x}{5x + \sin 5x}$

4.13 . $\lim\limits_{x \to +\infty} -3x^2 + 5x^3 - x$

4.14 . $\lim\limits_{x \to +\infty} \frac{1-x}{(1+x)^2}$

4.15 . $\lim\limits_{x \to -3} \frac{2x^2 + 12x + 18}{2x^2 + 8x + 6}$

4.16 . $\lim\limits_{x \to 2} \frac{3x^2 + 3x - 12}{3x^2 + 3x - 18}$

4.17 . $\lim\limits_{x \to 1} \frac{x^3 - x^2 + x - 1}{x - 1}$

4.18 . $\lim\limits_{x \to +\infty} \sqrt{\frac{3x+1}{x-2}}$

4.19 . $\lim\limits_{x \to -2} \frac{x^2 + 4x + 4}{x^3 + 8}$

4.20 . $\lim\limits_{x \to 3} \frac{\sqrt{x+1} - 2\sqrt{x-2}}{x-3}$

4.21 . $\lim_{x \to -3} \dfrac{|x+3|-2x}{4x-6-|x+3|}$

4.22 . $\lim_{x \to 1} \dfrac{\sqrt{3+x} - \sqrt{5-x}}{\sqrt{2x+7} - \sqrt{10-x}}$

4.23 . $\lim_{x \to 0} \dfrac{\sin x}{\sqrt{1+x}-1}$

4.24 . $\lim_{x \to +\infty} (\sqrt{x^2+2x-1}-(x+1))$

4.25 . $\lim_{x \to 1} \dfrac{\sqrt[3]{x}-1}{\sqrt{x}-1}$

4.26 . $\lim_{x \to -1} \dfrac{|2x^2-8x-10|}{2x+2}$

4.27 . $\lim_{x \to 0} \dfrac{3\sin 2x}{\sqrt{3x+1}\ -1}$

4.28 . $\lim_{x \to 0} \dfrac{\sin 5x}{\tan 3x}$

4.29 . $\lim_{x \to 0} \dfrac{\sin x - \tan x}{x^3}$

4.30 . $\lim_{x \to +\infty} (\sqrt{x^2+x+1}-x-2)$

4.31 . $\lim_{x \to \pm\infty} \left(\sqrt[3]{x^3+1} - x\right)$

4.32. Let f be a function defined on $\Re$ by :

$$\begin{cases} f(x)=x^2 & if \quad x \leq 0 \\ f(x)=x & if \quad x \succ 0 \end{cases}$$

Is f continuous at 0 ?

4.33. Let f be a function defined on $\Re^*$ by f(x) $= \frac{|x|}{x}$. Study the continuity of f at the point of abscissa 0.

4.34.

1- Show that for every x > 0 , $\frac{-1}{x} \leq \frac{\sin x}{x} \leq \frac{1}{x}$.

2- Deduce the $\lim_{x \to +\infty} \frac{\sin x}{x}$.

3- Calculate $\lim_{x \to +\infty} \frac{x + \sin x}{x}$.

4.35. Let f be a function defined by :

$$f(x) = \begin{cases} \frac{x^2 - 5x + 6}{x - 3} & when\ x \neq 3 \\ 1 & when\ x = 3 \end{cases}$$

Study the continuity of f at x = 3 .

In problems 4.36 – 4.37, are the given functions continuous at x = 0? Justify.

4.36 . $g(x) = \begin{cases} \frac{1-\tan x}{x} & , \ x \neq 0 \\ 0 & , \ x = 0 \end{cases}$

4.37. f(x) = $\begin{cases} \dfrac{\sin x}{x} & , x \neq 0 \\ 1 & , x = 0 \end{cases}$

4.38. Consider the function f(x) = $\dfrac{2x+7}{x+3}$.

a) Use long division to show that f(x) = $a + \dfrac{b}{x+3}$, where a and b are two real numbers to be determined.

b) Verify that, for every x > 0 : $0 < \dfrac{1}{x+3} < \dfrac{1}{x}$ and deduce $\lim\limits_{x \to +\infty} \dfrac{1}{x+3}$.

c) Calculate $\lim\limits_{x \to +\infty} f(x)$.

4.39. Consider the function f(x) = $\dfrac{\sqrt{1+x} - 1}{x}$.

a) Show that f(x) has an indeterminate form as $x \to +\infty$.

b) Verify that f(x) = $\dfrac{1}{\sqrt{1+x}+1}$ and calculate $\lim\limits_{x \to +\infty} f(x)$.

4.40. a) Show that for x > 1 we have : $1 - \dfrac{1}{x} < \dfrac{\sqrt{x^2-1}}{x} < 1$.

b) Deduce $\lim\limits_{x \to +\infty} \dfrac{\sqrt{x^2-1}}{x}$.

c) Calculate $\lim\limits_{x \to -\infty} \dfrac{\sqrt{x^2-1}}{x}$.

4.41. Consider the function f(x) = $\dfrac{x-1}{\sqrt{1-x^2}}$.

a) Determine the domain of definition of f.

b) Calculate $\lim\limits_{x \to 1^-} f(x)$ and $\lim\limits_{x \to 1^+} f(x)$.

4.42. Consider the function $f(x) = \dfrac{|x|\sqrt{x+1}}{x}$.

a) Calculate $\lim\limits_{x \to 0^+} f(x)$ and $\lim\limits_{x \to 0^-} f(x)$.

b) Does f has a limit at 0?

In problems 4.43 – 4.45, show that the given straight lines are asymptotes to their functions :

4.43. $f(x) = \dfrac{1}{x^3}$, $y = 0$ (x-axis)

4.44. $g(x) = \dfrac{1}{x-1}$, $x = 1$

4.45. $p(x) = \dfrac{2x+1}{x+2}$, $x = -2$ and $y = 2$

4.46. Consider the function $f(x) = \dfrac{6x-1}{2x+4}$.

a) Determine the limit of f at $+\infty$ *and at* $-\infty$. Then interpret this result graphically.

b) Calculate $\lim\limits_{x \to -2^+} f(x)$ and $\lim\limits_{x \to -2^-} f(x)$. Then interpret this result graphically .

4.47. Consider the function f defined over $\Re$ by:

$$f(x) = \begin{cases} \dfrac{|x-2|}{x-2} & \text{if} \quad x \neq 2 \\ -1 & \text{if} \quad x = 2 \end{cases}$$

Study the continuity of f at 2.

4.48. Consider the function g defined by :

$$g(x) = \begin{cases} \dfrac{x^2+x^3}{x^2-x^3} & if \quad x \neq 0 \\ 2 & if \quad x = 0 \end{cases}$$

Is g continuous at 0? Justify .

4.49. Consider the function g defined by :

$$g(x) = \begin{cases} x^2 - \dfrac{\sqrt{(x-1)^2}}{x-1} & if \quad x \neq 1 \\ 0 & if \quad x = 1 \end{cases}$$

Study the continuity of g over $\Re$.

4.50. Consider the function f defined by:

$$f(x) = \begin{cases} \dfrac{x^3-8}{x-2} & if \quad x \neq 2 \\ a & if \quad x = 2 \end{cases}$$

Determine a so that f is continuous at the point x = 2 .

4.51. Consider the function g defined by :

$$g(x) = \begin{cases} \dfrac{\sqrt{1+x}\,-\,2}{x-3} & if \quad x \succ 3 \\ \dfrac{1}{4}x - \dfrac{1}{2} & if \quad -1 \leq x \leq 3 \end{cases}$$

Show that g is continuous over [-1 , + ∞ [.

4.52. Consider the function f defined by :

$$f(x) = \begin{cases} \dfrac{x-1-|x-1|}{x-1} & \textit{for} \quad x \neq 1 \\ 2 & \textit{for} \quad x = 1 \end{cases}$$

Study the continuity of f at 1 .

4.53. Let h be a function defined by :

$$h(x) = \begin{cases} 3x^2 + ax + 1 & \textit{if} \quad x \prec 1 \\ \dfrac{3x-1}{x+2} & \textit{if} \quad x \geq 1 \end{cases}$$

Find a so that h is continuous over $\Re$.

4.54. For each of the following functions defined by their representative

curve state , if there exist , the discontinuous points.

a)

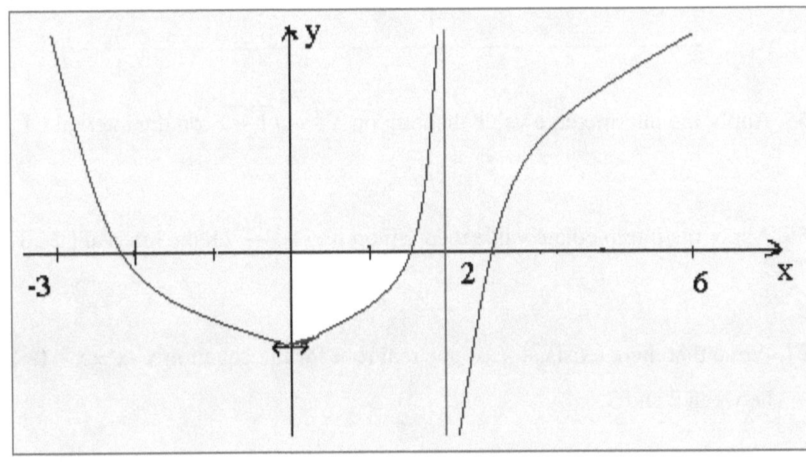

b)

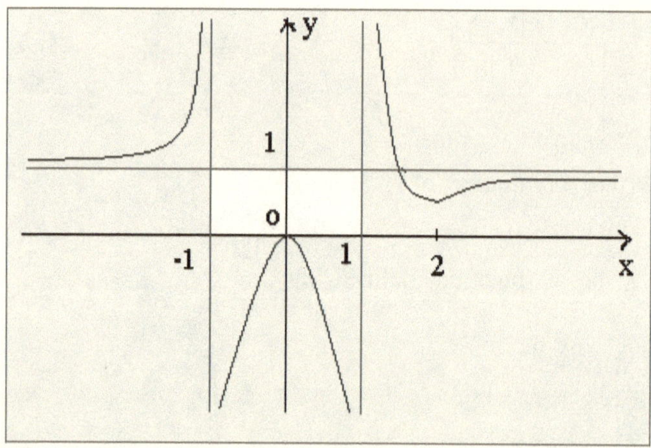

c)

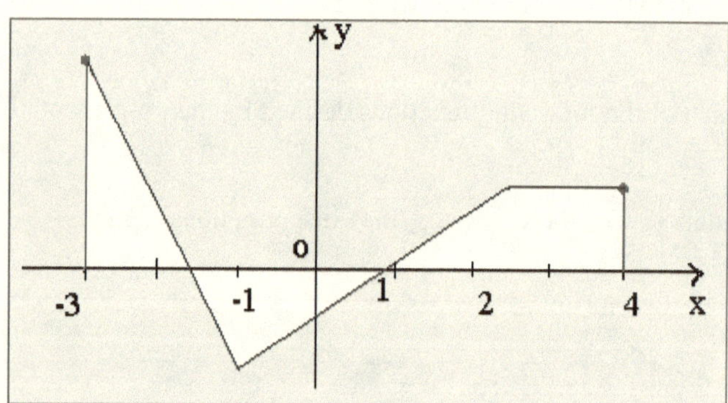

4.55. Apply the intermediate value theorem on $x^2 - \sqrt{1+x}$ on the interval [1 , 2].

4.56. Apply the intermediate value theorem on $f(x) = \frac{1}{x^2}$ on the interval [2 , 3].

4.57. Prove that there exists at least one real root for the equation $x^3 - x^2 + x = 0$ between 2 and 3.

4.58. Prove that there exists at least one real root for the equation $\cos x - x = 0$ on the interval (0 , 1).

4.59. The total cost of a manufacturer in producing x units per week is determined by $C(x) = 1000 + 10x$ for $x \geq 0$.

a. Find the average cost $\overline{C}(x)$ per unit.

b. Find the limit of $\overline{C}(x)$ as x approaches $+\infty$.

c. Give the horizontal asymptote for $\overline{C}$.

d. Find the vertical asymptote for $\overline{C}$.

e. For which values of $x > 0$ is the average cost per unit decreasing?

f. Sketch the graph of $\overline{C}$.

4.60. The average cost $\overline{C}(q)$ per unit a manufacturer in producing q units per week is determined by

$$\overline{C}(q) = \frac{200q + 20}{q^2} \quad \text{for } q \geq 0.$$

a. Find the total cost function C (q) per unit.

b. Find the limit of C (q) as q approaches $+\infty$.

c. Give the horizontal asymptote for C.

d. Find the vertical asymptote for C.

e. For which values of $q > 0$ is the cost per unit decreasing?

f. Sketch the graph of C.

4.61. A company produces a liquid detergent, its total cost is given by the following function $C(x) = 0.5x + \frac{8}{x}$ where $x > 0$. C (x) is expressed in thousands of dollars and x is expressed in hectoliters. Let f be the representative curve of C in an orthonormal system of axes x' o x , y' o y.

a. Determine the limit of C (x) as x tends to 0 and deduce an asymptote to f.

b. Determine the limit of C (x) as x tends to ∞ and show that the straight line (d) of equation $y = 0.5x$ is an asymptote to f.

c. Trace f over $]0, \infty)$.

d. The company sells all its production x, the revenue is given by $R(x) = -0.8x + 13$. Designate by g the graphical representation of R.

   1) Trace g in the same system as that of f.

   2) Determine the interval of production for the company to experience profit.

4.62. The total cost of production of x articles manufactured by a company is given by $C(x) = 0.3x^2 + x + 60$ where $x > 0$. The average unit cost of fabrication of an article is defined over $]0, \infty)$ by $\overline{C}(x) = \frac{C(x)}{x}$. Let g be the representative curve of $\overline{C}$ in an orthonormal system of axes x' o x , y' o y.

c. Determine the limit of $\overline{C}(x)$ as x tends to 0 and deduce an asymptote to g.

d. 1) Determine the limit of $[\overline{C}(x) - (0.3x + 1)]$ as x tends to ∞ and interpret this result graphically.

   2) Give an economical interpretation of this result specifying the variation of the unit average cost for big values of x.

Chapter 5 **Differentiation**

Introduction

The study of the notion of the limit has paved the way to rigorous calculations of velocities: namely the instantaneous velocity at a given time (instead of the average velocity between two given times) according to the laws of motion. Transforming the problem of discussing the variations of a function into a problem of determining the sign of its derived function has immensely simplified such a discussion.

Why ?

The scientist Sharafeddine Al Tousi has studied the equations of degree ≤ 3 in his book "قوام الحساب". He was treating the function without mentioning its name, but he came to another form of this concept which is known later on as derivative. To solve these equations, Al Tousi has studied the greatest value to algebraic statements and has taken the first derivative of these statements, without using its name, then eliminated and prove that the root of the obtained equation which has been substituted in the algebraic statement, gave the greatest value of the statement. Derivation has a major importance in the direction of the bombs, rockets, manufactured moons, astronomies and stars ....

Objectives

❖ After studying the material in this chapter, you should be able to:

- Define the derivative of a number and the derivative of a function.
- Find the average change of variation over an interval.
- Find the derivative of a function by using the definition.
- Examine the differentiability of the function.
- Identify the laws of derivation.
- Know the derivative of trigonometric functions.
- Practice geometric and physical applications.
- Know the law of the derivative of composite of functions.

Practice

A stone is thrown down from a building of 30 m height. We know, in physics, that the distance d held by the stone in terms of the time t is given by :

d (t) = $\frac{1}{2}9.81t^2$, where d is given in meters and t is given in seconds.

1) Calculate the time needed for the stone reaches the ground .

2) What is the average velocity of the stone:

   a) between the two times $t_1 = 1$ s and $t_2 = 2$ s?

   b) to cover all its target?

3) We are interested in velocity after one second of the beginning. So we take $t_1 = 1$ and $t_2 = 1 + h$. Show that the average velocity between t_1 and t_2 is $v(h) = 9.81\left(1+\frac{h}{2}\right)$.

4) Calculate lim v(h) as h tends to 0 (this limit is called the velocity at the instant t = 1).

5) Calculate the velocity at the instant t = 2 .

Section 5.1

What is the Derivative?

In calculus, the derivative is the slope of the tangent line to a graph of f(x), and is usually denoted by f ' (x).

In figure 5.1-1

* t_1 is the tangent to f(x) at the point of abscissa c_1
* t_2 is the tangent to f(x) at the point of abscissa c_2
* t_3 is the tangent to f(x) at the point of abscissa c_3

So as we see we have different tangent lines to f(x) at different values of c. Then, there are different derivatives at each point of c.

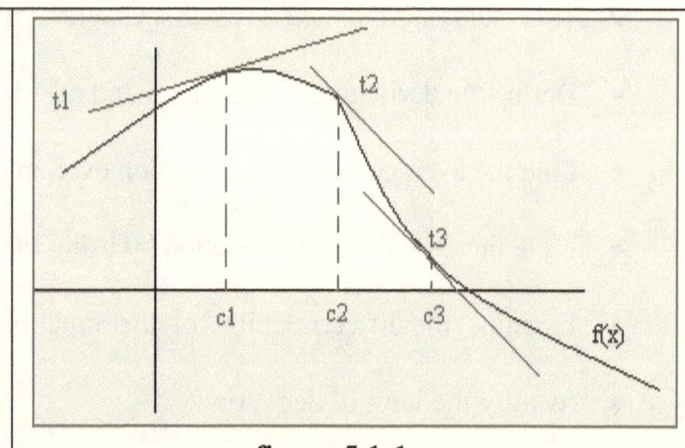

figure 5.1-1

The question now is how do we calculate the slope of a tangent line?

To calculate this we have to use the difference quotient.

Constructing the Difference Quotient

Here we need to find the slope of the tangent line (t) to f at x.

Construction

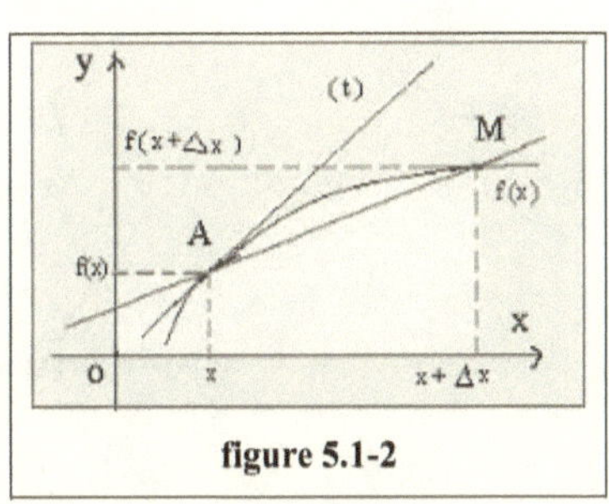

figure 5.1-2

* Move x to a small distant Δx to the right of x. So the new location is the point M (x +Δx , f (x + Δx)).(Figure5.1-2)

* The slope of the secant line (AM) through A (x , f (x)) and M is equal to $\frac{f(x+\Delta x)-f(x)}{x+\Delta x-x}=\frac{f(x+\Delta x)-f(x)}{\Delta x}$.

The fraction $\frac{f(x+\Delta x)-f(x)}{\Delta x}$ is part of the

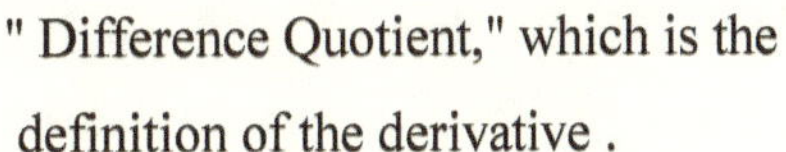

" Difference Quotient," which is the definition of the derivative .

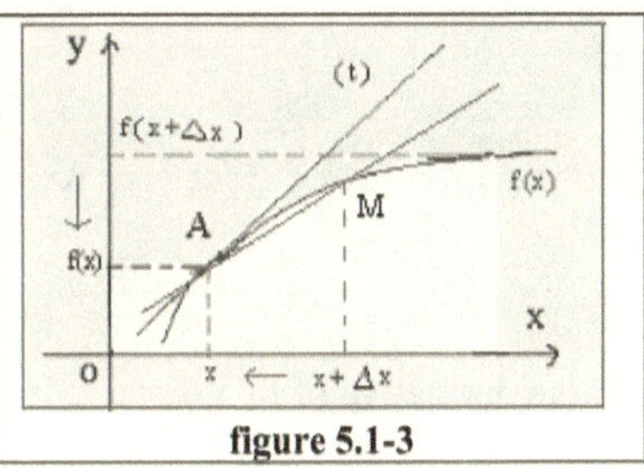

figure 5.1-3

Now, if M is moved close to A, the smaller the Δx , the more the secant and tangent lines resemble one another.(Figure 5.1- 3)

We want Δx to be as small as possible. $\frac{f(x+\Delta x)-f(x)}{\Delta x}$ is the seca

already found. So $\lim_{\Delta x \to 0} \frac{f(x+\Delta x)-f(x)}{\Delta x}$

is the difference quotient, the definition of the derivative (figure 5.1 – 4).

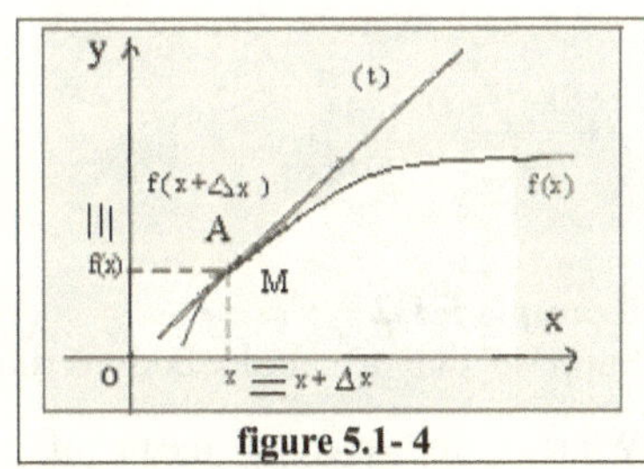

figure 5.1- 4

Example 5.1-1

Use the difference quotient to differentiate $f(x) = x^2$.

Solution

If $f(x) = x^2$, then $f'(x) = \lim_{\Delta x \to 0} \frac{f(x+\Delta x) - f(x)}{\Delta x}$

$$= \lim_{\Delta x \to 0} \frac{(x+\Delta x)^2 - x^2}{\Delta x}$$

$$= \lim_{\Delta x \to 0} \frac{x^2 + 2x\Delta x + (\Delta x)^2 - x^2}{\Delta x}$$

$$= \lim_{\Delta x \to 0} \frac{2x\Delta x + (\Delta x)^2}{\Delta x}$$

$$= \lim_{\Delta x \to 0} \frac{\Delta x\,(2x + \Delta x)}{\Delta x}$$

$$= \lim_{\Delta x \to 0} 2x + \Delta x$$

$$= \lim_{\Delta x \to 0} (2x + 0)$$

$$= 2x.$$

So the derivative of $f(x) = x^2$ is $f'(x) = 2x$.

Differentiability of functions

Definition 5.1-1

Let f be a function defined over the interval I and a and b are two elements of I. we call average change of variations of f between a and b the real number $\frac{f(b) - f(a)}{b-a}$.

Example 5.1-2

Suppose that a certain body is running on a straight line by the function $f(x) = x^2$. What is the average mean of velocity for a body in the interval 2 and $\frac{5}{2}$.

Solution

$f(2) = 2^2 = 4$

$f(\frac{5}{2}) = (\frac{5}{2})^2 = \frac{25}{4}$

Which means that the body cover the distance by $\frac{25}{4} - 4 = \frac{9}{4}$ km during the interval $\frac{5}{2} - 2 = \frac{1}{2}$ km. Therefore , the average mean of velocity $= \frac{\frac{9}{4}}{\frac{1}{2}} = \frac{9}{2}$ km /s.

Definition 5.1-2

Let f be a function defined over the interval I and a $\in$ I. we say that f is differentiable at x if $\lim_{h \to 0} \frac{f(x+h) - f(x)}{h} = l \in \Re$. This limit l is called the derivative number of f at x . It is denoted by f'(x).

Example 5.1-3

$f(x) = x^2$ is a differentiable function (has a derivative) at 1 because :

$$f'(1) = \lim_{h \to 0} \frac{f(x+h) - f(x)}{h} = \lim_{h \to 0} \frac{f(1+h) - f(1)}{h} = \lim_{h \to 0} \frac{(1+h)^2 - 1^2}{h} =$$

$$\lim_{h \to 0} 2 + h = 2 + 0 = 2 \in \Re .$$

Definition 5.1-3

Let f be a function defined over the interval I and a $\in$ I. we say that f is Differentiable at x_0 if $f'(x) = \lim_{x \to x_o} \frac{f(x) - f(x_o)}{x - x_o} = l \in \Re$.

Example 5.1-4

$f(x) = x^2$ is a differentiable function (has a derivative) at 1 because :

$$f'(1) = \lim_{x \to 1} \frac{f(x) - f(1)}{x-1} = \lim_{x \to 1} \frac{x^2 - 1^2}{x-1} = \lim_{x \to 1} \frac{(x-1)(x+1)}{x-1}$$

$$= \lim_{x \to 1} x+1 = 1+1 = 2 \in \Re .$$

Property 5.1-1

If f is a differentiable function at a, then the equation of the tangent to the graph of f at the point A(a , f(a)) is : y = f(a) + (x – a) f '(a) . (figure 5.1-5)

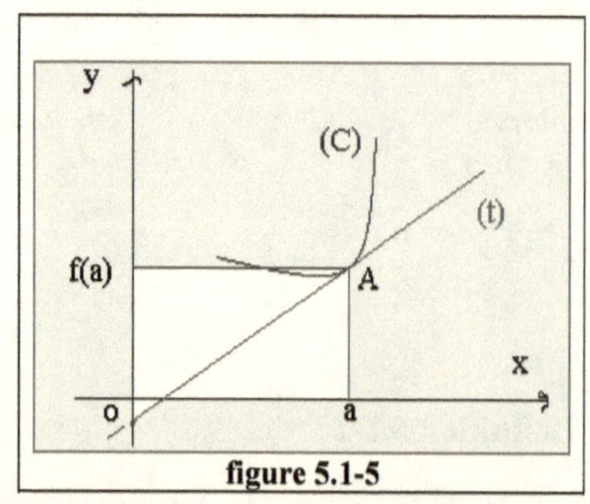

figure 5.1-5

Remark 5.1-1

1) In particular, if f '(a) = 0 , then the equation of the " horizontal " tangent is y = f(a) (figure 5.1 – 6).

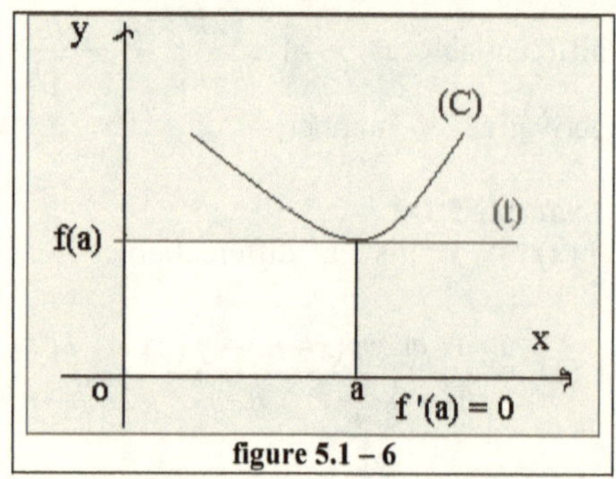

figure 5.1 – 6

2) If $\lim_{h \to 0} \frac{f(x+h)-f(a)}{h} = +\infty \ (or -\infty)$

The tangent is then a " vertical " line with equation x = a (figure 5.1 – 7).

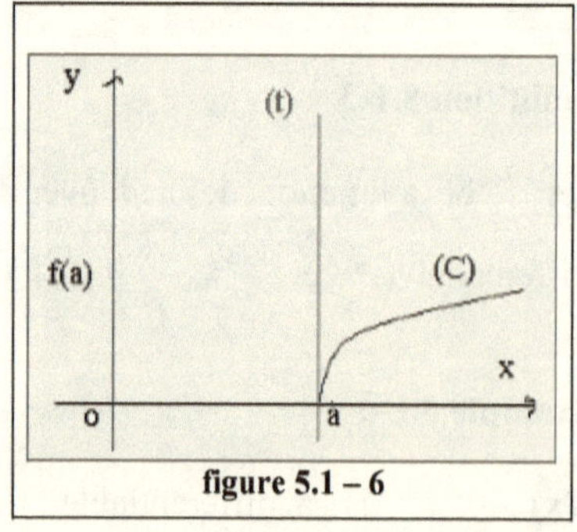

figure 5.1 – 6

Definition 5.1-4

Let f be a function defined on an interval I and that has a derivative f '(a) at every a in I . The function f '(x) defined on I is called the derivative function of f and is denoted by f ' .

Remark 5.1-2

The domain of definition of f ' is a subset of that of f .

Section 5.2

Rules For Derivations

* The **Power Rule** will make the derivative of $f(x) = x^2$ much easier!

$f(x) = x^2$ $f'(x) = 2x^{2-1}$ $= 2x$ Notice that the same answer is gotten with difference quotient	Steps to follow in the Power Rule 1. Pull a copy of the exponent out in front of them 2. Subtract 1 from the exponent

* What if there ' s **a coefficient**?

$f(x) = 5x^8$

$f'(x) = 5 . 8x^{8-1}$

$= 40x^7$

* **The Product Rule**

Right now, you can find the derivative of things like $6x^9$, a number multiplied by a variable raised to a power. However, The Power Rule doesn't work for a lot of derivatives. For example, look at this function: $f(x) = 4x^3 . \ln x$. This function f consists of two separate pieces: $4x^3$ and $\ln x$ multiplied together. What is the derivative of $f(x) = 4x^3 . \ln x$? Take the derivative of each piece, we get : $12x^2$ and $\frac{1}{x}$. Does $f'(x) = 12x^2 . \frac{1}{x}$? Of course no!

Therefore, we need a method to find derivatives when variables are multiplied. To find the derivative of $f(x) = 4x^3 . \ln x$, we follow the following steps of the Product Rule:

1. Imagine that f actually broken into two pieces. The first piece is $4x^3$ and the second piece is $\ln x$.
2. First term: Differentiate one piece $4x^3$ and leave the second piece $\ln x$, we get $12x^2 . \ln x$
3. Add the second term: Differentiate the second piece $\ln x$ and leave the first piece $4x^3$,we get $12x^2 . \ln x + \frac{1}{x} . 4x^3$

Hence, $f'(x) = 12x^2 . \ln x + 4x^2$.

***The Quotient Rule**

By using the product rule if $f(x) = \cos x + x^2$, then $f'(x) = -\sin x + 2x$. So what is the derivative of $g(x) = \frac{\cos x}{x^2}$?

$g(x) = \frac{\cos x}{x^2}$ is a quotient of two variable expressions: Top (cos x) and button (x^2) so it needs the Quotient Rule!

$f(x) = \frac{\cos x}{x^2}$ $f'(x) = \frac{x^2.(\cos x)' - (\cos x).(x^2)'}{(x^2)^2}$ $= \frac{-x^2 \sin x - 2x\cos x}{x^4}$ $= \frac{-x(x\sin x + 2\cos x)}{x^4}$ $= \frac{-x\sin x - 2\cos x}{x^3}$	<u>Steps to follow in the Quotient Rule</u> 1. Pull a copy of the button and multiply it by the derivative of the top 2. Pull a copy of the top and multiply it by the derivative of the button 3. Subtract 2 from 1 and divided by the square of the button

***The Chain Rule**

Here ' s the difference

$f(x) = \sin x$ has a derivative $f'(x) = \cos x$. However, if $g(x) = \sin 3x$, then $g(x) \neq \cos 3x$ (pretty close to the derivative of g , but not quite right. So if a function contains something other than plain old " x ", you must use the Chain Rule to find the derivative. That is, when a function is inside another function, use the Chain Rule to find the derivative. For instance, $f(x) = \sin 3x$ is a composite function.

As a conclusion, you will use the Chain Rule to find derivatives given:

- The composition of functions.
- A function containing something other then just plain x.

For f(x) = sin 3x , there are two functions here : the sine function and the 3 x variable expression. The Chain Rule tells you which function to derive first.

Between " sin " and " 3x " which is the inner function? and which is the outer function? The inner function is " 3 x " and the outer function is " sin " because Sin 3x = sin (3x).

outer

$$f(x) = \sin(3x)$$

inner

To apply the Chain Rule:

1. Take derivative of outer function, leaving the inner function alone : cos 3x
2. Multiply by the inner function derivative: 3

So $f'(x) = \cos 3x \,.\, 3 \quad \Rightarrow \quad f'(x) = 3\cos 3x$

Another Chain Rule derivative : $f(x) = (\ln x)^4$.

Solution plan:

inner

1) Identify inner / outer functions: $(\ln x)^4$ outer

2) Derive the outer function, leaving the inner function alone: $4(\ln x)^3$.

3) Derive the inner function: $\frac{1}{x}$ and multiply by the result obtained in part 2.

Finally, $f'(x) = 4(\ln x)^3 \,.\, \frac{1}{x} \Rightarrow f'(x) = \frac{4(\ln x)^3}{x}$.

Property 5.2-1

Let u, v be functions of x , k is constant and n is a natural integer.

1) If $f(x) = k$, then $f'(x) = 0$.
2) If $f(x) = u + v$, then $f'(x) = u' + v'$.
3) If $f(x) = k\,u$, then $f'(x) = k\,u'$.
4) If $f(x) = u \,.\, v$, then $f'(x) = u'v + v'u$.
5) If $f(x) = \frac{u}{v}$, then $f'(x) = \frac{u'v - v'u}{v^2}$.

6) If f(x) = $\sqrt{u}$, then f'(x) = $\frac{u'}{2\sqrt{u}}$.
7) If f(x) = u^n , then f'(x) = $nu^{n-1}.u'$.
8) If f(x) = ln u , then f'(x) = $\frac{u'}{u}$.
9) If f(x) = e^u , then f'(x) = $u'e^u$.
10) If f(x) = a^u , then f'(x) = $u'a^u \ln a$

Example5.2-1

1) If f(x) = 4, then f'(x) = 0.
2) If f(x) = $-\frac{1}{2}$, then f'(x) = 0.
3) If f(x) = 2x + 7, then f'(x) = 2 + 0 = 2.
4) If f(x) = -4x, then f'(x) = - 4.
5) If f(x) = x (2x+8), then f'(x) = 1 (2x+8) + 2 (x) = 4 x + 8.
6) If f(x) = $\frac{x+2}{5x-3}$, then f'(x) = $\frac{1(5x-3)-5(x+2)}{(5x-3)^2} = \frac{-13}{(5x-3)^2}$.
7) If f(x) = x^5, then f'(x) = $5x^4(1) = 5x^4$.
8) If f(x) = $\sqrt{x-4}$, then f'(x) = $\frac{1}{2\sqrt{x-4}}$.
9) If f(x) = ln (3 x + 1), then f' (x) = $\frac{3}{3x+1}$.
10) If f(x) = $e^{(-2x^2+x+1)}$, then f' (x) = $(-4x+1)e^{(-2x^2+x+1)}$.
11) If f(x) = $3^{(5x^5+x)}$, then f' (x) = $(25x^4+1)\,3^{(5x^5+x)}\ln 3$.

Remark 5.2-1

1)

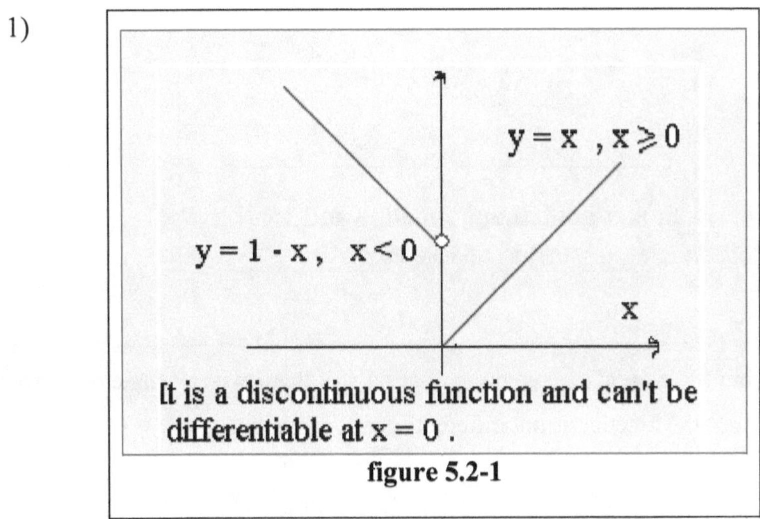

figure 5.2-1

2)

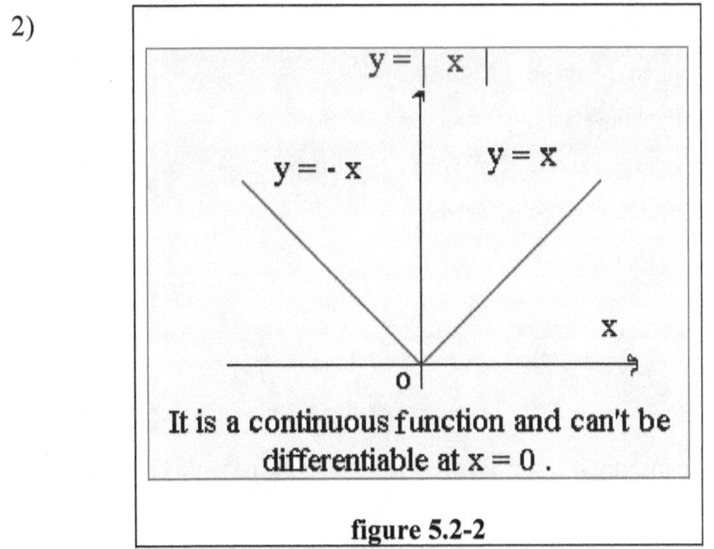

figure 5.2-2

3)

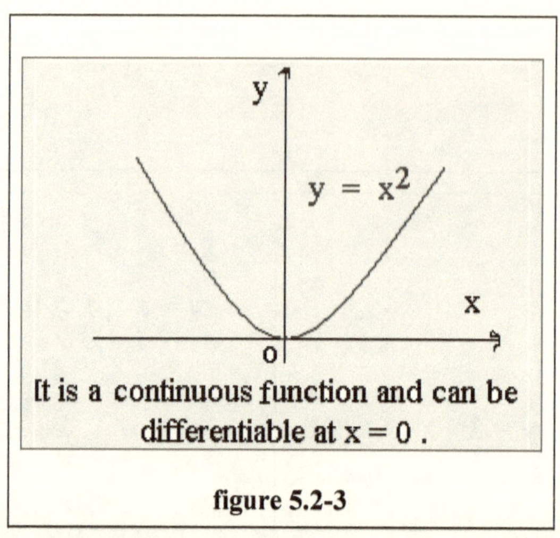

figure 5.2-3

4) In general, if at a given point of abscissa a there exist an angle or avertical tangent, then the function is not differentiable at this point.

Example 5.2-2 Economical Application

A study implemented on a certain article, lead to the following relation

$D(p) = \dfrac{1}{p(\ln p - 1)}$ where p represents the price in dollars and D(p) the demand function related to this product for the price p.

a. Calculate the elasticity of the demand E(p).

b. Calculate E(4). Is D elastic or inelastic at p = 4. Give an economical interpretation of your answer.

Solution

a. $E(p) = -p\dfrac{D'(p)}{D(p)} = p\dfrac{\ln p}{(p\ln p - p)^2} \times (p\ln p - p) = \dfrac{\ln p}{\ln p - 1}$.

b. $E(4) = \dfrac{\ln 4}{\ln 4 - 1} \approx 3.5 > 1$, then the demand is elastic for the price p = 4. If the price increases 1 % from the price of $ 4, the demand decreases 3.5 %.

Section 5.3

Differentiability of Trigonometric Functions

Property 5.3-1

1) If u is a differentiable function at a, then sin u is differentiable at a and its derivative at this point is u' cos u.

2) If u is a differentiable function at a, then cos u is differentiable at a and its derivative at this point is - u' sin u.

3) If u is a differentiable function at a, then tan u is differentiable at a and its derivative at this point is $u'(1+\tan^2 u) = \frac{u'}{\cos^2 u} = u' \sec^2 u$.

4) If u is a differentiable function at a, then cot u is differentiable at a and its derivative at this point is $\frac{-u'}{\sin^2 u} = -u' \csc^2 u$.

5) If u is a differentiable function at a, then sec u is differentiable at a and its derivative at this point is u' sec u tan u.

6) If u is a differentiable function at a, then csc u is differentiable at a and its derivative at this point is - u' csc u cot u.

Example5.3-1

1) If $y = \sin 3x$, then $y' = 3\cos 3x$.

2) If $y = \cos(-12x)$, then $y' = -(-12)\sin(-12x) = 12\sin(-12x)$.

3) If $y = \tan(x^2)$, then $y' = \frac{2x}{\cos^2 x^2} = 2x(1+\tan^2 x^2)$.

4) If $y = \cot(5x^4 + 3x)$, then $y' = \frac{-(20x^3+3)}{\sin^2(5x^4+3x)}$.

5) If $y = \frac{\sin x}{3+\sec x}$, then $y' = \frac{(3+\sec x)(\cos x) - \sin x(\sec x \tan x)}{(3+\sec x)^2}$.

Section 5.4

Geometric Applications

Property 5.4-1

Suppose that figure 5.4-1 represents the curve of the function y = f(x) and (x_1 , $f(x_1)$) is a point on this curve.

If $h \neq 0$, then (x_1+h , $f(x_1+h)$) represents another point on the curve, and it is clear that

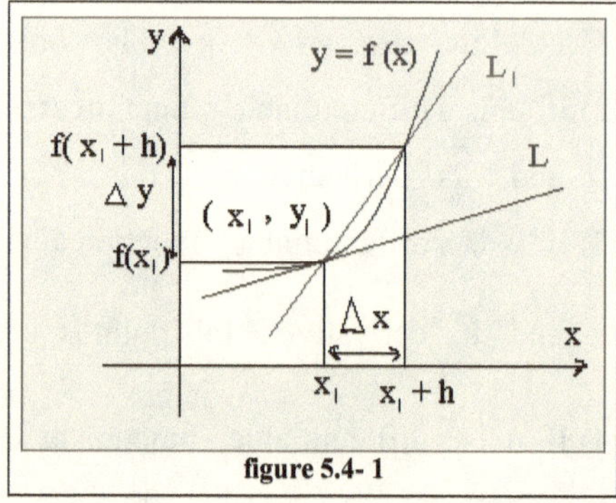

figure 5.4- 1

$\dfrac{\Delta y}{\Delta x}=\dfrac{f(x_1+h)-f(x_1)}{h}$ is the slope of the straight line L_1 which passes through these two points. As we mention before if $\lim\limits_{h\to 0}\dfrac{f(x_1+h)-f(x_1)}{h}$ exists, then it is a real number equals to the slope of the straight line L tangent to the curve y = f(x) at the point ($x_1,f(x_1)$) = (x_1, y_1), and it is called the derivative function at x_1 and is denoted by $f'(x_1)$. We conclude that the equation of the straight line L that passes through the point (x_1 , y_1) with slope $f'(x_1)$ is given by :

$$t: \frac{y-y_1}{x-x_1}=f'(x_1) \Leftrightarrow y-y_1=f'(x_1)(x-x_1).$$

It is the equation of the tangent line L at the point (x_1 , y_1).

If the straight line D is perpendicular to the tangent L and passes through the point of tangency (x_1 , y_1) as shown in figure 5.4-2, then the slope of the straight line D which is called normal line is equal to $-\dfrac{1}{f'(x_1)}$

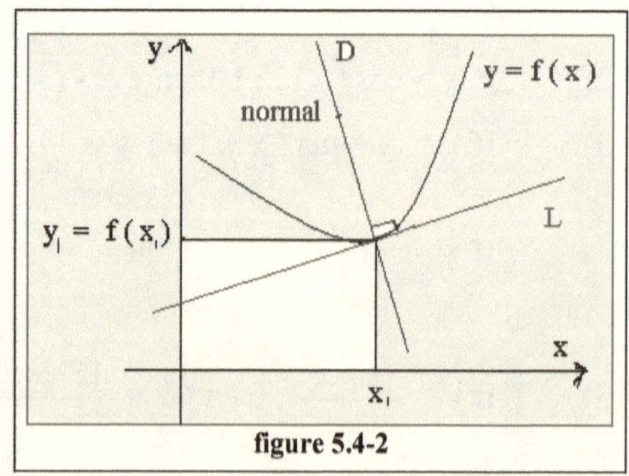

figure 5.4-2

if $f'(x_1) \neq 0$ then the equation of the normal becomes:

$$D: \frac{y-y_1}{x-x_1} = -\frac{1}{f'(x_1)} \quad \text{or} \quad y-y_1 = -\frac{1}{f'(x_1)}(x-x_1).$$

Remark 5.4-1

If $f'(x_1) = 0$, then the tangent line L becomes parallel to the x-axis as shown in figure 5.4-3 and its equation is $y = y_1$ and in this case the normal D becomes parallel to the y-axis and its equation is $x = x_1$.

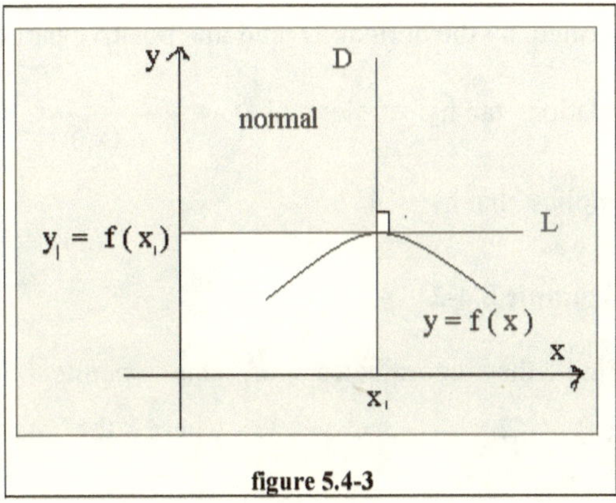

figure 5.4-3

Example 5.4-1

Find an equation of the tangent line t and the normal D to the curve $f(x) = \frac{1}{x}$ at $x_1 = 1$.

Solution

$f'(x) = \frac{-1}{x^2}$ then slope of t =

$f'(x_1) = f'(1) = \frac{-1}{1^2} = -1.$

$y_1 = f(1) = \frac{1}{1} = 1$

$t: y - y_1 = f'(1)(x - x_1)$ then

$y - 1 = (-1)(x - 1)$ so $t: y = -x + 2$

$$D: y - y_1 = -\frac{1}{f'(x_1)}(x - x_1)$$

then $y - 1 = -\frac{1}{-1}(x - 1)$ So D : $y = x$. (Figure 5.4- 4)

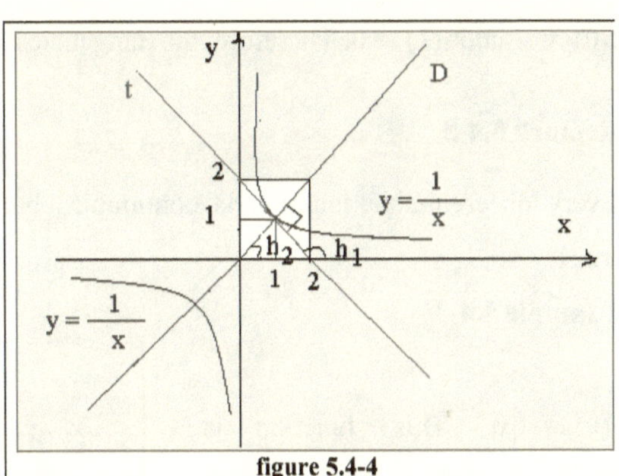

figure 5.4-4

Remark 5.4-2

We noticed that in figure 5.4-4 the angle h_1 formed by the tangent t and the positive part of the x-axis is determined by the relation tan h_1 = slope of t = f '(x_1) = -1, where $0 \le h_1 \le 180^o$ this implies that $h_1 = 135^o$. While the angle h_2 formed by the normal D and the positive part of the x-axis is determined by the relation $\tan h_2$ = slope of D $= -\dfrac{1}{f'(x_1)} = 1$, where $0 \le h_2 \le 180^o$ this implies that $h_2 = 45^o$.

Example 5.4-2

Find the coordinates of the points of tangency of the function $f(x) = 2x^3 - 3x^2 + 1$ where the tangent is parallel to the x-axis.

Solution

$f'(x) = 6x^2 - 6x$. The tangent is parallel to the x-axis if tan h = 0 that is its slope = 0 . So f '(x) = 0. Hence, $6x^2 - 6x = 0$ implies that $6x(x-1) = 0$, thus, x = 0 or x = 1.

f(0) = 1 and f(1) = 0. Therefore, the tangencies points are (0,1) and (1,0).

Remark 5.4-3

Every differentiable function is continuous, but the converse is not necessarily true .

Example 5.4-3

$f(x) = |x|$. This function is continuous at x = 0, but it is not differentiable at this point because the graph makes an angle at this point.

(Figure 5.4 – 5)

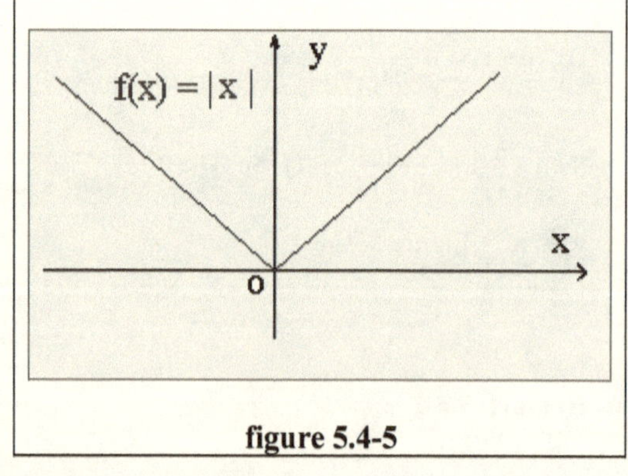

figure 5.4-5

Section 5.5

Physical Applications

Property 5.5-1

Suppose that a body is moving on a straight line so that its position (its distance from a fixed point on the straight line) is defined by the relation : x = f(t) . x is the directed displacement , t is the time and f is a differentiable function.

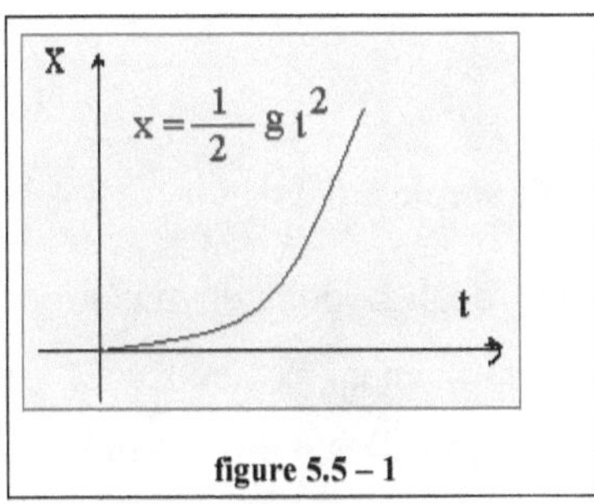

figure 5.5 – 1

It is known that $\frac{f(t_1+h)-f(t_1)}{h}$ is the average velocity of the body during the period t_1 to $t_1 + h$, and on the limit when h tends to zero the limit becomes $\lim_{h \to 0} \frac{f(t_1+h)-f(t_1)}{h}$. As we know, a definition to the initial velocity of the body at time t_1 is $f'(t_1)$. For instance, if a body is fell down at instant t = 0 under the effect of gravity for the earth, the experiments proved that the covered displacement by the body is $x = \frac{1}{2} g t^2$, where g = velocity of gravity = 9.80m / s. According to the preceding information, we can say that the body's velocity at instant t is :

$$v = \frac{dx}{dt} = \frac{1}{2} g \times 2t = gt.$$

This means that the velocity is increased normally with the increase of time as shown in figure 5.5- 2.

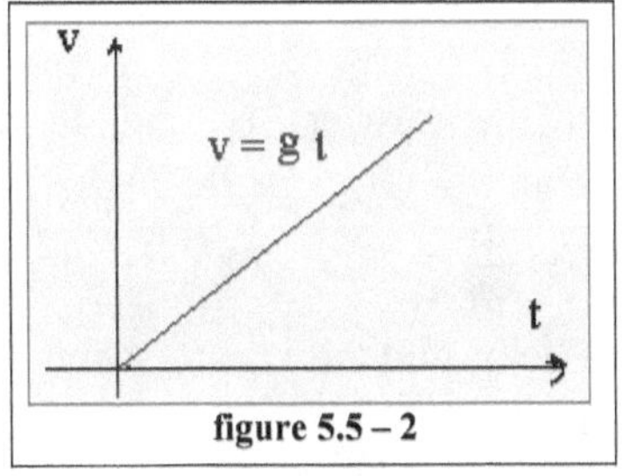

figure 5.5 – 2

Since the velocity is the average of the changing distance with respect to the time, then the acceleration γ is known as the average change of velocity with respect to the time.

That is $\gamma = \dfrac{dv}{dt} = g$, and it is known the acceleration of the body during its descend remains constant and it is equal to the acceleration of the gravity as shown in figure 5.5- 3.

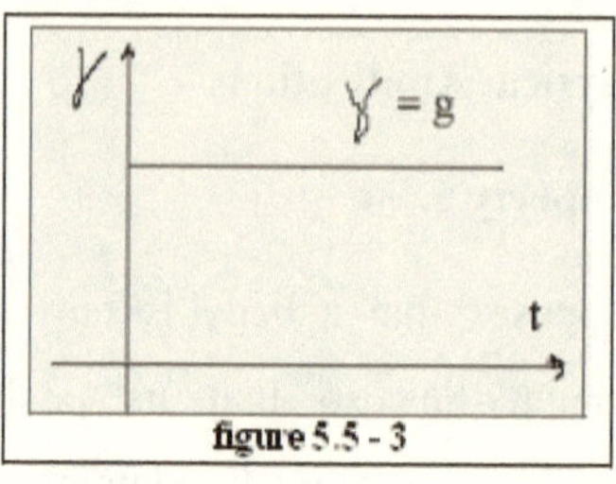

figure 5.5 - 3

Example 5.5-1

Through a point on the ground, a body is lunched up vertically and the displacement covered at t second from the beginning of the lunch is: $x = 24.5\,t - 4.9\,t^2$ m. Find :

1) the time along which the body come back to the starting point of lunch.
2) the velocity that the body lunched by.
3) the maximum distance that the body could reach.
4) the instant along which the velocity of the body is 14.7 m / s.
5) the acceleration of the body at each instant.

Solution

1) At t = 0 the body is lunched from its place on the ground where v = 0. The body will be back to the starting point on the ground when v = 0

$$\Rightarrow \quad 24.5\,t - 4.9\,t^2 = 0$$

$$\Rightarrow \quad t\,(24.5 - 4.9) = 0$$

$\Rightarrow \quad t = 0$ or $t = \dfrac{24.5}{4.9} = 5$ s . Since t = 0 is the time of the beginning,

then the answer is the instant of being back which is = 5 s.

2) $v(t) = \dfrac{dx}{dt} = 24.5 - 9.8\,t$ is the velocity at instant t . The velocity that the body start with is v(t) when t = 0. That is the initial velocity is then $v_0 = 24.5$ m / s.

3) The body reach the highest height when the velocity becomes 0; that

is $v(t) = 24.5 - 9.8\,t = 0 \Rightarrow t = \dfrac{24.5}{9.8} = 2.5$ s. At this instant, the height is $x(2.5) = 24.5(2.5) - 4.9(2.5)^2 = 30.625$ m.

4) The instant that the velocity of the body is 14.7 m / s is the value of t that satisfies the equation : $v = 24.5 - 9.8\,t = 14.7$

$\Rightarrow \quad 9.8\,t = 9.8$

$\Rightarrow \quad t = 1$ s.

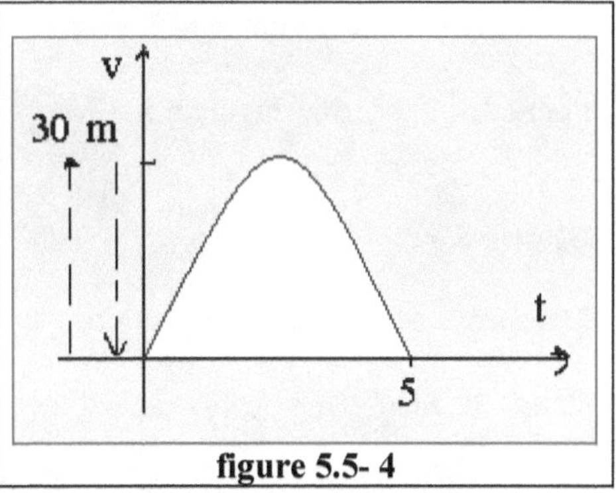

figure 5.5- 4

Notice that the body reaches this velocity after one second from its lunches that is while it is ascending. But if it is demand to find the instant that its velocity reaches 14.7 m / s during its descending, then the of the velocity becomes negative and the answer is to solve the equation :

$v = 24.5 - 9.8\,t = -14.7 \Rightarrow$

$9.8\,t = 39.2 \Rightarrow t = 4$ s. That is, the sign of the velocity is positive (upwards) from $t = 0$ to $t = 2.5$ s then it becomes negative (downwards) from $t = 2.5$ s to $t = 5$ s as shown in figure 5.5 - 4.

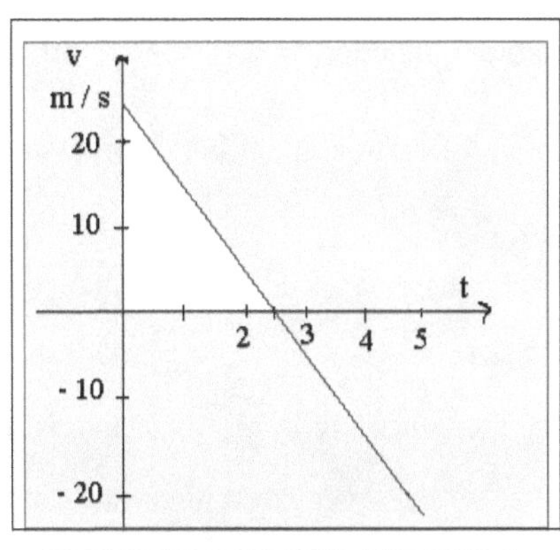

figure 5.5 – 5

5) $\gamma = v'(t) = -9.8 \text{ m / s}^2$.

That is the body moves in a constant acceleration which is the gravity acceleration of the earth that pull it down so that its velocity decreases during the period $[0 , 2\frac{1}{2}]$, but it increases during its descend during the period $[2\frac{1}{2} , 5]$.(Figure 5.5 – 5).

Section 5.6

Derivative of Composite Functions (Chain Rule)

Property 5.6-1

If f has derivative at a point x_0 of an interval I, and if g has derivative at $u_0 = f(x_0)$, then the composite function $h = g \circ f$ has a derivative at x_0 and we have : $h'(x_0) = g'(u_0) \times f'(x_0)$.

Example 5.6-1

The function $x \to \sin\frac{1}{x}$ is differentiable on $\Re^*$ since it is composite of two differentiable functions : $f(x) = \frac{1}{x}$, differentiable on $\Re^*$, and the function $g(x) = \sin x$, differentiable on $\Re$. Moreover, $f'(x) = -\frac{1}{x^2}$ and $g'(x) = \cos x$, hence $h'(x) = -\frac{1}{x^2}\cos\frac{1}{x}$.

Remark 5.6-1

1) If $y = f(x)$, then $dy = y'\,dx$; that is, the derivative of y is $y' = \frac{dy}{dx}$.

2) Let x, y, z be open real intervals such that :

$f : x \to y$ with $y = f(x)$

$g : y \to z$ with $z = g(x)$

If we assume that y is a function of x so that $y = g(f(x)) = g(y)$ and y = f(x), then the previous equation, which is known as chain rule, takes the form :

$$\frac{dz}{dx} = \frac{dz}{dy} \times \frac{dy}{dx}$$

Example 5.6-2

If $z = (x^2 - 1)^6$, then find $\frac{dz}{dx}$.

Solution

By assuming that $z = g(y) = y^6$ and $y = f(x) = x^2 - 1$, it is obvious that $z = g \circ f(x)$ and then we can apply the chain rule $\frac{dz}{dx} = \frac{dz}{dy} \times \frac{dy}{dx}$

$$\Rightarrow \frac{dz}{dx} = (6y^5)(2x)$$

$$= 12xy^5$$

$$= 12x(x^2 - 1)^5.$$

Property 5.6-2

Let f(x) be a continuous function at a point of abscissa x_0. f(x) is said to be differentiable at x_0 if and only if $f'(x_0) = \lim_{x \to x_0} \frac{f(x) - f(x_0)}{x - x_0} = k \in \Re$ or in another word $f'(x_0^-) = f'(x_0^+)$.

Example 5.6-3

Let f be a function defined by $f(x) = \frac{x + |x-1| + 2}{x+2}$. Study the differentiability of f at the point $x_0 = 1$.

Solution

The function f is the ratio of two continuous functions over $\Re$, then it is continuous over its domain of definition $D_f = \Re - \{-2\}$; in particular at 1.

$$* \; f'(1^+) = \lim_{x \to 1^+} \frac{f(x) - f(1)}{x-1} = \lim_{x \to 1^+} \frac{\frac{x + x - 1 + 2}{x+2} - 1}{x-1} = \lim_{x \to 1^+} \frac{1}{x+2} = \frac{1}{3} \in \Re$$

$$* \; f'(1^-) = \lim_{x \to 1^-} \frac{f(x) - f(1)}{x-1} = \lim_{x \to 1^-} \frac{\frac{x - x + 1 + 2}{x+2} - 1}{x-1} = \lim_{x \to 1^-} \frac{-1}{x+2} = \frac{-1}{3} \in \Re$$

Since $f'(1^+) \neq f'(1^-)$, then f is not differentiable at 1.

Example 5.6-4

Let D be the demand function defined by D(p) = $\frac{4000}{2+p}$ where p is the unit price, expressed in dollars, of a product .

a. Calculate the elasticity of the demand E(p).

b. Find E(10) and give an economical interpretation to the value found.

Solution

a. $E(p) = -p\frac{D'(p)}{D(p)} = -p\frac{\frac{4000}{(2+p)^2}}{\frac{4000}{2+p}} = \frac{p}{2+p}$.

b. $E(10) = \frac{10}{2+10} \approx 0.8$.

E(10) < 1, then the demand is elastic for p = 10.If the price increases by 1 % starting from a price of $ 10, the demand decreases by 0.8 %.

Section 5.7

Second Order Differentiation

Definition 5.7-1

Let f be a function having a derivative function f ' on an open interval I. If f ' has a derivative at a point a of I, we say that f has a second derivative at a, denoted by f '' (a):

$$\frac{d^2y}{dx^2} = f''(a) = \lim_{x \to a} \frac{f'(x) - f'(a)}{x - a}.$$

Example 5.7-1

Let y = f(x) = $7x^3 - x^2$.

$$f'(x) = \frac{dy}{dx} = 21x^2 - 2x$$

$$f''(x) = \frac{d^2y}{dx^2} = \frac{d}{dx}\left(\frac{dy}{dx}\right) = \frac{d}{dx}\left(21x^2 - 2x\right) = 42x - 2.$$

Higher-Order Derivatives

The derivative f ′ of the function f is often referred to as the first-order derivative of the function. The second-order derivative, f ″ (denoted at *x* as f ″(x) or d^2y/dx^2 measures the slope and the instantaneous rate of change of the first derivative with respect to a change in *x*, just as the first derivative measures the slope and the rate of change of the original or primitive function. The third-order derivative, f ‴(x) (or d^3y/dx^3), measures the slope and instantaneous rate of change of the second-order derivative, etc. Higher-order derivatives are found by applying the rules of differentiation to lower-order derivatives, as illustrated in the following example.

Example: If $f(x) = 2x^5 + 4x^3 + 2x$,

$f'(x) = 10x^4 + 12x^2 + 2$

$f''(x) = 40x^3 + 24x$

$f'''(x) = 120x^2 + 24$

$f^{(4)}(x) = 240x$

$f^{(5)}(x) = 240$

$f^{(6)}(x) = 0$

All additional higher-order derivatives will also equal zero.

Section 5.8

Implicit Differentiation

Definition 5.8-1

For an equation to be differentiate implicitly, we find $y' = \frac{dy}{dx}$ as follows :

1. Differentiate both sides of the equation with respect to x .

2. Collect all terms involving y ' on one side of the equation, and collect all other terms on the other side.
3. Factor y ' from the side involving the y ' terms.
4. Solve for y ' then replace y ' by $\frac{dy}{dx}$.

Example 5.8-1

Find $\frac{dy}{dx}$ by implicit differentiation for the equation: $x^2 y = x^3 + 3$.

Differentiate both sides: $2xy + x^2 y' = 3x^2$

$$x^2 y' = 3x^2 - 2xy$$

$$y' = \frac{3x^2 - 2xy}{x^2}$$

Then, $\frac{dy}{dx} = \frac{3x^2 - 2xy}{x^2}$.

Summary

☒ The average change of variations of f between a and b is the real number $\frac{f(b)-f(a)}{b-a}$

☒ f is differentiable at x if $\lim_{h\to 0} \frac{f(x+h)-f(x)}{h} = l \in \Re$

☒ f is differentiable at x_o if $f'(x) = \lim_{x\to x_o} \frac{f(x)-f(x_o)}{x-x_o} = l \in \Re$

☒ u, v are functions of x, k is constant and n is a natural integer, then:

- If $f(x) = k$, then $f'(x) = 0$
- If $f(x) = u + v$, then $f'(x) = u' + v'$
- If $f(x) = k u$, then $f'(x) = k u'$
- If $f(x) = u . v$, then $f'(x) = u'v + v'u$
- If $f(x) = \frac{u}{v}$, then $f'(x) = \frac{u'v - v'u}{v^2}$

- If $f(x) = \sqrt{u}$, then $f'(x) = \dfrac{u'}{2\sqrt{u}}$
- If $f(x) = u^n$, then $f'(x) = nu^{n-1}.u'$
- If $f(x) = \ln u$, then $f'(x) = \dfrac{u'}{u}$
- If $f(x) = e^u$, then $f'(x) = u'e^u$
- If $f(x) = a^u$, then $f'(x) = u'a^u \ln a$

☒ If $y = \sin u$ then $y' = \cos u$

☒ If $y = \cos u$ then $y' = -u' \sin u$

☒ If $y = \tan u$ then $y' = u'(1 + \tan^2 u) = \dfrac{u'}{\cos^2 u} = u' \sec^2 u$

☒ If $y = \cot u$ then $y' = \dfrac{-u'}{\sin^2 u} = -u' \csc^2 u$

☒ If $y = \sec u$ then $y' = u' \sec u \tan u$

☒ If $y = \csc u$ then $y' = -u' \csc u \cot u$

☒ the composite function $h = g \circ f$ has a derivative at x_0 defined by

$$h'(x_0) = g'(u_0) \times f'(x_0) \; ; \; \frac{dz}{dx} = \frac{dz}{dy} \times \frac{dy}{dx}$$

☒ f(x) is said to be differentiable at x_0 if and only if $f'(x_0) =$

$$\lim_{x \to x_0} \frac{f(x) - f(x_0)}{x - x_0} = k \in \Re \text{ or in another word } f'(x_0^-) = f'(x_0^+)$$

☒ f has a second derivative at a ,denoted by $f''(a) = \dfrac{d^2y}{dx^2} = \lim\limits_{x \to a} \dfrac{f'(x) - f'(a)}{x - a}$

Exercises

In problems 5.1 – 5.6, use the definition of the derivative of a function f at a point a to calculate the derivative in each case :

5.1. $f(x) = 3x^2 - 6x + 3$, $a = 2$

5.2. $f(x) = x^3$, $a = -1$

5.3. $f(x) = \sqrt{x+1}$, $a = 1$

5.4. $f(x) = x^2 + x$, $a = 2$

5.5. $f(x) = \dfrac{x+1}{x-1}$, $a = 2$

5.6. $f(x) = x - \sqrt{x}$, $a = 1$

In problems 5.7 – 5.29, determine the derived function of each of the given functions over their domain of definitions.

5.7. $f(x) = 2x^4 - 5x^3 + x^2 - 1$

5.8. $g(x) = -x^2 + x\sqrt{x} + 4x - 2$

5.9. $p(x) = \dfrac{2x}{x+1}$

5.10. $l(x) = \dfrac{\sqrt{4x}}{x-1}$

5.11. $A(x) = \sin\left(\dfrac{\pi}{6} + x\right)$

5.12. $B(x) = \cos\left(\dfrac{\pi}{4} - x\right)$

5.13. $C(x) = \sin 2x \tan 3x$

5.14. $R(x) = \dfrac{2}{\sin x} + \dfrac{1}{\sin 2x}$

5.15. $z(x) = \cos(\cos x)$

5.16. $s(x) = x \cdot \sqrt{\dfrac{x-1}{x+1}}$

5.17. $f(x) = x^2 \ln x$

5.18. $f(x) = \ln(\cos x)$

5.19. $f(x) = \dfrac{1-\ln x}{1+\ln x}$

5.20. $f(x) = \sqrt{1+\ln^2 x}$

5.21. $f(x) = e^{\frac{x}{x+1}}$

5.22. $f(x) = e^{\sqrt{\ln x}}$

5.23. $f(x) = x e^{\frac{1}{x}}$

5.24. $f(x) = e^{x \sin x}$

5.25. $f(x) = \dfrac{x^3 + 2^x}{\ln x}$

5.26. $f(x) = 2^{3x+1}$

5.27. $f(x) = \dfrac{2^x}{4^x}$

5.28. $f(x) = 6^{\frac{1}{x}}$

5.29. $f(x) = \pi^{\sin x}$

In problems 5.30 – 5.34, find the equation of the tangent to the graph of f at the point A(a , f(a)) in each of the given cases:

5.30. $f(x) = x^2 + 3x$, $a = 1$

5.31. $f(x) = \dfrac{2}{x-1}$, $a = 0$

5.32. $f(x) = \sqrt{3x+2}$, $a = -1$

5.33. $f(x) = e^{-x+1}$, $a = 1$

5.34. $f(x) = \ln(x-2)$, $a = 3$

In problems 5.35 – 5.43, determine the second derivative of the given functions:

5.35. $A(x) = \sqrt{1+x}$

5.36. $B(x) = \dfrac{x-2}{x-1}$

5.37. $C(x) = 2\sin x + 3x$

5.38. $D(x) = \tan x - 2x$

5.39. $E(x) = \sqrt{10x+1}$

5.40. $F(x) = \dfrac{5}{3+x^2}$

5.41. $G(x) = \sin x + \cos x$

5.42. $H(x) = \tan x + \dfrac{1}{x+3}$

5.43. $I(x) = x^3 - 2x\sqrt{x}$

5.44. Let f be the function defined by $f(x) = \left|\frac{2x-1}{x+1}\right|$.

a) Study the differentiability of f at $x_0 = \frac{1}{2}$.

b) Write the equations of the semi – tangents drawn through A of abscissa $\frac{1}{2}$ to the curve of f.

5.45. Let f be the function defined by $f(x) = x^3 - 3x^2 + 1$. Determine, if possible, the values of x such that:

a) $f'(x) = 0$.

b) $f'(x) = 1$.

In problems 5.46 – 5.56, determine the derived function of each of the given functions over their domain of definitions . Calculate the value of the derivative at each given point.

5.46. $k(x) = (x - \sqrt{x})(2x + 1)$; $a = 2$
5.47. $r(x) = (x - 3)^4$; $a = -3$
5.48. $s(x) = \frac{2x^2 + 3x + 1}{2x^2 - 3x - 1}$; $a = 1$

5.49. $z(x) = \sqrt{-x + 2}$; $a = \frac{1}{2}$

5.50. $y(x) = \frac{1}{\sqrt{2x+1}} - 4x$; $a = 4$

5.51. $u(x) = \sin x \cos x$; $a = \frac{\pi}{4}$

5.52. $v(x) = \sin^2 x$; $a = \frac{\pi}{2}$

5.53. $t(x) = \cot(-3x)$; $a = \frac{\pi}{4}$

5.54. $f(x) = \sin(\cos 3x)$; $a = \frac{\pi}{3}$

5.55. $B(x) = \frac{\sin 2x}{\cos^2 x}$; $a = \frac{\pi}{4}$

5.56. $A(x) = x^3 \sin(4x^2 - 1)$; $a = \frac{\pi}{2}$

In problems 5.57 – 5.61, The position of a particle M on a line is determined by the law of motion $t \to x(t)$. Calculate the instantaneous velocity of M at the given value of t :

5.57. $x(t) = \dfrac{4}{4t+3}$, $t = 2$

5.58. $x(t) = 3t^2 + 3t$, $t = 0$

5.59. $x(t) = \dfrac{t-2}{t+2}$, $t = 3$

5.60. $x(t) = 4t^3 + t^2$, $t = 1$

5.61. $x(t) = \sqrt{t+2}$, $t = 1$

5.62. Let f be the function defined by $f(x) = x^2 + |x-1|$. Study the differentiability of f at 1.

5.63. Let f be the function defined by $f(x) = \begin{cases} ax^2+1 & \text{if} \quad x \le 2 \\ bx & \text{if} \quad x > 2 \end{cases}$.
Calculate a and b so that f is continuous and differentiable at 2.

5.64. Let f be the function defined by $f(x) = \dfrac{bx+c}{2x+1}$. Calculate b and c so that the tangent at A (1 , $\dfrac{1}{3}$) of the curve (C), the representative graph of f in an orthonormal system, is parallel to the straight line (D) : $y = \dfrac{-11}{9}x + 4$. Write then the equation of this tangent.

5.65. A body moves in a straight line that cover a distance x ft. after t seconds so that $x = t^4 - 2t^3 + t^2 - 5$. Find the time that the velocity of the body is null.

5.66. A body moves vertically up and down in only one straight line so that its height from the ground surface after n seconds is $x = 128t - 16t^2$ feet . Find :

a) the velocity of the body at any instant.

b) the set of values of $t \ge 0$ such that the velocity is positive.

c) the highest altitude from the ground that the body can reach.

d) the initial velocity of the body.

e) draw the curve of each of the velocity and the distance function.

In problems 5.67 – 5.70, find $\frac{dy}{dx}$ **using chain rule :**

5.67. $y = u^3 - 2u + 1$ where $u = x^2$

5.68. $y = u^2$ where $u = x^2 + x - 1$

5.69. $y = u^3$ where $u = \sqrt{x} + 1$

5.70. $y = u^4 + 2u^2 + 1$ where $u = x + \frac{1}{x}$

In problems 5.71 – 5.74, find $\frac{dy}{dt}$ **using chain rule :**

5.71. $y = u^3 + u$ where $u = \sqrt{x} + 1$ and $x = 3t^2 + t$ when $t = 1$.

5.72. $y = u^4 - 2u^2$ where $u = \frac{1}{x}$ and $x = 3t$ when $t = 2$

5.73. $y = \ln(u + 1)$ where $u = x^3$ and $x = t^4$ when $t = 1$

5.74. $y = e^{u^3 - 4u}$ where $u = x^4$ and $x = 3t^2$ when $t = 0$

5.75. Determine the value of the positive constant k if :

a) $\frac{d}{dx}(2e^x - 4x^k) = 0$ when $x = 1$

b) $\frac{d}{dx}(e^x - x^k) = 0$ when $x = 1$

5.76. Consider the function f defined by $f(x) = \begin{cases} x^2 \sin x & \text{if} \quad x \neq 0 \\ 0 & \text{if} \quad x = 0 \end{cases}$

a) Prove that $0 \leq |x \sin x| \leq |x|$ and deduce $\lim\limits_{x \to 0} x \sin x$.

b) Study the differentiability of f at $x = 0$.

In problems 5.77 – 5.83, find the derivative of the given functions :

5.77. $y = \dfrac{\sec x}{1+\tan x}$

5.78. $y = \sec x \tan x$

5.79. $y = \sec^3 (3 x^3 - 2)$

5.80. $y = \sqrt{\csc x^3}$

5.81. $y = \dfrac{\cot x}{1+\csc x}$

5.82. $y = \sec x \csc 2x$

5.83. $y = \tan^2 3x$

5.84. Consider the function f defined by $f(x) = x^2 - 4x + 4$. Let (C) be its graph in an orthonormal system.

a) Determine the equation of the tangent to (C) with slope -2 after calculating the coordinates of the point of tangency.

b) Determine an equation of a tangent to (C) drawn through the origin O . How many solutions are there?

c) Let M be the point of (C) with abscissa a. Determine an equation of the tangent at M to (C) in terms of a. Find the result of part b) again.

d) Can we determine a so that the tangent at M passes through the point A (1 , 2)? Justify your answer.

In problems 5.85 – 5.90, find $\dfrac{dy}{dx}$ by implicit differentiation in each case:

5.85 . $x^2 + y^2 = 1$

5.86 . $\sqrt{x} + \sqrt{y} = 4$

5.87. $y \ln (x+1) = x e^y$

5.88. $\ln (x y) + x^2 = 6$

5.89. $x + xy + y^2 = 10$

5.90. $\ln(xy) = x e^y$

5.91. The total cost of producing q units of output is given by:

$$C(q) = 3q^2 + 6q + 100$$

a. Find an expression for the firm's marginal cost function

b. Find an expression for the firm's average cost function

c. Evaluate the firm's marginal and average costs when output equals 25 units.

d. What is the value of output at which marginal and average costs equal?

5.92. a. The total cost of producing q units of output is given by:

$$C(q) = q^2 + 7q + 23$$

1. Find an expression for the firm's marginal cost function

2. Find an expression for the firm's average cost function

b. Given the demand function $p = 30 - 2q$.

1. Find the total revenue function.
2. Find the marginal revenue function.

5.93. Let D be the demand function defined by $D(p) = 10 - \sqrt{p+1}$ where p is the unit price, expressed in dollars, of a product such that $0 \le p < 90$.

a. Calculate the elasticity of the demand E(p).

b. Find E(80) and give an economical interpretation to the value found.

5.94. The function of demand of a butcher is given by $P(q) = \frac{-120}{q+1} + 100$, where q is the quantity consumed in kg and P(q) is the price in dollars. The demand function is given by $D(q) = 100 - 10q$.

a. Determine quantity of equilibrium.

b. 1) Calculate the elasticity of the demand.

2) Determine q for the demand to be elastic.

5.95. Let D be the demand function defined by $D(p) = 200 - 2p$ where p is the unit price, expressed in dollars, of a product such that $10 \leq p < 100$.

a. How many units are sold for a price p = 25?

b. 1) Find the elasticity of the demand E(p).

2) Calculate E(25). Is the function D elastic for p = 25? Give an economical interpretation to the value found.

c. Find the unit price p for E(p) = 1.

d. Determine p if the demand is elastic.

5.96. The mayor of a city decides to promote bike riding in his struggle against pollution. He buys 1000 bicycles for rent. The study certifies that the demand function of price is given by $D(p) = \dfrac{5 \ln p}{p^2}$ where p is the price in dollars with $p \in]\,0, 10\,]$, and D (p), the demand expressed in thousands of bicycles.

a. For what price is the demand less than 500 bicycles?

b. Suppose that the rent charge \$ 3, calculate the percentage of variations of the demand to the nearest 10^{-2} as the rent charge increases \$ 0.03.

c. Calculate the elasticity of the demand.

d. Calculate E(3) to the nearest 10^{-2} and interpret this result.

5.97. A study was implemented on back to school items specifically on the price of files. The function $D(p) = 4 \ln\left(\dfrac{6}{p}\right)$ is the demand function of files, expressed in thousands, in terms of a unit price of a file expressed in dollars. The function $S(p) = 4 \ln(p - 1)$ is the supply function expressed in thousands of files. D and S are defined over [2 , 6].

a. Find the price of equilibrium.

b. What is the total revenue of producers for the price of equilibrium?

c. Calculate the elasticity of the demand.

d. Calculate E(2) and give an economical interpretation o the result.

5.98. A factory produces x machines per month. The total cost of production , in millions of dollars, is given by: $C(x) = \ln(2x + 4)$ where $0 \leq x \leq 10$.

a. Study the variations of the function C and draw its table of variations over [0 , 10].

b. What is the total cost of producing five machines.

c. Each machine is sold by \$ 500 000.

1) Determine the profit, in millions of dollars.

2) What is the profit obtained by selling five machines?

5.99. Consider a product, in p units of price in dollars, and whose demand is given by $D(p) = (2.5 + p)\, e^{-0..5p+1}$ and its supply by $S(p) = 0.3p + 1$ with $p \in [0, 10]$. The equation $f(p) = D(p) - S(p) = 0$ admits a unique solution α between 4.1 and 4.2 take $\alpha = 4.15$.

a. Find the market – equilibrium price.

b. Calculate the elasticity of the demand E(p).

c. Calculate E(4) and interpret its result.

5.100. In a store , the supply, expressed in dozens of articles of this item, is defined over $[0, +\infty[$ by $S(p) = \dfrac{e^{ap}-1}{4}$ where a is a real positive number and p represents the unit selling price expressed in terms of hundreds of dollars.

a. Determine a when a unit selling price of $ 400 corresponds to a supply of 745 dozens of articles.

b. in the same store and for the same article demand expressed in dozens of articles of this item is defined over $[0, +\infty[$ by $D(p) = \dfrac{12}{e^{2p}+1}$ where p is expressed in terms of hundreds of dollars.

1) Calculate $\lim\limits_{p \to +\infty} D(p)$ and deduce an asymptote to the graph of D.

2) Calculate D ' (p) and draw the table of variations of g.

3) Determine the coordinates of the point of intersection of S and D and give an economical interpretation of the value obtained.

4) Determine the elasticity of the demand E(p).

5) Calculate E(2) and give an economical interpretation of the value obtained.

Chapter 6 Application on Derivatives

Introduction

In the nineteenth century, economists developed the concept of " marginal analysis " such as marginal cost, marginal product and others .....

It was developed that the derivative of the total cost function is called the marginal cost. Similarly, the derivatives of the revenue and the profit are called marginal revenue and marginal profit, respectively.

In addition, when a quotient $\dfrac{f(x)}{g(x)}$ has the indeterminate form at a point of abscissa a where a is real constant, the concept of the derivative may help us find its limit by using the so-called " L'Hopital 's Rule ".

Why?

The ability of a function to be differentiable on an interval provides us with a big capacity to discover its behavior on this interval and this has an important issue to predict the scientific results when using the arithmetic differentiation in its different applications in physics, astronomy, sciences, geometry, economics, ...

To know the first derivative of a function is good, knowing the second derivative is even better. If the first derivative gives information on the velocity of a moving body, the tangent to a curve.... the second derivative gives the acceleration of the moving body, the curvature (concavity) ....The second derivative brings additional information about the behavior of the observed phenomenon.

Objectives

❖ After studying the material in this chapter, you should be able to:

- Understand the mean value theorem
- Study the sense of variation of functions by using derivatives and limits
- Find the local extreme values
- Determine the absolute extreme values

- Study the concavity of a function and recognize the inflection point
- Graph certain functions
- Apply problems on maximum and minimum values
- Solve problems in finance.

Section 6.1

Minimum and Maximum Values

Definition 6.1-1

* We say that a function f has absolute maximum on an interval I if there exists $c \in I$ such that $f(c) \geq f(x)$ for every $x \in I$. f(c) is called the absolute maximum of f.
* We say that a function f has absolute minimum on an interval I if there exists $c \in I$ such that $f(c) \leq f(x)$ for every $x \in I$. f(c) is called the absolute minimum of f.
*Maximum and minimum values are called extremes or extrema.

Example 6.1-1

In figure 6.1 -1, the function f admits a maximum at a and a minimum at b and d.

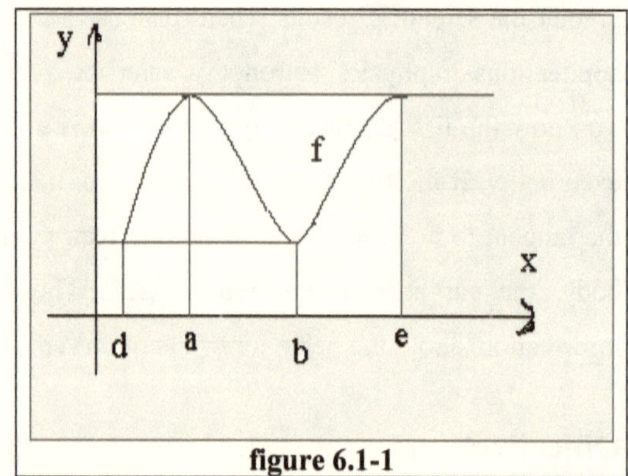

figure 6.1-1

Definition 6.1- 2

* f is said to have a local maximum at $c \in (a, b)$ if $\forall x \in (a, b)$ we have $f(x) \leq f(c)$. f(c) is called the local maximum of f over (a , b).
* f is said to have a local minimum at $c \in (a, b)$ if $\forall x \in (a, b)$ we have $f(x) \geq f(c)$. f(c) is called the local minimum of f over (a , b).

Example 6.1-2

In figure 6.1-2, f(a) = M is a local maximum, and f(b) = m is a local minimum on the interval [d , e].

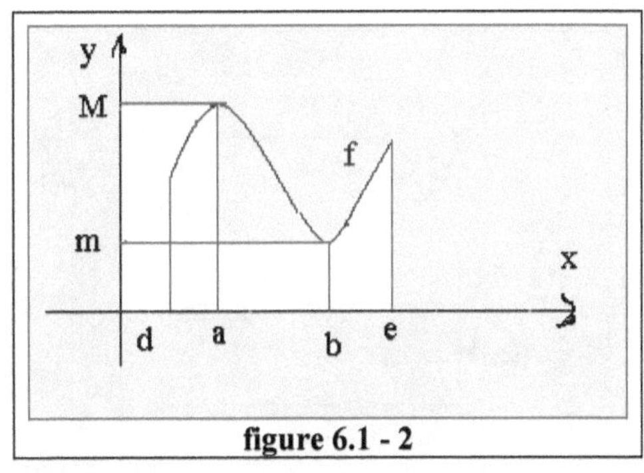

figure 6.1 - 2

Section 6.2

Increasing and Decreasing Functions

Definition 6.2-1

* A function f is said to be increasing over an interval I if for every x and x + 1 in I we have $f(x+1) \geq f(x)$.
* A function f is said to be decreasing over an interval I if for every x and x + 1 in I we have $f(x+1) \leq f(x)$.

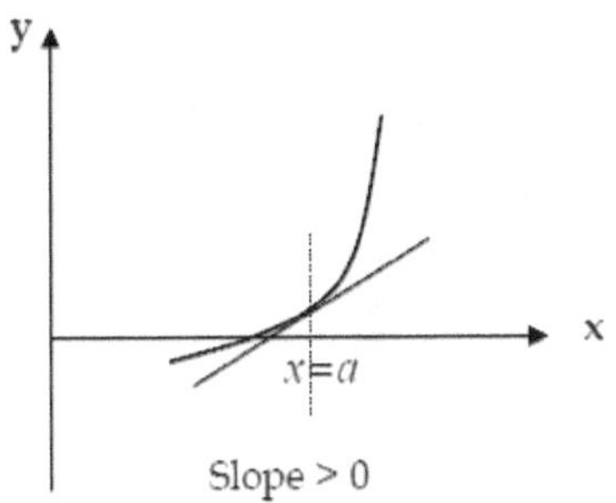

Increasing function at $x=a$

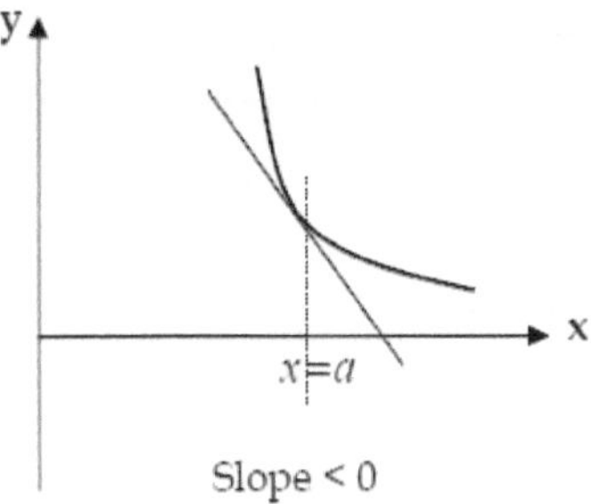

Decreasing function at $x=a$

Note: A function that increases (or decreases) along its entire domain is called a monotonic function.

Example 6.2-1

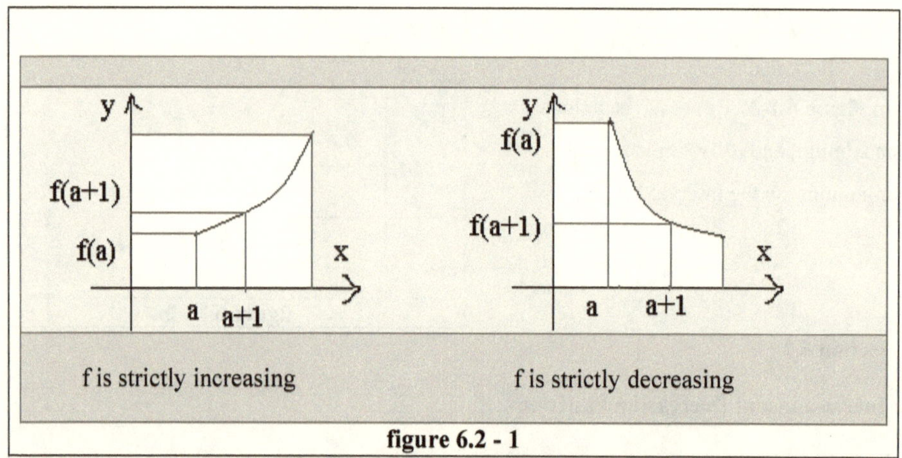

figure 6.2 - 1

Property 6.2- 1

* If $f' > 0$ on I, then f is strictly increasing on I.
* If $f' < 0$ on I, then f is strictly decreasing on I.
* If $f' = 0$ on I, then f is constant on I.

Example 6.2-2

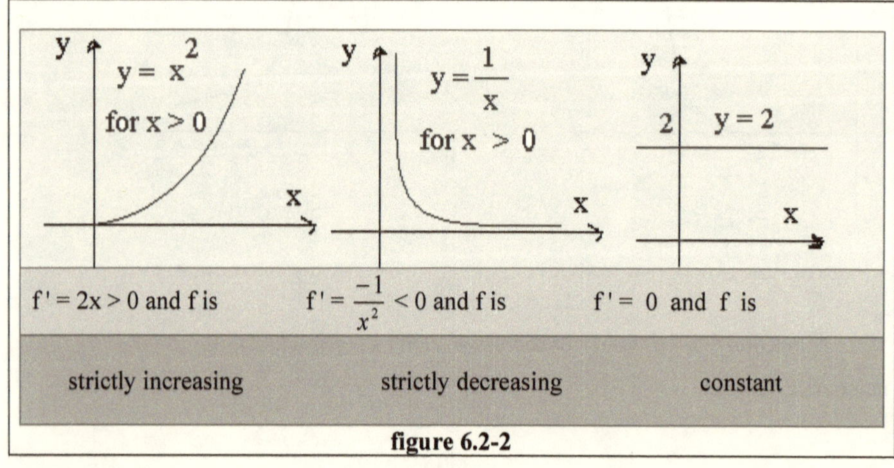

figure 6.2-2

Remark 6.2-1

If $f'(x)$ is zero at a point and keeps a constant sign everywhere on I, then f is strictly increasing or strictly decreasing on I.

Example 6.2-3

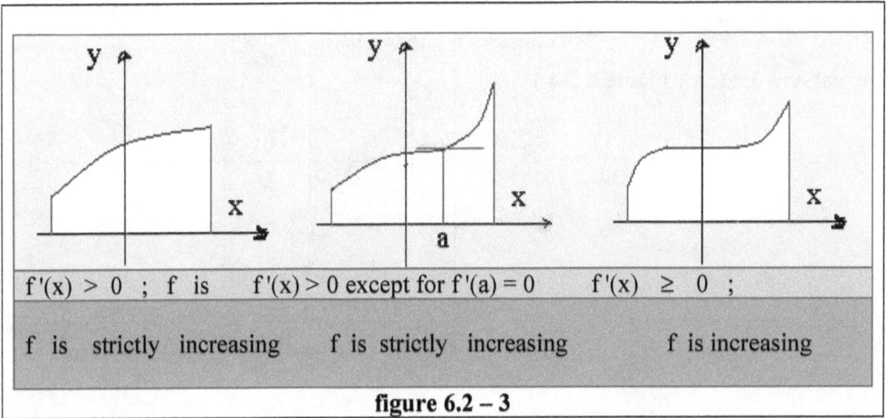

figure 6.2 – 3

Property 6.2-2

* If f ' becomes zero at a with changing signs from negative to positive, then f(a) is a local minimum of f .
*If f ' becomes zero at b with changing signs from positive to negative, then f(b) is a local maximum of f .

Example 6.2-4

Consider the function defined on $\Re$ by $f(x) = x^3 - 3x^2 + 1$. It is differentiable on $\Re$ and $f'(x) = 3x^2 - 6x = 3x(x-2)$.

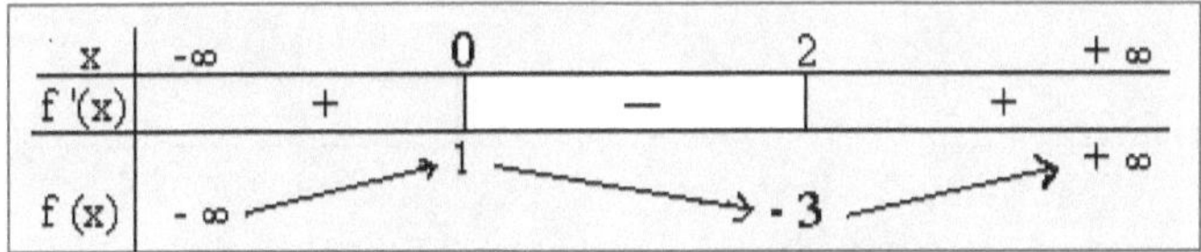

x	$-\infty$		0		2		$+\infty$
f '(x)		+		−		+	
f (x)	$-\infty$	↗	1	↘	-3	↗	$+\infty$

f has a local maximum equal to 1 at 0 and a local minimum equal to - 3 at 2. (Figure 6.2-4)

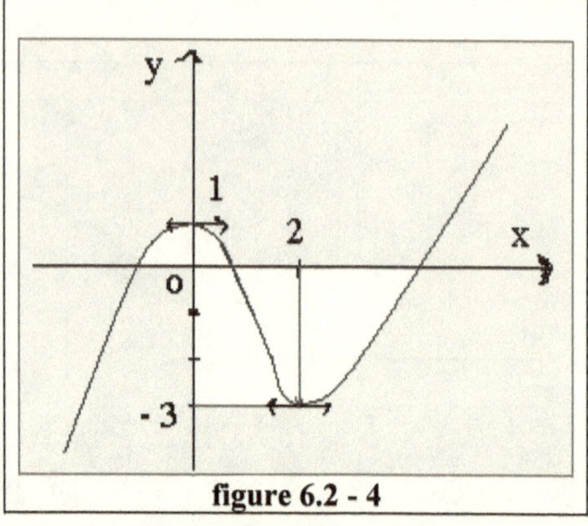

figure 6.2 - 4

Remark 6.2-2

a) A function can be continuous and have an extremum at a point without being differentiable at this point.

b) If f ' equals zero without changing signs, f has no local extreme.

c) If the interval I is closed at the endpoint a, then f can have an extremum at a without having f ' zero.(Figure 6.2 – 5).

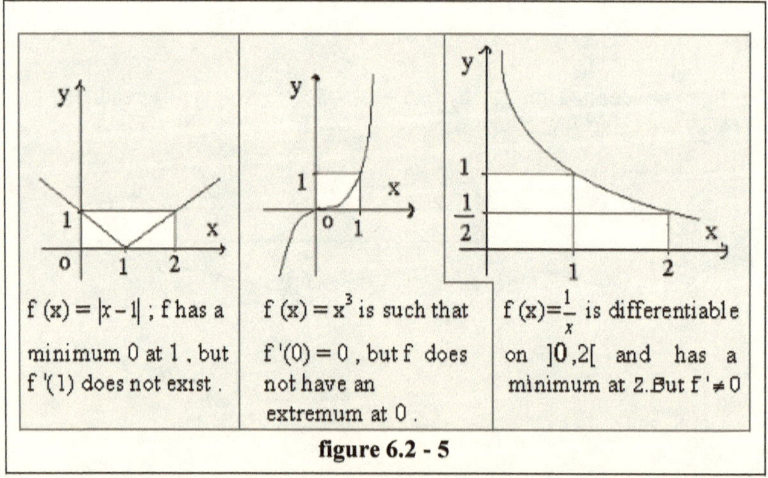

figure 6.2 - 5

Section 6.3

Critical Number

Definition 6.3-1

The number $c \in D_f$ is called critical number of the function f if one of the following conditions are satisfied:

a) $f'(c) = 0$.

b) $f'(c)$ does not exist.

Remark 6.3-1

If the function f has a local maximum or a local minimum at the number c, then c is a critical number. $(c, f(c))$ is a critical point.

Property 6.3-1(a method to find the absolute maximum and minimum)

1. Find the value of f at the critical numbers on the open interval (a,b), that is find all numbers $c \in (a, b)$ so that $f'(c) = 0$ or $f'(c)$ does not exist, then find $f(c)$.
2. Find the value of f on the boundaries of $[a, b]$; that is find $f(a)$ and $f(b)$.
3. The biggest values in (1) and (2) are absolute maximum.
4. The smallest values in (1) and (2) are absolute minimum.

Example 6.3-1

1) Find the critical point of $f(x) = x^2 - 2x$.

2) Find the critical point of $g(x) = \dfrac{3x+1}{x}$.

Solution

1) $f'(x) = 2x - 2$. Then $f'(x) = 0$ if $2x - 2 = 0$; $x = 1$, so the critical point is at $x = 1$.

2) $g'(x) = \dfrac{3x - (3x+1)(1)}{x^2} = -\dfrac{1}{x}$. So $g'(0)$ does not exist. Hence, the critical point is at $x = 0$.

Section 6.4

The Mean Value Theorem

Property 6.4-1(Rolle's Theorem)

This theorem gives sufficient conditions to the existence of at least one critical point over a defined interval.

If f satisfies the following conditions:

1) f is continuous over $[a, b]$.
2) f is differentiable over (a, b).
3) $f(a) = f(b)$.

Then there exists at least one number $c \in (a, b)$ satisfying $f'(c) = 0$.

* If $f(x) = k \in \mathfrak{R}$ is a constant function on $[a, b]$, then $f'(x) = 0$ for all $x \in (a, b)$. So any number $c \in (a, b)$ satisfies $f'(c) = 0$ (figure 6.4 - 1).

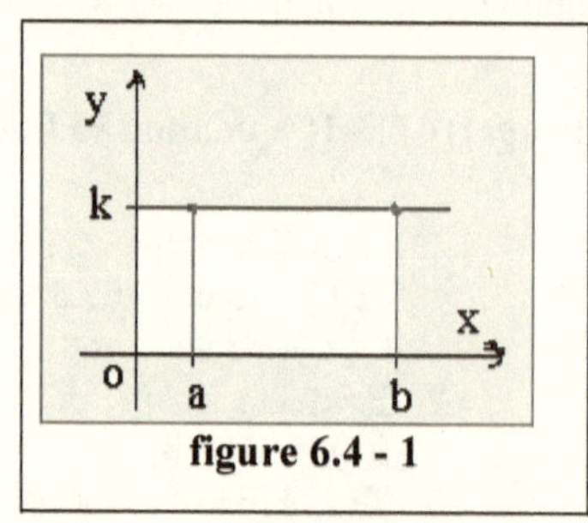

figure 6.4 - 1

* f is not a constant function so that $x \in (a, b)$ satisfies $f(x) > f(c)$. From the extremum value theorem, there exists for f a maximum value in the interval $[a, b]$. Since $f(a) = f(b)$, it is necessary that this maximum value should be on $c \in (a, b)$. So f has a local maximum at c, and since f is differentiable over (a, b), then f is differentiable at c and $f'(c) = 0$. (Figure 6.4 – 2).

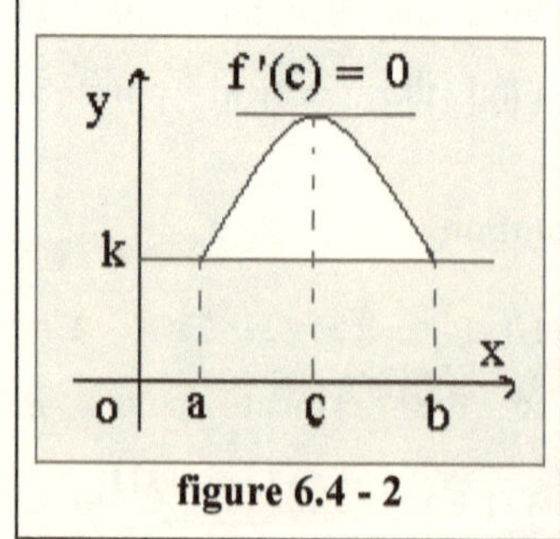

figure 6.4 - 2

* f is not a constant function so that $x \in (a, b)$ satisfies $f(x) < f(c)$. From the extremum value theorem, there exists for f a minimum value in the interval $[a, b]$. Since $f(a) = f(b)$ it is necessary that this minimum value should be on $c \in (a, b)$. So f has a local minimum at c and since f is differentiable over (a, b), then f is differentiable at c and $f'(c) = 0$. (Figure 6.4 – 3).

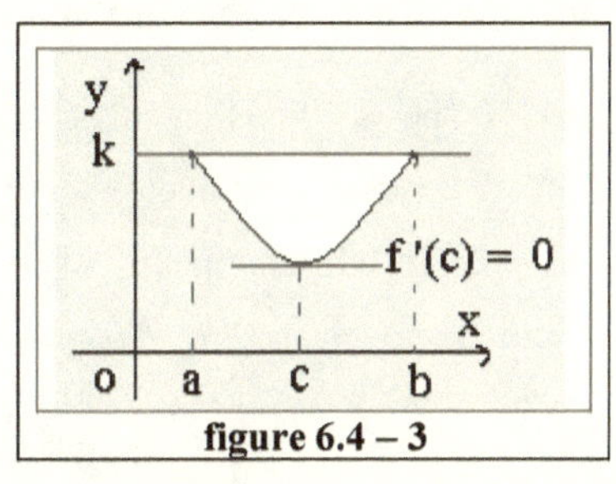

figure 6.4 – 3

Remark 6.4-1

Rolle's theorem includes at least one point $(c, f(c))$ so that the tangent at this point is horizontal and $f'(c) = 0$.(Figure 6.4 – 4).

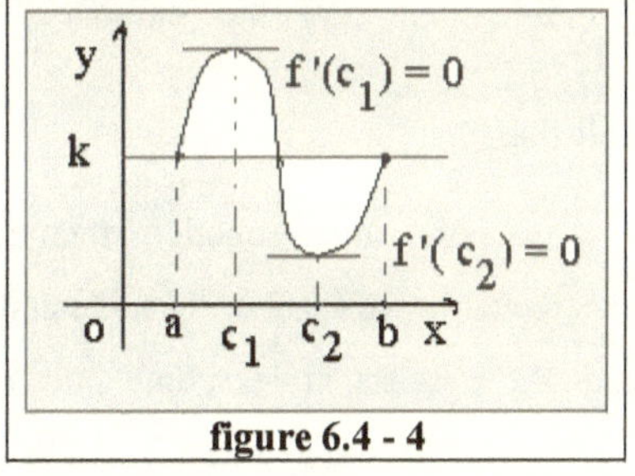

figure 6.4 - 4

Example 6.4-1

Prove that the equation $2x^5 - 1 = 0$ admits only one real root in $[0, 1]$.

Proof

Since f is a continuous function and $f(0) = -1 < 0$ and $f(1) = 1 > 0$, then there exists at least on root $c \in (0, 1)$ such that $f(c) = 0$. We need to prove the uniqueness of this root. Suppose that there exist two distinct real roots a, b for the given equation. Then $f(a) = f(b) = 0$; and since f is a polynomial function then it is :

1. continuous over $[a, b]$.
2. differentiable over (a, b).
3. $f(a) = f(b) = 0$.

Therefore, f satisfies Rolle 's theorem. Thus, there exists at least one number $c \in (a,b)$ such that $f'(c) = 0$. However, $f'(x) = 10x^4 \neq 0$ for all $x \in (a, b)$; this contradicts Rolle 's theorem and then there is only one real root.

Property 6.4-2(The Mean Value Theorem)

Let f be a function that satisfies the following conditions :

1. f is continuous over [a , b].
2. is differentiable over (a , b).

Then there exists at least one number $c \in (a, b)$ satisfying

$f'(c) = \dfrac{f(b)-f(a)}{b-a}$ that is $f(b) - f(a) = f'(c)(b-a)$.

Example 6.4-2

Show that $f(x) = x^3 - x + 2$ satisfies the mean value theorem on the interval [0 , 2], then find the numbers that satisfy this theorem.

Solution

f is a polynomial function , then it is continuous on the interval [0 , 2] and differentiable on (0, 2). Then f satisfies the conditions of the mean value theorem and there exists at least one number $c \in (0, 2)$ such that

$$f'(c) = \frac{f(b)-f(a)}{b-a}.$$

$f'(x) = 3x^2 - 1 \Rightarrow f'(c) = 3c^2 - 1$

$f(a) = f(0) = 2$

$f(b) = f(2) = 8 - 2 + 2 = 8$

$$\frac{f(b)-f(a)}{b-a} = \frac{8-2}{2-0} = 3$$

$$f'(c) = \frac{f(b)-f(a)}{b-a} \Rightarrow 3c^2 - 1 = 3$$

$$\Rightarrow c^2 = \frac{4}{3} \Rightarrow c = \pm\frac{2}{\sqrt{3}}.$$

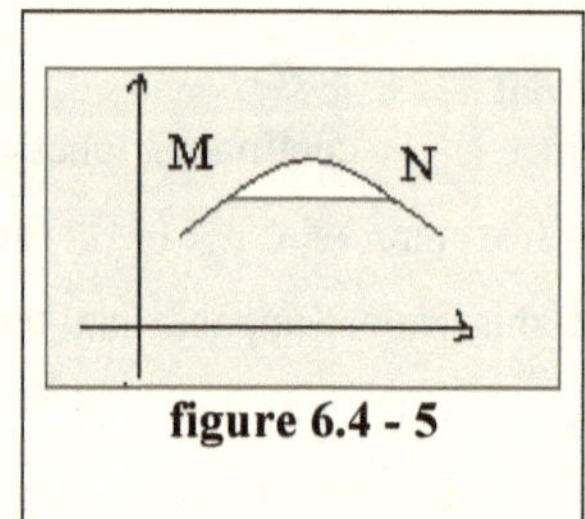

figure 6.4 - 5

so $c = \dfrac{2}{\sqrt{3}} \in (0, 2)$ satisfies the mean value theorem. (Figure 6.4 – 5).

Section 6.5

Concavity

Definition 6.5-1

Let f be a function defined on I = [a , b] whose curve is (c).

* f or (c) is said to be concave upward or (convex) on I if and only if every chord [M N] of (c) subtends an arc M N located below this chord.

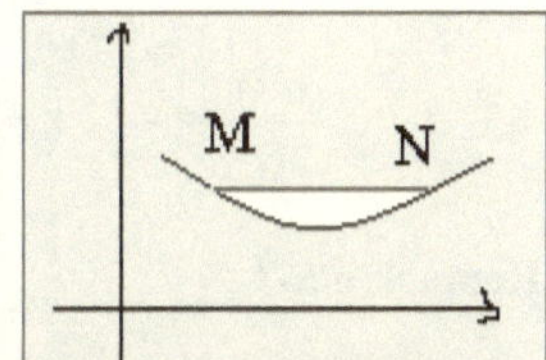

figure 6.5 - 1

* f or (c) is said to be concave downward on I if and only if every chord [M N] of (c) subtends an arc M N located above this chord.

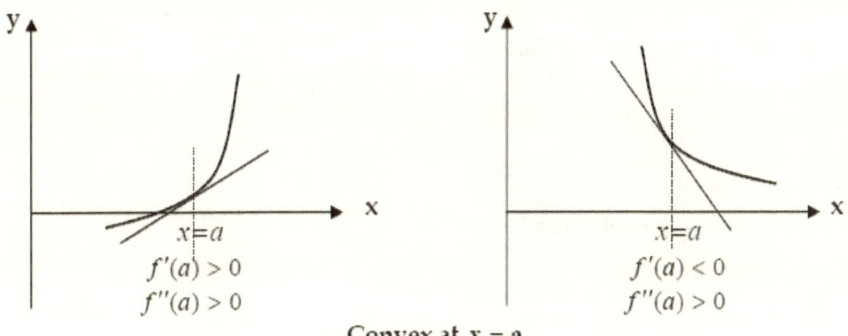

Convex at x = a

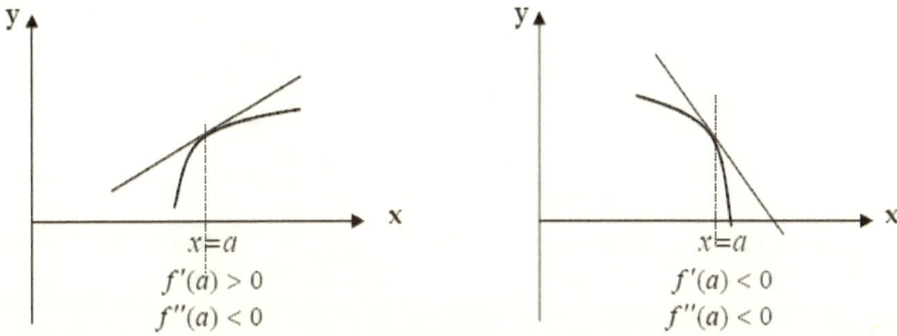

Concave at x = a

Note: If f" (x) > 0 for all x in the domain, f(x) is strictly convex. If f" (x) < 0 for all x in the domain, f(x) is strictly concave.

Example 6.5-1

Test whether the function $y = -2x^3 + 4x^2 + 9x - 15$ is concave or convex at $x = 3$.

$\frac{dy}{dx} = -6x^2 + 8x + 9.$

What we need to test for concavity / convexity is the sign of the second derivative at $x = 3$.

$$\frac{d^2y}{dx^2} = -12x + 8 = -12(3) + 8 = -28 < 0; \text{ therefore, concave.}$$

Remark 6.5-1

* If f is concave downward, then - f is concave upward.

* If f is concave upward, then - f is concave downward.

Property 6.5-1

Let f be a function having a second derivative function f".

* If $f'' \geq 0$ on an interval $]a, b[$, then f is concave upward.

* If $f'' \leq 0$ on an interval $]a, b[$, then f is concave downward.

x	$-\infty$	a	$+\infty$
f"(x)	+	0	−
f	∪ concave upward		∩ concave downward

Example 6.5-2

1) The function $f(x) = \sqrt{x}$ is concave downward on $]0, +\infty[$, since for $x > 0, f''(x) = \frac{-1}{4x\sqrt{x}} < 0$. (Figure 6.5 – 2).

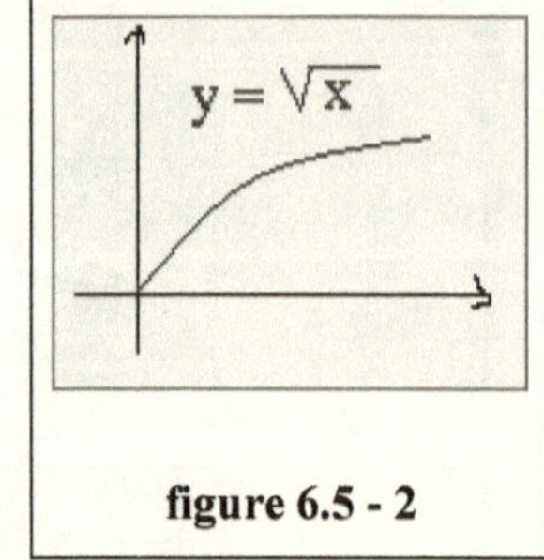

figure 6.5 - 2

2) The function $f(x) = \sin x$ is concave upward on the interval $[-\pi, 0]$, Since $f''(x) = -\sin x > 0$ on this interval. (Figure 6.5 – 3).

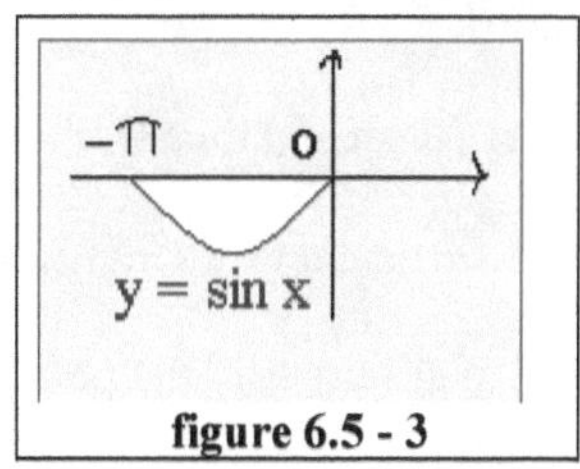

figure 6.5 - 3

Definition 6.5-2

We say that a function f has a point of inflection at $x = a$ on its domain of definition, if f changes concavity at a.

Example 6.5-3

The function $f(x) = x^3$ has a point of inflection at $x = 0$ since $f''(x) > 0$ for $x > 0$ and $f''(x) < 0$ for $x < 0$; hence the function f is concave downward on the interval $]-\infty, 0[$, and is concave upward on the interval $]0, +\infty[$. (Figure 6.5 - 4).

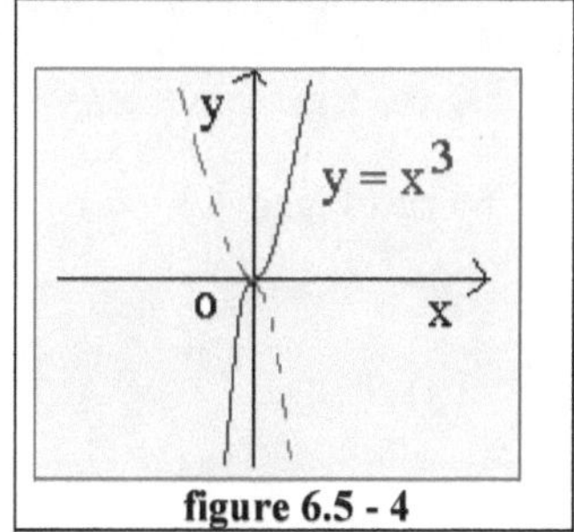

figure 6.5 - 4

Property 6.5-2

Let f be a function having a second derivative f". f has a point of inflection at a if and only if f" vanishes at this point and changes sign at this point.

Example 6.5-4

$f(x) = \sqrt{x}$ does not have a point of inflection, since f" is never 0.

Remark 6.5-2

1) The condition $f''(a) = 0$ is not sufficient to guarantee that the point of abscissa a is a point of inflection. For example, $f(x) = x^4$ is concave upward on $\Re$, nevertheless $f''(0) = 0$. (Figure 6.5 – 5).

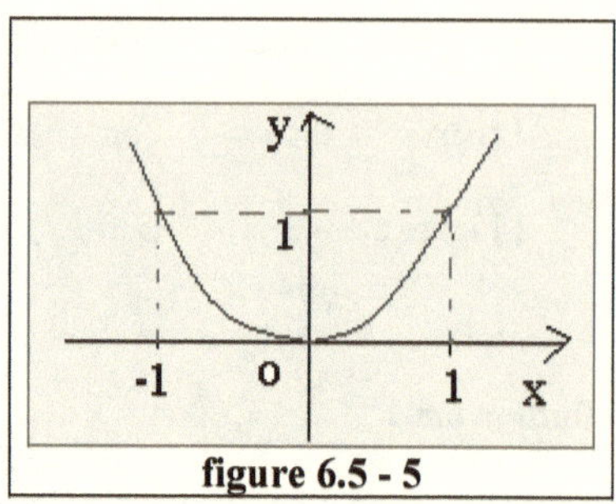

figure 6.5 - 5

2) A polynomial function of degree 1 or 2 has no inflection point. While a polynomial of degree 3 or above may have inflection points.(Figure 6.5 – 6).

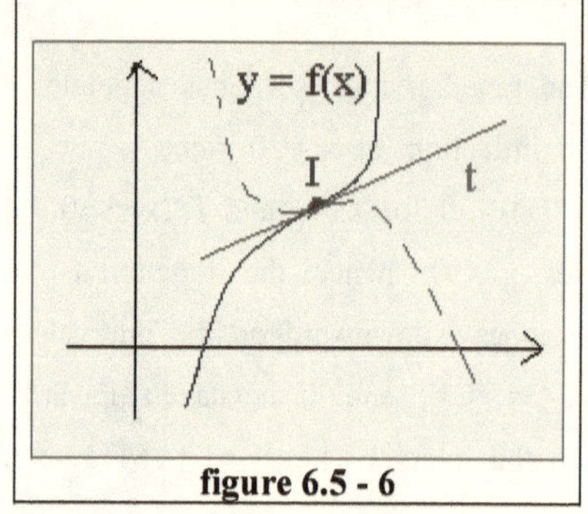

figure 6.5 - 6

3) There is a tangent that cut the curve on the inflection point.

Property 6.5-3

Let f and g be differentiable functions in the neighborhood of the real number a. If $f(a) = g(a) = 0$, if $g'(x) \neq 0$ at $x \neq a$, and if $\lim_{x \to a} \frac{f'(x)}{g'(x)} = l$, then $\lim_{x \to a} \frac{f(x)}{g(x)} = l$. This rule, called L' Hopital Rule, l could be finite or infinite.

Remark 6.5-3

The condition $f(a) = g(a) = 0$ explain the label " indeterminate form " of $\lim_{x \to a} \frac{f(x)}{g(x)} = l$.

Example 6.5-5

Show that

a. $\lim_{x \to 0} \dfrac{\sin x}{x} = 1$

b. $\lim_{x \to 0} \dfrac{\tan x}{x} = 1$

c. $\lim_{x \to 0^+} x \ln x = 0$

d. $\lim_{x \to +\infty} \dfrac{\ln x}{x} = 0$

e. $\lim_{x \to +\infty} \dfrac{e^x}{x} = +\infty$

Solution

a. $\lim_{x \to 0} \dfrac{\sin x}{x} = \dfrac{0}{0}$ indeterminate form, apply L' Hopital Rule:

$\lim_{x \to 0} \dfrac{\sin x}{x} = \lim_{x \to 0} \dfrac{\cos x}{1} = 1.$

b. $\lim_{x \to 0} \dfrac{\tan x}{x} = \dfrac{0}{0}$ indeterminate form, apply L' Hopital Rule:

$\lim_{x \to 0} \dfrac{\tan x}{x} = \lim_{x \to 0} \dfrac{1 + \tan^2 x}{1} = 1.$

c. $\lim_{x \to 0^+} x \ln x = 0(-\infty)$ indeterminate form,

$\lim_{x \to 0^+} x \ln x = \lim_{x \to 0^+} \dfrac{\ln x}{\frac{1}{x}} = \dfrac{-\infty}{+\infty}$ indeterminate form, then apply L' Hopital

Rule : $\lim_{x \to 0^+} x \ln x = \lim_{x \to 0^+} \dfrac{\frac{1}{x}}{\frac{-1}{x^2}} = \lim_{x \to 0^+} \dfrac{1}{\frac{-1}{x}} = \lim_{x \to 0^+} -x = 0$.

d. $\lim_{x \to +\infty} \dfrac{\ln x}{x} = \dfrac{\infty}{\infty}$ indeterminate form, then apply L' Hopital Rule:

$$\lim_{x \to +\infty} \frac{\ln x}{x} = \lim_{x \to +\infty} \frac{\frac{1}{x}}{1} = 0.$$

e. $\lim_{x \to +\infty} \frac{e^x}{x} = \frac{\infty}{\infty}$ indeterminate form, then apply L' Hopital Rule:

$$\lim_{x \to +\infty} \frac{e^x}{x} = \lim_{x \to +\infty} \frac{e^x}{1} = +\infty.$$

Section 6.6

Variation and Graph of Polynomial Functions

The following conditions are needed to study the variation and drawing the graph of polynomial functions :

1. Domain of definition.
2. limits on the boundaries of the domain of definition.
3. Finding the zero derivative.
4. Constructing the table of variation.
5. Concavity (polynomials of degree 3 and 4).
6. Particular points (when needed).
7. Graph .

Example 6.6-1

Study the variation of the following functions and draw their representative curves in an orthonormal system of axes x' o x, y 'o y:

1) $f(x) = x^2 - 4x + 3$.

2) $g(x) = x^3 - 3x^2$.

3) $h(x) = x^4 + 2x^3$.

Solution

1) f(x) is a polynomial function, then f is defined and continuous for every $x \in \Re$.

$$\lim_{x \to \pm\infty} f(x) = \lim_{x \to \pm\infty} x^2 - 4x + 3 = \lim_{x \to \pm\infty} x^2 = +\infty.$$

$f'(x) = 2x - 4$, then $f'(x) = 0$ if $2x - 4 = 0 \Rightarrow 2x = 4 \Rightarrow x = 2$. for $x = 2$, $f(x) = 2^2 - 4(2) + 3 \Rightarrow$ $f(x) = 4 - 8 + 3 = -1$. So the point s (2 , -1) is the vertex of the given parabola.

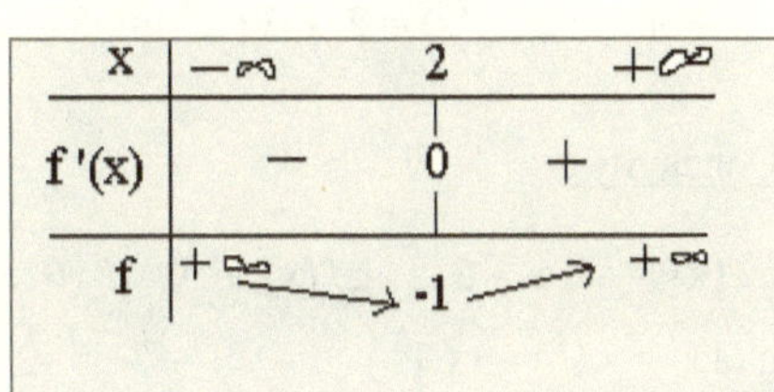

x	$-\infty$	2	$+\infty$
f '(x)	−	0	+
f	$+\infty$ ↘	-1	↗ $+\infty$

Particular points :

X	0	1 or 3
Y	3	0

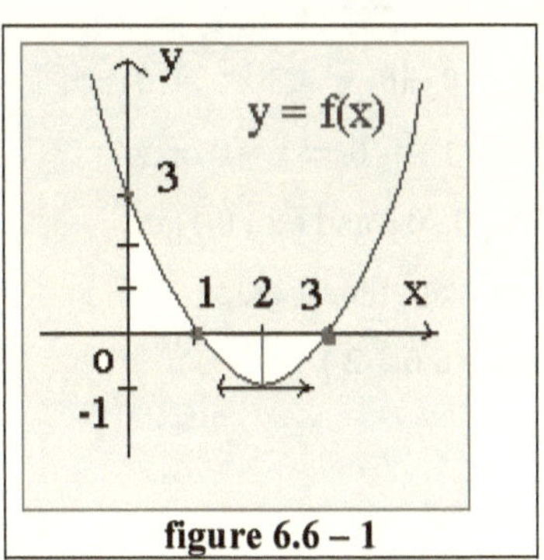

figure 6.6 – 1

2) g(x) is a polynomial function, then g is defined and continuous for every $x \in \Re$.

$$\lim_{x \to +\infty} g(x) = \lim_{x \to +\infty} x^3 - 3x^2 = \lim_{x \to +\infty} x^3 = +\infty.$$

$$\lim_{x \to -\infty} g(x) = \lim_{x \to -\infty} x^3 - 3x^2 = -\infty - \infty = -\infty.$$

$g'(x) = 3x^2 - 6x$.

$g'(x) = 0$ if $3x^2 - 6x = 0$

$\Rightarrow 3x(x-2) = 0 \Rightarrow x = 0$

or $x = 2$.

For $x = 0$, $g(x) = 0$ then $(0, 0)$

is a vertex.

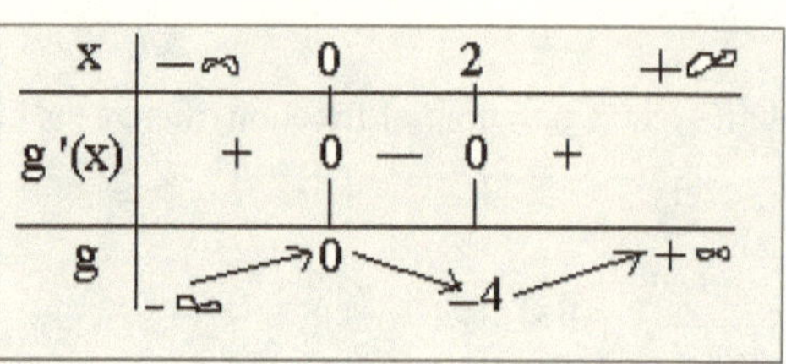

x	$-\infty$		0		2		$+\infty$
g '(x)		+	0	—	0	+	
g	$-\infty$	↗	0	↘	−4	↗	$+\infty$

For $x = 2$, $g(x) = 2^3 - 3(2^2) = 8 - 12 = -4$ then $(2, -4)$ is a vertex.

Concavity :

$g''(x) = 6x - 6$. $g''(x) = 0$ if $6x - 6 = 0 \Rightarrow x = 1$.

$g(1) = 1^3 - 3(1^2) = 1 - 3 = -2$.

Since $g''(x) = 0$ with changing of sign, then the point $I(1, -2)$ is an inflection point.

x	$-\infty$	1	$+\infty$
g "(x)	—	0	+
g	∩		∪

Particular points

$g(x) = 0$ if $x^3 - 3x^2 = 0 \Rightarrow$

$x^2(x-3) = 0 \Rightarrow x = 0$ or $x = 3$.

Then $(0, 0)$ and $(3, 0)$ are points of the graph of g.

(Figure 6.6 – 2).

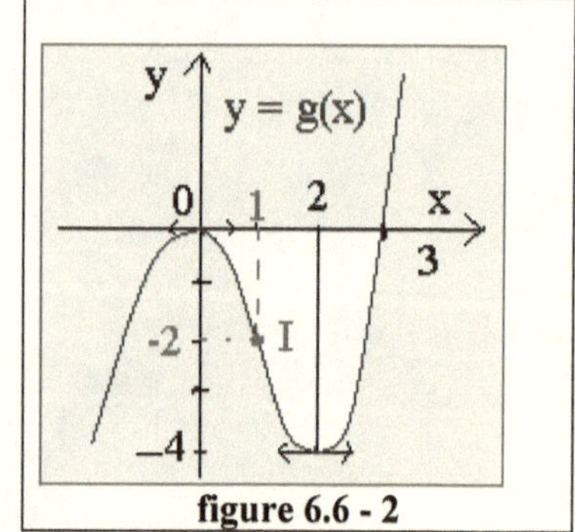

figure 6.6 - 2

3) h(x) is a polynomial function, then h is defined and continuous for every $x \in \Re$.

$$\lim_{x \to \pm\infty} h(x) = \lim_{x \to \pm\infty} x^4 + 2x^3 = \lim_{x \to \pm\infty} x^4 = +\infty.$$

$h'(x) = 4x^3 + 6x^2$. $h'(x) = 0$ if $4x^3 + 6x^2 = 0$

$\Rightarrow 2x^2(2x+3) = 0 \Rightarrow x = 0$ or $x = -1.5$.

For $x = 0$, $h(x) = 0$ then $(0, 0)$ is a vertex.

For $x = -1.5$, then $h(x) = (-1.5)^4 + 2(-1.5)^3 = 5.0625 - 6.75 \approx -1.7$

Then $(-1.5, -1.7)$ is a vertex.

Sign of h '(x) :

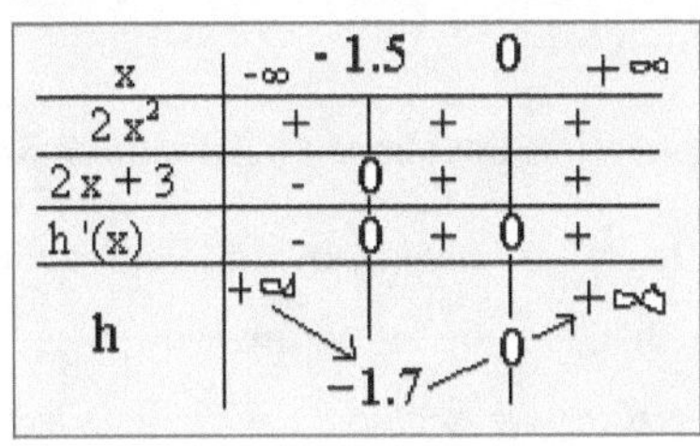

x	$-\infty$		-1.5		0		$+\infty$
$2x^2$		+		+		+	
$2x+3$		-	0	+		+	
h '(x)		-	0	+	0	+	
h	$+\infty$	↘	−1.7	↗	0	↗	$+\infty$

Concavity :

$h''(x) = 12x^2 + 12x$; $h''(x) = 0$ if

$12x^2 + 12x = 0$

x	$-\infty$		-1		0		$+\infty$
h"(x)		+	0	−	0	+	
h		∪		∩		∪	

$\Rightarrow 12x(x+1) = 0 \Rightarrow x = 0$ or $x = -1$

For $x = 0$, $h(x) = 0$ and for $x = -1$, $h(x) = (-1)^4 + 2(-1)^3$

$\Rightarrow h(x) = 1 - 2 = -1.$

Since h"(0) and h"(- 1) are zeros with changing of sign, then the points $I_1(0, 0)$ and $I_2(-1, -1)$ are inflection points.

Particular points :

$h(x) = 0$ if $x^4 + 2x^3 = 0$,then $x^3(x+2) = 0$

$\Rightarrow x = 0$ or $x = -2$. then(0 , 0) and (- 2 , 0) are points of the graph of h. (Figure 6.6 – 3)

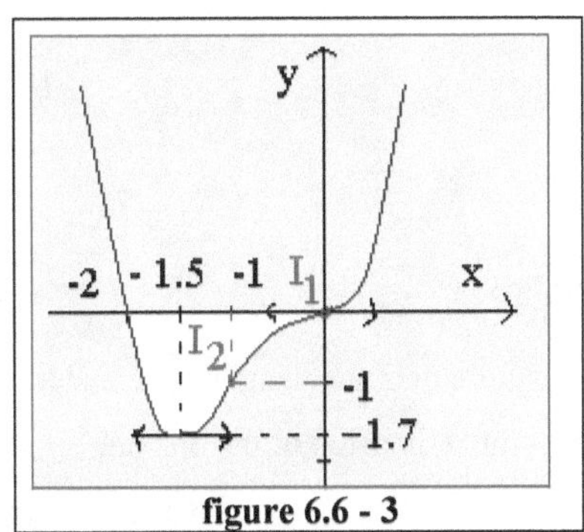

figure 6.6 - 3

Section 6.7

Application Problems on Maximum and Minimum Values

There are infinitely many problems about the applied problems of extremum along which there are no permanent rules followed to reach the needed results. Therefore, we will just have certain examples about it.

Example 6.7-1

1) We would like to make a box with rectangular base from strong papers with dimensions 30 cm and 16 cm ; this will be done by cutting equal squares from each corner then folding the boundaries. What will be the length of the side of the cutter square so that the volume of the obtained box will be maximum?

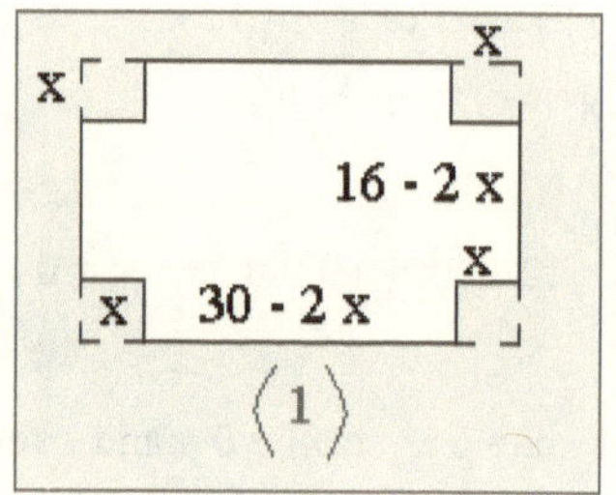

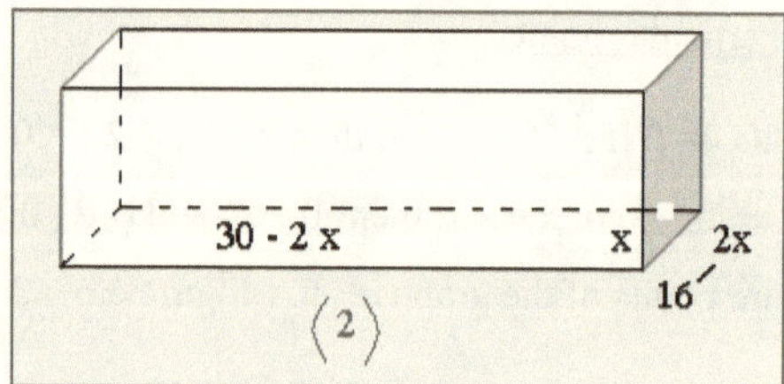

Solution

Suppose that the length of the side of the cutter square is x then it is obvious that the dimension of the box are x, $30 - 2x$, and $16 - 2x$. If the volume is y then $y = x(30 - 2x)(16 - 2x)$. To have a box, its height must not be negative; that is, $x \geq 0$, and that its width must not be negative; that is, $16 - 2x \geq 0$; so in this case we notice that the length becomes also not negative; we then see that $0 \leq x \leq 8$. Hence, mathematically, we need to find where the function $f(x) = x(30 - 2x)(16 - 2x)$ is maximum in the interval $[0, 8]$. The needed value occurs at zero, or 8, or at a critical point of the function f in $(0, 8)$. Since f is a polynomial function, its critical points are the points where the derivative f' is zero.

$f(x) = 4(x^3 - 23x^2 + 120x)$ then $f'(x) = 4(3x^2 - 46x + 120)$
$= 4(3x - 10)(x - 12)$

$f'(x) = 0$ if $4(3x - 10)(x - 12) = 0 \Rightarrow x = \frac{10}{3}$ or $x = 12$. So the critical points are $\frac{10}{3}$ and 12. $12 \notin (0, 8)$ then it is rejected. Hence, the needed maximum value occurs at zero, 8 or $\frac{10}{3}$. Now, $f(0) = 0$, $f(8) = 0$ and $f(\frac{10}{3}) > 0$; then to have a box with maximum volume, we take the length of the cutter side to be $\frac{10}{3}$ cm.

Example 6.7-2

Find the dimension of a cylindrical box opened up with volume 1 ft^3 and which use a least amount of quantity of iron of fixed thickness.

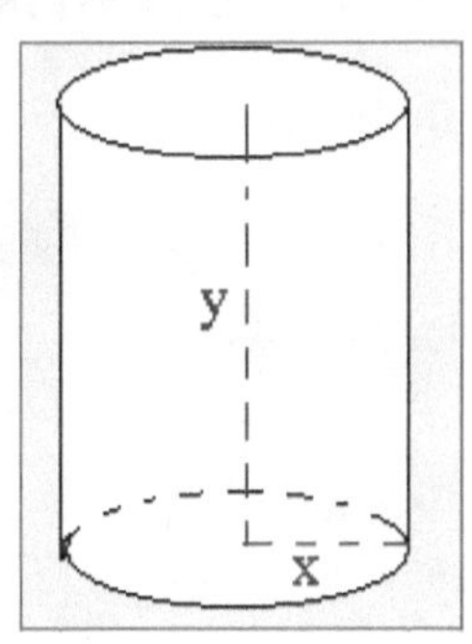

Solution

Suppose that the radius of the circular base is x and that the height is y then the area of the upper surface is $z = 2\pi xy + \pi x^2$. The volume is $v = \pi x^2 y = 1$;

then $y = \frac{1}{\pi x^2}$. So $z = \frac{2}{x} + \pi x^2$. So z becomes now with only one unknown x,

and it is obvious that $x \in (0, \infty)$. Therefore, we need to find the value of x in $(0, \infty)$ which allows z as small as possible. Notice here that the interval is not finite and not closed; to get out of this problem we apply the fact that the local extremum is being absolute extremum if the function has only one critical point in the interval. $f(x) = \frac{2}{x} + \pi x^2$ then $f'(x) = \frac{2(\pi x^3 - 1)}{x^2}$. So

$\frac{1}{\sqrt[3]{\pi}}$ is the only critical point of the function in $(0, \infty)$.

Now, $f''(x) = \frac{4}{x^3} + 2\pi$, then $f''(\frac{1}{\sqrt[3]{\pi}}) > 0$. This means that $f(\frac{1}{\sqrt[3]{\pi}})$ is a local minimum. Since there exists no critical point other than $(\frac{1}{\sqrt[3]{\pi}})$ in $(0, \infty)$, then $f(\frac{1}{\sqrt[3]{\pi}})$ is the smallest value of the function in $(0, \infty)$.

Thus, the radius $x = \frac{1}{\sqrt[3]{\pi}}$ and the height $y = \frac{1}{\pi.(\frac{1}{\sqrt[3]{\pi}})^2} = \frac{1}{\sqrt[3]{\pi}}$.

Section 6.8

Finance

Rate of Change of Price with Respect to Quantity

Definition 6.8.1

If p is the demand function for a manufacturer's product and q is the quantity, then the rate of change of price p with respect to quantity q is $\frac{d_p}{d_q}$.

Example 6.8-1

Let $p = q^3 - 2q + 100$ be the demand function for a manufacture's product. Determine the rate of change of price p per unit with respect to quantity q. How fast is the price changing with respect to q when q = 4? Assume that p is in dollars.

Solution

The rate of change of p with respect to q is: $\frac{d_p}{d_q} = 3q^2 - 2$

Then, $\frac{d_p}{d_q}\Big|_{q=4} = 3 \times 4^2 - 2 = 46$.

This means that when four units are demanded , an increase of one extra unit demanded corresponds to an increase of approximately \$ 46 in the price per unit that consumers are willing to pay .

Marginal Cost

Definition 6.8.2

If c = f (q) is the total cost function of a manufacturer, q is the marketing units of product, and $\bar{c}$ is the average cost per unit, then the marginal cost is the rate of change of c with respect q. That is, marginal cost = $\frac{d_c}{d_q}$ where $c = q . \bar{c}$

Example 6.8-2

If a manufacturer's average - cost equation is: $\bar{c} = 0.001q^2 - 0.1q + \frac{4000}{q} + 6$. Find the marginal – cost function. What is the marginal cost when 30 units are produced?

Solution

$$c = q\,\bar{c}$$

$$= q\,(0.001q^2 - 0.1q + \frac{4000}{q} + 6\,)$$

$$= 0.001q^3 - 0.1q^2 + 4000 + 6\,q$$

the marginal – cost function is: $\frac{dc}{dq} = 0.003\,q^{2} - 0.2\,q + 6$

Then the marginal cost when 30 units are produced is:

$$\frac{d_c}{d_q}\Big|_{q=30} = 0.003(\,30\,)^{2} - 0.2\,(\,30\,) + 6$$

$$= 2.7.$$

This means that when production is increased by one unit and c is in dollars, then the cost of the additional unit is approximately \$ 2.7.

Marginal Revenue

The revenue r received for selling q units when the price per unit is p is given by:

Revenue = Price × Quantity = p . q .

Definition 6.8.3

If r = f (q) is the total - revenue function for a manufacturer. q is the marketing units of product, then the marginal revenue is the rate of change of the total dollar value received (r) with respect to the total number of units sold (q). That is, marginal revenue = $\frac{d_r}{d_q}$.

Example 6.8-3

For a certain manufacturer, the revenue obtained from the sale of q units of a product is given by r = 20 q - 0.2 q ^{2}. Find the marginal revenue when q = 10.

Solution

The marginal revenue function is: $\frac{dr}{dq} = 20 - 0.4\,q$.

When q = 10, the marginal revenue is: $\left.\frac{dr}{dq}\right|_{q=10} = 20 - 0.4\,(10)$

$= 16.$

Relative Rate of Change

Definition 6.8.4

The relative rate of change of y = f (x) is: $\frac{y'}{y}$.

Percentage Rate of Change

Definition 6.8.5

The percentage rate of change of y = f (x) is: $\frac{y'}{y} \times 100$.

Example 6.8-4

Determine the relative and percentage rates of change of $y = f(x) = 4x^2 - 6x + 15$ when x = 4.

Solution

Since $y' = 8x - 6$, then the relative of change of y is: $\frac{y'}{y} = \frac{8x-6}{4x^2 - 6x + 15}$.

When x = 4, then the relative of change of y is: $\left.\frac{y'}{y}\right|_{x=4} = \frac{8(4)-6}{4(4)^2-6(4)+15}$

$\approx 0.427.$

and the percentage rate of change of y is: $\left.\frac{y'}{y}\right|_{x=4} \times 100 = 0.427 \times 100$

$= 42.7\,\%.$

Taylor Series

A function, f(x), may be approximated at some point, x_0, by the following:

$$f(x) = f(x_0) + f^1(x_0)(x - x_0) + \frac{1}{2!} f^2(x_0)(x - x_0)^2 +$$

$$\frac{1}{3!} f^3(x_0)(x - x_0)^3 + \ldots\ldots + \frac{1}{n!} f^n(x_0)(x - x_0)^n + \text{Remainder}$$

where $f(x_0)$ is the value of f(x) at the point x_0 and n! reads n factorial where

n! = n (n-1)(n-2)…..3.2.1.

Note: 0! = 1! = 1.

Maclaurin Series

A special case of the Taylor Series is when $x_0 = 0$, in which case

$$f(x) = f(0) + f^1(0)\,x + \frac{1}{2!} f^2(0)\,x^2 + \frac{1}{3!} f^3(0)\,x^3 + \ldots + \frac{1}{n!} f^n(0)\,x^n + \text{Remainder}$$

Example 6.8-5

Let $f(x) = (1 + x)^3$

We know that $(1 + x)^3 = 1 + 3x + 3x^2 + x^3$. Use Maclaurin Series to prove it.

$f(x) = (1 + x)^3$ $\quad$ $f(0) = 1$

$f^1(x) = 3(1 + x)^2$ $\quad$ $f^1(0) = 3$

$f^2(x) = 6(1 + x)$ $\quad$ $f^2(0) = 6$

$f^3(x) = 6$ $\quad$ $f^3(0) = 6$

$f^4(x) = 0$ $\quad$ $f^4(0) = 0$

Therefore, $\quad f(x) = 1 + 3x + \frac{1}{2!} 6x^2 + \frac{1}{3!} 6x^3$

Thus, $\quad (1 + x)^3 = 1 + 3x + 3x^2 + x^3$

Example 6.8-6

Find the value of e^x.

Let $f(x) = e^x$ then

$$f(x) = e^x \qquad f(0) = e^0 = 1$$
$$f^1(x) = e^x \qquad f^1(0) = e^0 = 1$$
$$f^2(x) = e^x \qquad f^2(0) = e^0 = 1$$
$$f^3(x) = e^x \qquad f^3(0) = e^0 = 1$$
$$f^4(x) = e^x \qquad f^4(0) = e^0 = 1$$

. .

. .

. .

Therefore, $e^x = 1 + x + \frac{x^2}{2!} + \frac{x^3}{3!} + \ldots.. + \frac{x^n}{n!} + \ldots..$

Another example of Taylor Series

Taylor series expansion is used to measure the sensitivity of bond prices with respect to changes in interest rates. Using the first two terms of a Taylor series expansion to approximate the price change we can write

$$dP = \frac{dP}{dy}dy + \frac{1}{2}\frac{d^2P}{dy^2}(dy)^2 + error$$

Dividing both sides of the equation by P to find the percentage price change results in

$$\frac{dP}{P} = \frac{dP}{dy}\frac{1}{P}dy + \frac{1}{2}\frac{d^2P}{dy^2}\frac{1}{P}(dy)^2 + \frac{error}{P}$$

The duration of the bond is defined as $D = -\frac{1}{P}\frac{dP}{dy}$ The convexity of the bond is

defined as $C = \frac{d^2P}{dy^2}$

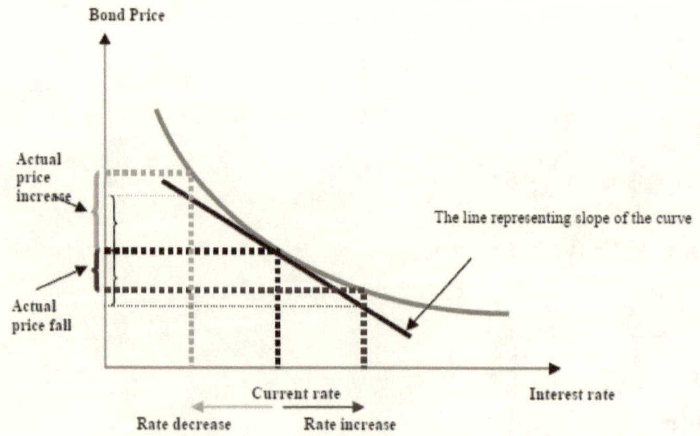

Optimization of Economic Functions

Profit Maximization:

Let's suppose an economist is asked to maximize the profits of his firm, given a total revenue function $R = 4000\,q - 33\,q^2$, where q represents the level of output, and a total cost function $C = 2\,q^3 - 3\,q^2 + 400\,q + 5000$, assuming $q > 0$.

(i) Set up the profit function: $P = R - C$

$$P = 4000\,q - 33\,q^2 - (2\,q^3 - 3\,q^2 + 400\,q + 5000)$$
$$= -2\,q^3 - 30\,q^2 + 3600\,q - 5000$$

(ii) Take the first derivative, set it equal to zero, and solve for q to find the critical points.

$$P' = -6\,q^2 - 60\,q + 3600 = 0$$
$$= -6\,(q^2 + 10\,q - 600) = 0$$
$$= -6\,(q + 30)(q - 20) = 0$$

Therefore, $q = -30$ and $q = 20$.

(iii) Take the second derivative, evaluate it at the critical point and ignore the negative critical point which has no economic significance and will prove mathematically to be a relative minimum. Then check the sign of the second derivative to be sure of a relative maximum.

$$P'' = -12\,q - 60$$

$P''(20) = -12\,(20) - 60 = -300 < 0$ concave, relative maximum.

Therefore, profit is maximized at $q = 20$ where

$$P(20) = -2\,(20)^3 - 30\,(20)^2 + 3600\,(20) - 5000 = 39000.$$

Optimization in Finance

A. Utility Functions

Utility theory postulates that an investor would choose between competing Investments by ranking them according to his preference (or utility) function. A number of brokerage firms and banks have developed programs to extract the utility functions of investors by confronting them with a choice between a series of simple investments. These have not been particularly successful. This is because many investors do not obey the rationality postulates when faced with a series of choice situations. Moreover, when faced with more complicated choice situations, they encounter aspects of the problem that were not of concern to them in the simple choice situations. However, utility functions remain of major importance as they provide insight into the process of rational choice. A risk-averse investor would reject a fair gamble because the disutility of loss is greater than the utility of an equivalent gain. Functions that exhibit this property must have a negative second derivative. In the figure below the utility function in utility of wealth space of a risk-averse investor is exhibited.

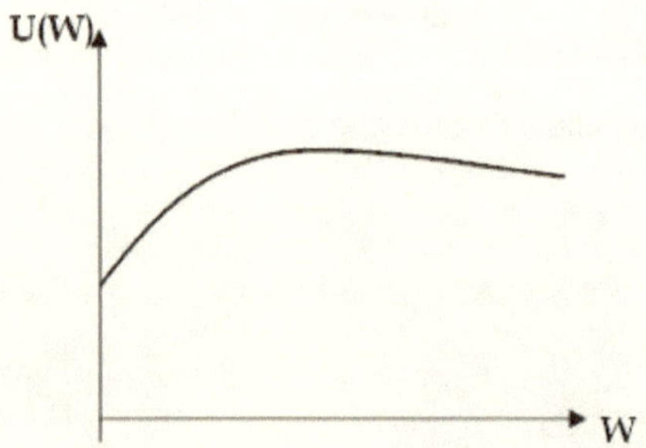

If $U(W)$ is the utility function of a risk-averse investor, and $U''(W)$ is the second derivative, then $U''(W) < 0$. Let's see why this is so. Suppose that an investor is faced with the choice of either investing \$ 1 in an investment which will yield a profit of \$ 2 with probability 0.5 or \$ 0. The alternative is not to invest at all and keep the dollar with certainty. The investment has an expected value of $(0.5)(2) + (0.5)(0) = 1$, the expected value of the gamble is equal to the cost.

As mentioned before, a risk-averse investor would reject this fair gamble. If so, the expected utility from not investing must be higher than the expected utility of investing, or

$$U(1) > (0.5)\,U(2) + (0.5)\,U(0)$$

Multiply both sides by two and rearranging we have

$$U(1) - U(0) > U(2) - U(1)$$

The preceding expression tells us that a one-unit change from 0 to1 is more valuable than a one – unit change from 1 to 2. A function where an additional unit increase is less valuable than the last unit increase is a function with a negative second derivative, such as the one shown in the above figure. Higher curvature, measured by $U''(W)$, implies higher risk aversion. An important property of utility functions rests on how the investor's preferences change with a change in wealth. Absolute risk aversion (ARA) measures the amount an individual is willing to expose to risk as wealth changes. If the investor increases the amount invested in risky assets as wealth increases, then the investor is said to exhibit decreasing absolute risk aversion (e.g. U = lnW). If the investor's investment in risky assets is unchanged as wealth increases, then the investor is said to exhibit constant absolute risk aversion. Finally, if he invests less money as wealth increases, he exhibits increasing absolute risk aversion. If U′(W) and U″(W) are the first and second derivatives of the utility function respectively at wealth level W, then

$$A(W)\frac{-U''(W)}{U'(W)}$$

can be used to measure an investor's absolute risk - aversion. Then A′(W), the derivative of A(W) with respect to wealth, is an appropriate measure of how absolute risk aversion behaves with respect to changes in wealth

Increasing ARA	$A'(W) > 0$
Constant ARA	$A'(W) = 0$
Decreasing ARA	$A'(W) < 0$

There is another measure of risk-aversion, known as relative risk-aversion, which refers to the change in the percentage investment in risky assets as wealth changes. The measure of relative risk-aversion (R R A or simply R (W)) is

$$R(W)=\frac{-W\,U''(W)}{U'(W)}=A(W)\,.\,W$$

Increasing RRA	$R'(W) > 0$
Constant RRA	$R'(W) = 0$
Decreasing RRA	$R'(W) < 0$

B. Portfolios

Optimization can also be used to pick the appropriate weights of a portfolio so as to minimize its variance, and thus choose the less risky investment.

Matrix Operations to Calculate Return and Variance of a Portfolio

Let X denote the vector of returns on the n assets we include in our investment portfolio:

$$X = \begin{bmatrix} X_1 \\ X_2 \\ . \\ . \\ . \\ X_n \end{bmatrix}$$

Where E [X $_i$] = μ_i, var (X$_i$) = σ_i^2 and cov (X$_i$, X$_j$) = σ_{ij}. Then we can write

$$E[X] = \begin{bmatrix} E[X_1] \\ E[X_2] \\ . \\ . \\ . \\ E[X_n] \end{bmatrix} = \begin{bmatrix} \mu_1 \\ \mu_2 \\ . \\ . \\ . \\ \mu_n \end{bmatrix} = \mu$$

Similarly, we can define the covariance matrix for the random vector X as follows

cov (X) = E [(X – E [X])(X – E [X])']

$$= E\begin{bmatrix} (X_1 - \mu_1)^2 & (X_1 - \mu_1)(X_2 - \mu_2) & ..... & (X_1 - \mu_1)(X_n - \mu_n) \\ (X_2 - \mu_2)(X_1 - \mu_1) & (X_2 - \mu_2)^2 & ..... & (X_2 - \mu_2)(X_n - \mu_n) \\ . & . & . & . \\ . & . & . & . \\ . & . & . & . \\ (X_n - \mu_n)(X_1 - \mu_1) & (X_n - \mu_n)(X_2 - \mu_2) & . & (X_n - \mu_n)^2 \end{bmatrix}$$

$$= \begin{bmatrix} \sigma_1^2 & \sigma_{12} & ..... & \sigma_{1n} \\ \sigma_{21} & \sigma_2^2 & ..... & \sigma_{2n} \\ . & . & . & . \\ . & . & . & . \\ . & . & . & . \\ \sigma_{n1} & \sigma_{n2} & . & \sigma_n^2 \end{bmatrix}$$

The covariance matrix of X is often denoted by $\sum_x$ and is a symmetric matrix with variances on the main diagonal and covariances as off-diagonal elements.

If a ' = (a $_1$, a $_2$, …., a $_n$) is a vector of weights, then the expected return of the portfolio can be calculated as:

$$E[a'X]=a'\mu=a_1\,\mu_1+a_2\,\mu_2+....+a_n\,\mu_n$$

And the variance of the portfolio as:

$$\text{var}[a'X]=a'\textstyle\sum_x a=\sum a_i^2\sigma_i^2+2\sum_{i<j}\sum a_i a_j \sigma_{ij}$$

Example 6.8-7

Let us consider a portfolio of five stocks with the following weights:

a = [0.35, 0.25, 0.15, 0.15, 0.10]

Mean Weekly Returns

Ret1	Ret 2	Ret 3	Ret 4	Ret 5
0.002786	0.000232	0.002572	0.002835	- 0.0011155

Correlation Matrix

	Ret1	Ret 2	Ret 3	Ret 4	Ret 5
Ret 1	1				
Ret 2	0.212307	1			
Ret 3	0.343187	0.245859	1		
Ret 4	0.155277	0.086078	0.120997	1	
Ret 5	0.078979	0.129544	0.048159	- 0.00148	

Variance- Covariance Matrix

	Ret1	Ret 2	Ret 3	Ret 4	Ret 5
Ret 1	0.000969	0.000124	0.000358	0.000187	0.000111
Ret 2	0.000124	0.000350	0.000154	0.000062	0.000110
Ret 3	0.000358	0.000154	0.001122	0.000157	0.000073
Ret 4	0.000187	0.000062	0.000157	0.001502	-0.000003
Ret 5	0.000111	0.000110	0.000073	-0.000003	0.002050

We know that

$$E[a'X]=a'\mu=a_1\,\mu_1+a_2\,\mu_2+....+a_n\,\mu_n$$

$$E[a'X]=a'\mu=[\,0.35,\;\;0.25,\;\;0.15,\;\;0.15,\;\;0.10\,]\begin{bmatrix}0.002786\\0.000232\\0.002572\\0.002835\\-0.001116\end{bmatrix}=0.001733$$

$$\text{var}[a'X] = a'\sum\nolimits_x a =$$

$$[\,0.35, 0.25,\ 0.15, 0.15, 0.10\,]\begin{bmatrix} 9.69\times10^{-4} & 1.24\times10^{-4} & \ldots\ldots & 1.11\times10^{-4} \\ 1.24\times10^{-4} & 3.50\times10^{-4} & \ldots\ldots & 1.10\times10^{-4} \\ . & . & . & . \\ . & . & . & . \\ . & . & . & . \\ 1.11\times10^{-4} & . & . & 20.5\times10^{-4} \end{bmatrix}\begin{bmatrix} 0.35 \\ 0.25 \\ 0.15 \\ 0.15 \\ 0.10 \end{bmatrix} = .000338$$

Optimization of Multivariate Functions

For a multivariable function such as y = f (x, z) to be at a relative maximum or minimum, three conditions must be met:

1. The first-order partial derivatives must equal zero simultaneously. This indicates that at the given point (a,b), called a critical point, the function is neither increasing nor decreasing with respect to the principal axes but is at a relative plateau. In other words, this is the first-order (**necessary**) condition for a stationary value.
2. The second-order direct partial derivatives, when evaluated at the critical point (a, b), must both be positive for a minimum and negative for a maximum. This ensures that from a relative plateau at (a, b) the function is moving upward in relation to the principal axes in the case of a minimum, and downward in relation to the principal axes in the case of a maximum.
3. The product of the second-order direct partials evaluated at the critical point must exceed the product of the cross partials evaluated at the critical point.

   Points 2 and 3 above together describe the second-order (**sufficient**) conditions for a relative extremum.

In summary

Turning Point To Be	First Order Necessary Condition	* Second Order Sufficient Condition
Maximum	$\frac{\partial y}{\partial x} = 0 \ and \frac{\partial y}{\partial z} = 0$	$\frac{\partial^2 y}{\partial x^2} < 0 \ and \frac{\partial^2 y}{\partial z^2} < 0$ $\frac{\partial^2 y}{\partial x^2} \times \frac{\partial^2 y}{\partial z^2} > \left(\frac{\partial^2 y}{\partial z \partial x}\right)^2$
Minimum	$\frac{\partial y}{\partial x} = 0 \ and \frac{\partial y}{\partial z} = 0$	$\frac{\partial^2 y}{\partial x^2} > 0 \ and \frac{\partial^2 y}{\partial z^2} > 0$ $\frac{\partial^2 y}{\partial x^2} \times \frac{\partial^2 y}{\partial z^2} > \left(\frac{\partial^2 y}{\partial z \partial x}\right)^2$
Saddle Point	$\frac{\partial y}{\partial x} = 0 \ and \frac{\partial y}{\partial z} = 0$	$\frac{\partial^2 y}{\partial x^2} > 0 \ and \frac{\partial^2 y}{\partial z^2} < 0$ Or vice versa

* Applies only if the first order necessary condition is met .

Note the following:

(a) Since $f_{xy} = f_{yx}$ by Young's theorem, $f_{xy} \times f_{yx} = (f_{xy})^2$ (see step 3)

(b) If $f_{xx} \times f_{yy} < (f_{xy})^2$, when f_{xx} and f_{yy} have the same signs, the function is at an *inflection point*. When f_{xx} and f_{yy} have different signs, the function is at a *saddle point*, i.e. the function is at maximum when viewed from one axis but a minimum when viewed from the other axis.

(c) $f_{xx} \times f_{yy} = (f_{xy})^2$ the test is inconclusive.

Example 6.8-8

(i) Find the critical points of the function $z = 2y^3 - x^3 + 147x - 54y + 12$

(ii) Test whether the function is at a relative maximum or minimum.

Solution

(i) Take the first-order partial derivatives, set them equal to zero, and solve for y and x.

$z_x = -3x^2 + 147 = 0$ $\qquad$ $z_y = 6y^2 - 54 = 0$

$x^2 = 49$ $\qquad$ $y^2 = 9$

$x = \pm 7$ $\qquad$ $y = \pm 3$

With $x = \pm 7$ and $y = \pm 3$, there are four distinct sets of critical points:

(7,3), (7, − 3), (−7, 3), (−7, − 3).

(ii) Take the second-order direct partials, evaluate them at each of the critical points, and check their signs.

$z_{xx} = -6x$ $\qquad\qquad$ $z_{yy} = 12y$

(1) $z_{xx}(7,3) = -6(7) = -42 < 0$ $\qquad$ $z_{yy}(7,3) = 12(3) = 36 > 0$

(2) $z_{xx}(7,-3) = -6(7) = -42 < 0$ $\qquad$ $z_{yy}(7,-3) = 12(-3) = -36 < 0$

(3) $z_{xx}(-7,3) = -6(-7) = 42 > 0$ $\qquad$ $z_{yy}(-7,3) = 12(3) = 36 > 0$

(4) $z_{xx}(-7,-3) = -6(-7) = 42 > 0$ $\qquad$ $z_{yy}(-7,-3) = 12(-3) = -36 < 0$

Since there are different signs for each of the direct partials in (1) and (4), the function cannot be at a relative maximum or minimum at (7 , 3) or (-7 , -3). When z_{xx} and z_{yy} are of different signs, $z_{xx} \times z_{yy}$ cannot be greater than $(z_{xy})^2$,and the function is at a saddle point.

With both signs of the second direct partials negative in (2) and positive in (3), the function *may* be at a relative maximum at (7 , -3) and at a relative minimum at (-7 , 3), but the third condition must be tested first to ensure against the possibility of an inflection point.

So, we need to take the cross partial derivatives and make sure that

$z_{xx}(a,b) \times z_{yy}(a,b) > [z_{xy}(a,b)]^2$.

$z_{xy} = 0$ $\qquad\qquad$ $z_{yx} = 0$

$$z_{xx}(a,b) \times z_{yy}(a,b) > [z_{xy}(a,b)]^2$$

From (2), $\qquad (-42)(-36) = 1512 > (0)^2$

From (3), $\qquad (42)(36) = 1512 > (0)^2$

The function is maximized at (7 , - 3) and minimized at (- 7 , 3).

Optimization of Multivariate Functions in Economics

Example 6.8-9

A firm producing two goods *x* and *y* has the following profit function:

$P = 64x - 2x^2 + 4xy - 4y^2 + 32y - 14$.

Find the profit maximizing level of output for each of the two goods and test to be sure profits are maximized:

1. Take the first-order partial derivatives, set them equal to zero, and solve for x and y simultaneously.

$$P_x = 64 - 4x + 4y = 0$$

$$P_y = 4x - 8y + 32 = 0$$

When solved simultaneously, x = 40 and y = 24 .

2. Take the second-order partial derivatives since both must be negative for the function to be maximum

$$P_{xx} = -4 \qquad\qquad P_{yy} = -8$$

3. Take the cross-partials to make sure

$$P_{xx}\,P_{yy} > (P_{xy})^2 .\ P_{xy} = P_{yx} = 4.$$

Thus ,

$$P_{xx}\,P_{yy} > (P_{xy})^2$$

$$(-4)(-8) > (4)^2$$

$$32 > 16$$

Profits are indeed maximized at x = 40 and y = 24 . At the point , $\pi = 1650$.

Example 6.8-10

A firm's total costs are related to its work force (L) and capital equipment (K) by the function, $TC = 10L^2 + 10K^2 - 25L - 50K - 5LK + 2000$. Find the combination of K, L to minimize TC.

Solution

<u>FOC:</u>

$$\frac{\partial(TC)}{\partial L} = 20L - 25 - 5K = 0 \qquad (1)$$

$$\frac{\partial(TC)}{\partial K} = 20K - 50 - 5L = 0 \qquad (2)$$

This is a system of two equations with two unknowns, K and L.

4 x (1) + (2) gives : 75 L – 150 = 0 , implies that L = 2

L = 2 implies that K = 3.

To verify it is a minimum take the SOC’s

SOC:

$$\frac{\partial^2(TC)}{\partial L^2} = 20 > 0 \quad \text{and} \quad \frac{\partial^2(TC)}{\partial K^2} = 20 > 0$$

Now test the existence using the product of the cross partial derivatives:

$$\frac{\partial^2(TC)}{\partial K \partial L} \equiv \frac{\partial^2(TC)}{\partial L \partial K} = -5 \Rightarrow \left(\frac{\partial^2(TC)}{\partial K \partial L}\right)^2 = 25$$

$$\frac{\partial^2(TC)}{\partial L^2} \times \frac{\partial^2(TC)}{\partial K^2} = 400 > 25$$

Therefore, there is a turning point at $K = 3$, $L = 2$, and it is a minimum.

At the above point, TC is:

$TC = 10\times2^2 + 10\times3^2 - 25\times2 - 50\times3 - 5\times2\times3 + 2000 = 1900.$

Summary

- ☒ f has absolute maximum on an interval I if there exists $c \in I$ such that $f(c) \geq f(x)$ for every $x \in I$. f(c) is called the absolute maximum of f
- ☒ f has absolute minimum on an interval I if there exists $c \in I$ such that $f(c) \leq f(x)$ for every $x \in I$. f(c) is called the absolute minimum of f
- ☒ f is said to have a local maximum at $c \in (a, b)$ if $\forall x \in (a, b)$ we have $f(x) \leq f(c)$. f(c) is called the local maximum of f over (a, b)
- ☒ f is said to have a local minimum at $c \in (a, b)$ if $\forall x \in (a, b)$ we have $f(x) \geq f(c)$, f(c) is called the local minimum of f over (a, b)
- ☒ A function f is said to be increasing over an interval I if for every x and x + 1 in I we have $f(x+1) \geq f(x)$.
- ☒ A function f is said to be decreasing over an interval I if for every x and x + 1 in I we have $f(x+1) \leq f(x)$
- ☒ If $f' > 0$ on I, then f is strictly increasing on I.
- ☒ If $f' < 0$ on I, then f is strictly decreasing on I.
- ☒ If $f' = 0$ on I, then f is constant on I.
- ☒ If f' becomes zero at a with changing signs from negative to positive, then f(a) is a local minimum of f.
- ☒ If f' becomes zero at b with changing signs from positive to negative, then f(b) is a local maximum of f

☒ The number $c \in D_f$ is called critical number of the function f if one of the following conditions are satisfied :

a) $f'(c) = 0$

b) $f'(c)$ does not exist

☒ **Rolle ' s Theorem**

1) f is continuous over $[a, b]$

2) f is differentiable over (a, b)

3) $f(a) = f(b)$

Then there exists at least one number $c \in (a, b)$ satisfying $f'(c) = 0$

☒ **The Mean Value Theorem**

1. f is continuous over $[a, b]$
2. is differentiable over (a, b)

Then there exists at least one number $c \in (a, b)$ satisfying

$f'(c) = \frac{f(b) - f(a)}{b - a}$ that is $f(b) - f(a) = f'(c)(b - a)$

☒ If $f'' \geq 0$ on an interval $]a, b[$, then f is concave upward

☒ If $f'' \leq 0$ on an interval $]a, b[$, then f is concave downward

☒ Let f and g be differentiable functions in the neighborhood of the real number a .If $f(a) = g(a) = 0$, if $g'(x) \neq 0$ at $x \neq a$, and if

$\lim_{x \to a} \frac{f'(x)}{g'(x)} = l$, then $\lim_{x \to a} \frac{f(x)}{g(x)} = l$. This rule, called L' Hopital Rule

☒ The rate of change of price p with respect to quantity q is $\frac{d_p}{d_q}$

☒ The marginal cost is the rate of change of c with respect q. That is, marginal cost = $\frac{d_c}{d_q}$ where $c = q \cdot \bar{c}$

☒ The marginal revenue is the rate of change of the total dollar value received (r) with respect to the total number of units sold (q). That is, marginal revenue = $\frac{d_r}{d_q}$

☒ The relative rate of change of $y = f(x)$ is: $\frac{y'}{y}$

☒ The percentage rate of change of $y = f(x)$ is: $\frac{y'}{y} \times 100$

Exercises

In problems 6.1 – 6.5, find the critical numbers and then find the maximum and the minimum values of the given functions on their respective intervals:

6.1. $f(x) = x^3 - 2x^2$; $[-1, 2]$

6.2. $f(x) = x^3 - 4x^2$; $[-1, 3]$

6.3. $f(x) = x^{\frac{2}{3}}$; $[-1, 2]$

6.4. $f(x) = 1 - \tan^2 x$; $[-\frac{\pi}{4}, \frac{\pi}{4}]$

6.5. $f(x) = 2\sin x - \cos 2x$; $[0, 2\pi]$

6.6. Prove that the equation $x^3 + 3x^2 + 6x + 1$ admits only one real root in the interval $[-1, 0]$.

In problems 6.7 – 6.9, find the number c, when possible, that makes the given functions satisfy Rolle 's theorem on their respective intervals:

6.7. $f(x) = x^2 - 4x + 1$; $[0, 4]$.

6.8. $g(x) = \sqrt{x} - x$; $[0, 1]$.

6.9. $h(x) = \begin{cases} x^2 & \text{for } 0 \le x < 1 \\ 2 - x & \text{for } 1 \le x < 2 \end{cases}$

In problems 6.10 – 6.12, show that the given functions satisfy the conditions of the mean value theorem on their intervals, then find the numbers that satisfy this theorem:

6.10. $f(x) = x^2 + 2x - 3$; $[-3, 0]$.

6.11. $g(x) = \cos x$; $[\frac{\pi}{4}, \frac{7\pi}{4}]$.

6.12. $h(x) = \frac{x+1}{x}$; $[\frac{1}{2}, 2]$

In problems 6.13 – 6.19, find the intervals along which the given functions increase or decrease on their domain of definitions, then describe the results by drawing their graphs in an orthonormal system:

6.13. $f(x) = x^2 - 2x - 3$

6.14. $g(x) = x^3 - 9x^2 + 24x + 1$

6.15. $h(x) = x^3(4 - x)$

6.16. $f(x) = x + \dfrac{1}{x}$

6.17. $g(x) = \cos x$ where $x \in [0, \pi]$

6.18. $h(x) = \sin x$ where $x \in [0, \pi]$

6.19. $p(x) = \dfrac{x}{x^2+1}$

In problems 6.20 – 6.26, study the concavity of the given functions and determine the inflection points if they exist:

6.20. $f(x) = \dfrac{x^4}{4} + x^3 + 2.$

6.21. $g(x) = \dfrac{x^2}{4} + \cos x$ where $x \in [0, \pi].$

6.22. $h(x) = \dfrac{1}{x}.$

6.23. $f(x) = (x + 2)^3.$

6.24. $g(x) = \tan x,$ where $x \in \left[-\dfrac{\pi}{2}, \dfrac{\pi}{2}\right].$

6.25. $h(x) = x^3(4 - x).$

6.26. $p(x) = \dfrac{x+1}{x-2}.$

In problems 6.27 – 6.33, find the local extremum of the given functions :

6.27. $f(x) = -x^4 + 2x^2 + 12.$

6.28. $g(x) = x^5 - \dfrac{5x^3}{3}.$

6.29. $f(x) = \sqrt{9-x^2}$ *where* $x \in [-1, 2].$

6.30. $g(x) = x^3 - 6x^2 + 9x - 5.$

6.31. $h(x) = \dfrac{x^2}{x^2+3}$ *where* $x \in [-1, 1].$

6.32. $p(x) = \cos x + \sin x$ where $x \in [0, 2\pi].$

6.33. $k(x) = 2\sec x - \tan x$ where $x \in \left[\dfrac{-\pi}{4}, \dfrac{\pi}{4}\right].$

In problems 6.34 – 6.44, study the variation and draw the graph of each of the given functions in an orthonormal system of axes x ' o x , y 'o y:

6.34. $f(x) = x^3 - 3x + 2.$

6.35. $g(x) = (x^2 - 1)^2.$

6.36. $h(x) = 2x^4 + 4x^2.$

6.37. $f(x) = 2x^3 - 6x^2 + 6x - 2.$

6.38. $g(x) = \frac{1}{3}x^3 - 2x^2 + 3x + 1.$

6.39. $h(x) = (5 - x)(x - 2)^2.$

6.40. $f(x) = \frac{x}{\ln x}$

6.41. $f(x) = x - \ln x$

6.42. $f(x) = \frac{\ln x}{x+1}$

6.43. $f(x) = e^x - e^{-x}$

6.44. $f(x) = e^x e^{-x}$

In problems 6.45 – 6.58, find the given limits if they exist:

6.45. $\lim_{x \to 0^+} \frac{\ln x}{x^2}$

6.46. $\lim_{x \to 0} (\ln x - x)$

6.47. $\lim_{x \to +\infty} x \ln(1 + \frac{1}{x})$

6.48. $\lim_{x \to +\infty} \frac{\ln x - 1}{\ln x + 1}$

6.49. $\lim_{x \to 0} \frac{\ln(\cos x)}{1 - \cos x}$

6.50. $\lim_{x \to 0^+} x^2 (1 + \ln x)$

6.51. $\lim_{x \to e} \frac{\ln x - 1}{x - e}$

6.52. $\lim_{x \to +\infty} (e^{2x} - 2e^x - x)$

6.53. $\lim\limits_{x \to +\infty} \dfrac{e^x}{e^x + 1}$

6.54. $\lim\limits_{x \to +\infty} \dfrac{2^x}{x^{10}}$

6.55. $\lim\limits_{x \to +\infty} e^{-x\sqrt{3}} \sin x$

6.56. $\lim\limits_{x \to -\infty} (x-1)e^{x+1}$

6.57. $\lim\limits_{x \to 0} \dfrac{2^x + 5^x - 2}{x}$

6.58. $\lim\limits_{x \to 0} \dfrac{e^{x^3} - 1}{\sin^3 x}$

6.59. Show that whether the function f defined by

$$f(x) = \begin{cases} x^2 & ; \quad x \in [0,1) \\ 2 - x & ; \quad x \in [1,2] \end{cases}$$

satisfies Rolle 's theorem on the interval [0 , 2].

6.60. Prove , using Rolle 's theorem, that the equation $2x^5 - 3x^3 + 1 = 0$ admits only one real root in the interval [0 , 1].

6.61. Show that the graph of the function f (x) = 4 sin 2x admits at least one tangent in the interval [0 , π] , then find the value of x, abscissa of the point of tangency.

6.62. Show that the following functions do not satisfy neither Rolle 's theorem nor the mean value theorem on their given intervals:

a) $f(x) = 1 - x^{\frac{2}{3}}$; [- 1 , 1].

b) $g(x) = \begin{cases} \dfrac{3-x}{2} & ; \quad x < 1 \\ x^{-1} & ; \quad x \geq 1 \end{cases}$; [0 , 2].

6.63. Find the value of a ,b ,c so that the graph (c) of $y = a x^3 + b x^2 + c x$ satisfies the following conditions:

a) (c) admits an inflection point at $x = \frac{1}{2}$.

b) (c) admits a horizontal tangent at $x = -1$.

c) (c) passes through the point (1 , 13) .

6.64. Using the function $f(x) = x^4$, prove that $f''(c)$ could be equal to zero but the point $(c, f(c))$ is not an inflection point.

In problems 6.65 – 6.68, let p be the demand function for a manufacturer ' s product. Find the rate of change of price p per unit with respect to quantity q. Assume that p is in dollars.

6.65. $p = 10 q^3 + 100 q$ when $q = 10$

6.66. $p = 20 q^2 + 100 q + 1$ when $q = 20$

6.67. $p = q \ln q$ when $q = 100$

6.68. $p = 12 e^{-0.01 q}$ when $q = 200$

In problems 6.69 – 6.72, c is the cost of producing q units of a product. In each of the given cases, find the marginal - cost function, then deduce the value of the marginal - cost function at the given value of q.

6.69. $c = 10 q + 200$ when $q = 100$

6.70. $c = 0.2 q^2 + 3 q + 540$ when $q = 2$

6.71. $c = 24 \ln (q + 2) + 8$ when $q = 6$

6.72. $c = 100 e^{q^2 + 2}$ when $q = 4$

In problems 6.73 – 6.77, $\bar{c}$ = f (q) is the average cost per unit, which is a function of the number q of units produced. Find the marginal- cost function, then deduce the value of the marginal- cost function for the indicated value of q.

6.73. $\bar{c} = 0.02q + 3 + \dfrac{400}{q}$ when $q = 50$

6.74. $\bar{c} = 0.003q^2 - 0.1q + 5 + \dfrac{10000}{q}$ when $q = 100$

6.75. $\bar{c} = \dfrac{240}{\ln(q+2)}$ when $q = 30$

6.76. $\bar{c} = \dfrac{430}{q} + 3000\,\dfrac{e^{\frac{(2q+4)}{400}}}{q}$ when $q = 200$

6.77. $\bar{c} = 0.01q^2 - 0.03q + 3 + \dfrac{200}{q}$ when $q = 20$

In problems 6.78 – 6.83, r = f (q) represents total revenue , which is a function of the number q of units sold. Find the marginal-revenue function, then deduce the value of the marginal-revenue function for the indicated value of q.

6.78. $r = 0.4q$ when $q = 50$

6.79. $r = 24q^2 + 230q - q^3$ when $q = 25$

6.80. $r = \dfrac{q-1}{q+1}$ when $q = 20$

6.81. $r = \sqrt{q^2 + q - 1}$ when $q = 10$

6.82. $r = \ln(q + 2) - q\ln q$ when $q = 30$

6.83. $r = e^{2q} + qe^{q}$ when $q = 15$

In problems 6.84 – 6.87, find the marginal – revenue function if the demand function p with respect to q units is:

6.84. $p = 10e^{q}\ln q$

6.85. $p = \dfrac{12}{\ln(q+1)}$

6.86. $p = 12e^{-0.02q}$

6.87. $p = (q+2)\ln q$

6.88. For the cost function $c = 0.1q^2 + 1.1q + 3$, how fast does c change with respect to q when $q = 4$? Determine the percentage rate of change of c with respect to q when $q = 4$.

In problems 6.89 – 6.93, suppose that p is a demand equation for a manufacturer's product , find:

1) The rate of change of p with respect to q.

2) Find the relative rate of change of p with respect to q at the given value of q.

3) The percentage rate of p at the given value of q.

6.89. $p = q + 4$ when $q = 5$

6.90. $p = 5 - 3q^3$ when $q = 1$

6.91. $p = 100 + \sqrt{q^2 + 20}$ when $q = 2$

6.92. $p = 10^{-q} + \ln(8+q) + 0.001e^{q-2}$ when $q = 0$

6.93. $p = (2q+3)(q+1)$ when $q = 4$

6.94. The profit of selling a high technology product is a function defined by $P(x) = -x^2 + 60x - 400$ where x is the number of units sold and P(x) is in thousands dollars.

a. Represent this function graphically for $x \geq 0$.

b. Determine the number of units that must be sold to achieve the maximum profit.

c. What is the maximum profit?

6.95. The total cost of producing a certain product is a function defined by: $C(x) = 100x^2 + 1300x + 1000$ where x is the number (in thousands) of units produced, an C (x) is in thousands of dollars. The selling price of each unit is $ 2000.

a. Determine in terms of x, the revenue function defined by R(x) in thousands of dollars.

b. What must the production be so that he profit is zero?

c. What must the production be so that he profit is minimum?

d. Calculate this profit.

6.96. The formula that relates the produced quantity x of kerosene to produce quantity y of fuel from crude oil is given by $y = \dfrac{125000 - 25x}{125 + 2x}$, x and y are in liters.

a. Represent graphically the given function for $x \geq 0$ and $y \geq 0$.

b. What are the maximum quantities that can be produced of these products?

c. Determine graphically the quantity of kerosene that can be produced when 100 liters of fuel are produced. Verify this result by calculation .

6.97. The revenue and the cost functions of a firm are given respectively by:

$R(x) = \dfrac{-11}{54}x^2 + \dfrac{235}{108}x - \dfrac{28}{27}$ with $2 \leq x \leq 5$,

$C(x) = \dfrac{3}{4}x + 1$ with $2 \leq x \leq 5$,

where x is the number of units in thousands.

a. Find the quantity that minimizes the cost.

b. Find the quantity that maximizes the revenue.

c. For what quantity the profit is maximum?

6.98. Suppose that the weekly revenue function from the sale of x units per week of a certain product is defined by R (x) = 20 ln (x + 2), while the cost function C is defined by $C(x) = \dfrac{x}{2}$. Find the number of units that should be produced to maximize the profit.

6.99. If the demand function D for cameras at a photo shop is defined by:

$D(x) = \dfrac{x^2}{3} - \dfrac{25x}{2} + 150$, where x is the number of cameras sold per week.

a. Find the values of x that maximizes revenue.

b. What is the maximum revenue?

6.100. When producing x units of its products weekly, the total cost and revenue functions of a manufacturer in thousands of dollars are determined by $C(x) = 10 + 0.2x$, and $R(x) = 0.3x$ respectively.

a. Sketch these functions on the same coordinate system.

b. Find the break even point.

c. Find the total weekly revenue at the break even point.

6.101. Let D be the demand function defined by $D(p) = \frac{4}{p^2}$.

a. Verify that $D'(p) < 0$ for all $p > 0$.

b. Find the elasticity of demand $E(p)$.

c. Find the values of p for which D is elastic.

d. Give an economical interpretation for the result in part (c).

6.102. The Total cost function $C(q) = q^3 - 21q^2 + 500q$. Find:

a. The average cost function.

b. The critical value at which average cost is minimized.

c. The minimum average cost.

6.103. A utility of wealth function is given by $U(w) = \alpha \ln(w+1)$, $w \geq 0$. Where α is a positive constant.

a. Show that $U'(w) > 0$.

b. that U is risk averse.

c. Find an expression for the relative risk aversion.

6.104. A portfolio P of two investments A and B is such that the returns are uncorrelated, so that $\sigma_P^2 = \alpha^2 \sigma_A^2 + (1-\alpha)^2 \sigma_B^2$, where σ_P^2, σ_A^2 *and* σ_B^2σ are the variances of the returns and *a* measures the proportion of the portfolio invested in A. Determine the value of *a* corresponding to minimum risk, and show that minimum risk is $\frac{\sigma_A \sigma_B}{\sqrt{\sigma_A^2 + \sigma_B^2}}$.

6.105. Historical data indicates that the average monthly returns for the Emerging Markets and S&P indices are 15 % and 10 %, respectively. The volatilities are found to be 20 % and 12 % for the two indices , respectively, while the correlation between the two is found to be - 0.2. As an employee at a fund management firm you are required to :

a. Find the optimum level of investment on each index, which minimizes the variance of the portfolio and verify the minimum variance attained.

b. Compute the minimum variance and the expected return on the optimum portfolio.

6.106. A fund manager is considering a diversified portfolio with domestic and foreign stocks and tries to decide on the allocation of funds. On average foreign portfolios (I N T) have been shown to be less risky than domestic portfolios (S&P), so he reasonably assumes that the former has a standard deviation of 14 % and the latter has a standard deviation of 14.9 %. In addition, the average correlation for the last five years between a domestic mutual fund and a foreign mutual fund is 0.33. At the time of the analysis, the fund manager estimated an average return for the domestic market of 12.5 % and was more pessimistic about foreign markets estimating returns 2 % lower.

a. Find appropriate weights for the domestic and foreign stock portfolios which minimizes the variance of the overall portfolio.

b. Compute the minimum variance and the expected return on the optimum portfolio.

6.107. A monopolistic firm has the following demand and profit functions for each of its two products:

$$p = 12 - 2x$$
$$q = 20 - y$$
$$P = 12x - 3x^2 + 20y - 3y^2 - 2xy,$$

where p and q are the respective prices for each product and x and y are the respective amounts of each sold.

a. Find the amounts x and y that maximize the profit and confirm the existence of a unique maximum.

b. Find also the profit – maximizing prices and the maximum profit.

6.108. For the following utility functions find expressions for the absolute and relative risk aversion and determine their properties

a. (i) $U(W) = \ln W$

(ii) $U(W) = aW^a \quad (a > 1)$

b. Given the following two investments:

A		B	
Outcome	Probability	Outcome	Probability
5	0.2	6	0.3
10	0.5	12	0.4
16	0.3	18	0.3

Which investment is preferred on the basis of expected return if the utility function is $U(W) = \ln(W)$?

6.109. A utility of wealth function is given by $U(W) = aW - bW^2 e^{cW}$, where a, b and c are positive constants.

a. Show that the investor is risk averse.

b. Find an expression for the relative risk-aversion.

6.110. A fund manager is trying to construct a diversified portfolio, which contains Stock and Bond indices. Collecting and analyzing monthly data for each index for the past five years has revealed the following information:

Index	Stocks	Bonds
Return	8.2 %	5.6 %

The variances of stock and bond returns along with their correlation are

	Stocks	Bonds
Variance	0.0064	0.0036
Correlation	- 0.5208	

a. Find appropriate weights for a portfolio of Stocks and Bonds, which minimize the variance of the portfolio.

b. Verify that the weights found above yield the minimum variance. Evaluate the minimum – variance and the expected return on the optimum portfolio.

In problems 6.111 – 6.114, find the Taylor polynomials of orders 0, 1, 2 and 3 generated by f at a.

6.111. $f(x) = \ln x$; $a = 1$

6.112. $f(x) = \dfrac{1}{x+2}$; $a = 0$

6.113. $f(x) = \cos x$; $a = \dfrac{\pi}{4}$

6.114. $f(x) = \sqrt{x}$; $a = 4$

In problems 6.115 – 6.118, find the Maclaurin series for the given functions.

6.115. $f(x) = \sin x$

6.116. $f(x) = (x+1)^2$

6.117. $f(x) = x^4 - 2x^3 - 5x + 4$

6.118. $f(x) = e^{-x}$

6.119. factory produces a quantity of plastic felt pens. The total cost of production of a quantity x is given, in dollars, by $C(x) = 2590 + 2\sqrt{900+x}$ where $x \in [0,+\infty[$.

a. What is the fixed cost?

b. Calculate the total cost of fabrication of 4000 units? deduce the average cost of production of one of these 4000 units.

c. Study the variation of C and draw its table of variations.

d. Justify that if the quantity produced exceeds 2700 units, the total cost of production exceeds \$ 2710.

e. Solve $C(x) \leq 3000$. deduce the maximal quantity to be produced in order to get a total cost less than or equal to \$ 3000.What is then the average cost of one of these produced items?

6.120. The total cost of production of a product is given by $C(x) = x^3 - 20x^2 + 300x$, with $x \in [0,300]$, x is the quantity produced in units and C is expressed in dollars.

a. Find the average cost function $\overline{C}$ in terms of x.

b. Determine the quantity to be produced for the average cost to be minimum then determine this minimum.

c. 1) Determine the marginal cost function $\frac{d_c}{dx}$ in terms of x.

2) What quantity should be produced for the marginal cost to be equal to the average cost?

6.121. A factory produces q units with $q \in [0, 55]$. The total cost of production of these q units, expressed in dollars, is given by $C(q) = q^2 - 20q + 200$.

a. Study the variations of C over [0 , 55] and draw its table of variations.

b. Each unit is sold for $ 34.

1) Study the variations of the function P(q) defined over $\Re$ by $P(q) = -q^2 + 54q - 200$.

2) Show that the profit obtained on selling q units is given by P(q). Deduce the quantity to be produced in order to get a maximum profit.

3) Solve P(q) = 0. Deduce the quantity to be produced in order to get a positive or zero profit.

6.122. The total cost of production of x hundreds of articles is given by $C(x) = x^3 - 12x^2 + 50x, \forall x \in [0,8]$ and C(x) is expressed in thousands of dollars.

a. 1) Express the marginal cost function $M = \frac{d_c}{dx}$ in terms of x .

2) Study the variations of M and draw its table of variations over [0 , 8].

3) Deduce that the total cost function C is increasing.

b. 1) Express the average cost function $\overline{C}$ in terms of x.

2) Study the variations of $\overline{C}$ and draw its table of variations over [0 , 8].

3) Show that the average cost is minimum when the average cost equals the marginal cost.

c. Designate by (f) and (g) the representative curves of $\overline{C}$ and M in an orthogonal system of axes x'ox, y'oy.

1) Trace (f) and (g) in the same system.

2) For what quantity does the marginal cost exceed the average cost?

6.123. Part A : The total cost of production of an item is given by

$$C(x) = \frac{1}{30}x^3 - 15x^2 + 2500x \text{ , } for\ x \in [0,300].$$

a. Express the marginal cost M (x) in terms of x.

b. Study the variations of M and draw its table of variations over [0 , 300].

c. Deduce that the function C is increasing over [0 , 300].

Part B : The equation that relates the sale unit price to the demanded quantity q , is given by $P(q) = \frac{-45}{8}q + 2750.$

a. Calculate the total revenue R(q) obtained on selling q units.

b. Calculate the marginal revenue and determine the values of q for which the marginal revenue is equal to the marginal cost.

c. 1) Show that the profit is given by $P(q) = -\frac{1}{30}q^3 + \frac{75}{8}q^2 + 250q$.

2) Calculate P'(q) and deduce that the profit is maximal when the marginal revenue is equal to the marginal cost. Find this maximal profit.

6.124. The marginal cost M_c, expressed in dollars, of producing a certain article is defined by: $M_c(q) = 10q + 200$ where q is the number of article produced.

a. Knowing that the total cost of production of 30 articles is $ 10 000, express the total cost C (q) in terms of q.

b. Calculate the total cot of production of the first 50 articles.

6.125. The mayor of a city decides to promote bike riding in his struggle against pollution. He buys 1000 bicycles for rent. The study certifies that the demand function of price is given by $D(p) = \frac{5\ln p}{p^2}$ where p is the price in dollars with $p \in]0, 10]$, and D (p), the demand expressed in thousands of bicycles.

a. Calculate the revenue for a rent charge equal to $ 3.

b. Express, in dollars, the revenue R (p) in terms of the price p.

c. Study the variations of R over [2,10].

d. Deduce the charge rent for which the revenue is maximum then determine this maximum.

6.126. A jeweler notes that the total cost of production, in thousands of dollars, of q necklaces is given by C(q) = $\frac{q^2}{2}+4-2\ln(q+1)$ where $0\le q\le 10$.Each necklace is sold for 6.75 thousands dollars.

a. Show that the profit function, in thousands of dollars, from selling q necklaces is given by P (q) = 6.75 q - $\frac{q^2}{2}-4+2\ln(q+1)$.

b. Study the variations of the function P and determine the minimum quantity to be produced for the company to experience profit.

6.127. A firm manufactures a quantity q (expressed in thousands of tons) of a certain product. The total cost of this product is given by

C (q) = $\frac{q^2}{4}+\frac{9}{2}\ln(q+1)$, where $0\le q\le 5$ and C (q) is expressed in millions of dollars.

a. Designate by $\overline{C}$ the average cost function for $0\le q\le 5$.Verify that

$$\overline{C}'(q)=\frac{f(q)}{2q^2} \text{ where } f(q)=\frac{q^2}{2}+\frac{9q}{q+1}-9\ln(q+1).$$

b. Show that the equation $f(q)=0$ admits a unique solution $\alpha \in]3.6, 3.7[$ .Then deduce the sign of f over [0 , 5].

c. Study the variations of $\overline{C}$ for $0\le q\le 5$.

d. For what production, expressed in tons to the nearest unit, is the average cost minimum? Find this minimum cost.

6.128. Let q be the number of articles, expressed in hundreds, produced by a factory, C(q) = 0.4 q + $e^{-0.4q+1}$ the total cost of production, expressed in thousands of dollars. Suppose that $0\le q<\infty$. Each article is sold for $ 5. Assume all the items are sold.

a. Express the revenue R(q), in thousands of dollars, in terms of q.

b. Verify graphically that the respective graphs of C and R intersect at a unique point of abscissa α such that $4.49<\alpha<4.5$.

c. Give an economical interpretation to the value α .

d. Show that the profit P(q), is expressed in thousands of dollars by P(q) = 0.1 q – $e^{-0.4q+1}$.

e. Study the variations of P over $[\,0\,,\,\infty\,[$.

f. Deduce the minimum number of articles produced in order to realize profit.

6.129. A factory produces an article , the total cost of production is given by $C(q) = 3q + 4 + (q-3)e^{q}$ where q is expressed in hundreds of articles and C is expressed in thousands of dollars, suppose that $0 < q \leq 3$.

a. Find the average cost function $\overline{C}(q)$.

b. Each article of this product is sold for \$ 30 ; all the production is sold.

1) Show that the profit function is given by $P(q) = (3-q)e^{q} - 4$.

2) Study the variations of the function P over $]\,0\,,\,3\,]$.

3) For what quantity produced is the profit maximal?

c. Will the factory realize profit for a production of 200 articles? 40 articles? Explain.

Chapter 7 Integration

Introduction

We have seen in the previous chapters the concept of derivative and its applications in calculating the differentiation of a function; in this chapter we are going to deal with the inverse operation of differentiation; that is, if you are given a function f, then how can you find a function F such that the derivative F ' of F = f ? This process is called " anti-differentiation ", and F is called the anti-derivative of f.

Objectives

- ❖ After studying the material in this chapter, you should be able to:
- Find anti-derivatives.
- Calculate indefinite integration.
- Calculate definite integration.
- Understand the main theory in differentiation and integration.
- Calculate integration by substitution.
- Understand the notion of integration by parts.
- Apply integration in finance.

Why ?

One of the most important mathematicians that helped in initializing the integral calculus was the famous Muslim Scientist " Al Hassan Bin Alhaytham", and the scientists " Yoskovitch" and " Ronfild " proved that Bin Alhaytham found the two chains sum of the third and the fourth power for real numbers, when he was calculating the volume of the rotating body obtained from rolling a conic piece around a center vertically, and these works has helped to discover differentiation and integration.

Section 7.1

Anti-derivative (Primitive) of a Function

Definition 7.1-1

f is a defined function over an interval I. F is said to be the anti-derivative of f if :

a) F is differentiable over I.

b) $F'(x) = f(x)$ for all $x \in I$.

Example 7.1-2

1) Suppose that f is defined by $f(x) = 7x^6$ for all $x \in \Re$, then

$F(x) = x^7 \Rightarrow F'(x) = 7x^6 = f(x)$

$F(x) = x^7 - 4 \Rightarrow F'(x) = 7x^6 = f(x)$

$F(x) = x^7 + \frac{1}{3} \Rightarrow F'(x) = 7x^6 = f(x)$

Hence, $\boxed{F(x) = x^7 + c\ , c \in \Re \Rightarrow F'(x) = 7x^6 = f(x)}$

Suppose that f is defined by $f(x) = \cos x$ for all $x \in \Re$, then $F(x) = \sin x \Rightarrow$
$F'(x) = \cos x = f(x)$

$F(x) = \sin x + 12 \Rightarrow F'(x) = \cos x = f(x)$

$F(x) = \sin x - \pi \Rightarrow F'(x) = \cos x = f(x)$

Hence, $\boxed{F(x) = \sin x + c\ , c \in \Re \Rightarrow F'(x) = \cos x = f(x)}$

Thus, we can deduce the following property:

Property 7.1-1

If F is the anti-derivative of f over the interval I, then the general signification f of F on I is: $f(x) = F(x) + c$ where c is any constant $\in \Re$.

Remark 7.1-1

1) The following table describes the anti-derivatives of some remarkable functions; c is any constant $\in \Re$:

Function	Anti-derivative
$K f(x)$	$K F(x) + c$
$f(x) + g(x)$	$F(x) + G(x) + c$
$f(x) = \cos x$	$F(x) = \sin x + c$
$f(x) = \sin x$	$F(x) = -\cos x + c$
$f(x) = \sec^2 x$	$F(x) = \tan x + c$
$f(x) = \sec x \tan x$	$F(x) = \sec x + c$
$f(x) = \csc x \cot x$	$F(x) = -\csc x + c$
$f(x) = \csc^2 x$	$F(x) = -\cot x + c$
$f(x) = x^n, n \in Z - \{-1\}, x \neq 0$	$F(x) = \dfrac{x^{n+1}}{n+1} + c$

2) Not all functions have anti-derivatives. The following is a counter-example about it :

$$f(x) = \begin{cases} 1 & ; \quad x>0 \\ 0 & ; \quad x=0 \\ -1 & ; \quad x<0 \end{cases}$$

There is no anti-derivative to f in a way that $x \in \Re$, $F'(x) = f(x)$.

To prove this, suppose that F ' exists.

1 . When $x > 0$, we have $F'(x) = 1 \Rightarrow F(x) = x + c_1$; $c_1 \in \Re$

2 . When $x < 0$, we have $F'(x) = -1 \Rightarrow F(x) = -x + c_2$; $c_2 \in \Re$

That is, $F(x) = \begin{cases} x + c_1 & ; \quad x>0 \\ x + c_2 & ; \quad x<0 \end{cases}$

And this function is not differentiable at $x = 0$ for all $c_1, c_2 \in \Re$. It is only continuous when $c_1 = c_2$.

Section 7.2

Indefinite Integral

Definition 7.2-1

The indefinite integral of f over an interval I is the set of all anti-derivatives of over I and it is denoted by $\int f(x)\,dx = F(x)+c$, where $F'(x) = f(x)$ for all $x \in I$, and c is any constant $\in \Re$.

" $\int$ " is read " integral "
" f(x) " is read " integrand "

" dx " is read " integration variable "

" c " is read " integration constant "

Basic Integration Form

Let f and g be two functions that have respective anti-derivatives F and G over an interval I, a, c $\in \Re$. Then:

$\int 0\,dx = c$
$\int a\,f(x)\,dx = a \int f(x)\,dx = a\,F(x) + c$
$\int [f(x) \pm g(x)]\,dx = \int f(x)\,dx \pm \int g(x)\,dx = F(x) \pm G(x) + c$
$\int a\,dx = a \int dx = ax + c$
$\int x^n\,dx = \dfrac{x^{n+1}}{n+1} + c$
$\int \cos x\,dx = \sin x + c$
$\int \sin x\,dx = -\cos x + c$
$\int \cos ax\,dx = \dfrac{\sin ax}{a} + c$
$\int \sin ax\,dx = \dfrac{-\cos ax}{a} + c$
$\int \sec^2 x\,dx = \int \dfrac{1}{\cos^2 x}\,dx = \int (1+\tan^2 x)\,dx = \tan x + c$

$\int \csc^2 x \, dx = \int \frac{1}{\sin^2 x} dx = -\cot x + c$
$\int \sec x \tan x \, dx = \sec x + c$
$\int \csc x \cot x \, dx = -\csc x + c$
$\int \frac{1}{2\sqrt{x}} dx = \sqrt{x} + c$

Example 7.2-1

1) $\int x^5 \, dx = \frac{x^{5+1}}{5+1} + c = \frac{x^6}{6} + c$

2) $\int x^{-3} \, dx = \frac{x^{-3+1}}{-3+1} + c = \frac{x^{-2}}{-2} + c = \frac{-1}{2x^2} + c$

3) $\int 3x^7 \, dx = 3 \int x^7 \, dx = 3 \frac{x^8}{8} + c$

4) $\int (x^2 + \frac{2}{3} x - 5x^3 - 7) \, dx = \frac{x^3}{3} + \frac{2}{3} \times \frac{x^2}{2} - 5\frac{x^4}{4} - 7x + c$

$= \frac{x^3}{3} + \frac{x^2}{3} - 5\frac{x^4}{4} - 7x + c$

5) $\int \cos 3x \, dx = \frac{\sin 3x}{3} + c$

6) $\int \sin \frac{2x}{5} dx = \frac{-\cos \frac{2x}{5}}{\frac{2}{5}} + c = -\frac{5}{2} \cos \frac{2x}{5} + c$

7) $\int \tan^2 x \, dx = \int \tan^2 x \, dx = \int (\tan^2 x + 1 - 1) \, dx =$

$\int [\tan^2 x + 1) - 1] \, dx = \int (\tan^2 x + 1) \, dx - \int dx = \tan x - x + c$

Property 7.2-1 (Integration by Substitution)

Let g be a function defined on an interval I, and f is a continuous function over I. If g is differentiable over its domain of definition D_g and F is the anti-derivative of f over I such that F ' = f, then

$\int f(g(x)) g'(x) \, dx = F(g(x)) + c.$

* If we assume that u $= g(x)$ then $\frac{du}{dx} = g'(x) \Rightarrow du = g'(x)\,dx$ and

$\int f(u)\,du = F(u) + c.$

Remark 7.2-1

These are some differentiation functions used in substitution integration

$u = g(x)$	$du = g'(x)\,dx$
$u = x^n$	$du = n x^{n-1}\,dx$
$u = \sin x$	$du = \cos x\,dx$
$u = \cos x$	$du = -\sin x\,dx$
$u = \tan x$	$du = \sec^2 x\,dx$
$u = \sec x$	$du = \sec x \tan x\,dx$
$u = \cot x$	$du = -\csc^2 x\,dx$
$u = \csc x$	$du = -\csc x \cot x\,dx$

Example 7.2-2

Calculate:

1) $I = \int (2x+1)(x^2+x)\,dx$

Let $u = x^2 + x$ then $du = (2x+1)\,dx$

Then $I = \int u\,du = \frac{u^2}{2} + c = \frac{(x^2+x)^2}{2} + c.$

2) $I = \int (3x^2-1)\cos(x^3-x)\,dx$

Let $u = x^3 - x$ then $du = (3x^2-1)\,dx$

Then $I = \int \cos u\,du = \sin u + c = \sin(x^3 - x) + c.$

3) $I = \int \frac{\sin 2x}{\sqrt{1+\cos 2x}}\,dx$

Let $u = 1 + \cos 2x$, then $du = -2\sin 2x\,dx \Rightarrow \sin 2x\,dx = \frac{-1}{2}\,du$

So $I = \int \frac{-1}{2\sqrt{u}}\,du = \frac{-1}{2}\int \frac{du}{u^{\frac{1}{2}}} = \frac{-1}{2}\int u^{\frac{-1}{2}}\,du = \frac{-1}{2} \times \frac{u^{\frac{-1}{2}+1}}{\frac{-1}{2}+1} + c$

$$= \frac{-1}{2} \times \frac{u^{\frac{1}{2}}}{\frac{1}{2}} + c = -\sqrt{u} + c = -\sqrt{1+\cos 2x} + c.$$

4) $I = \int \frac{\sin x \cos x}{(1+\sin^2 x)^2} dx$

Let $u = 1 + \sin^2 x$, $du = 2 \sin x \cos x\ dx \Rightarrow \sin x \cos x\ dx = \frac{du}{2}$

$$I = \int \frac{du}{2u^2} = \frac{1}{2} \int u^{-2} du = \frac{1}{2} \times \frac{u^{-1}}{-1} + c = \frac{-1}{2u} + c = \frac{-1}{2(1+\sin^2 x)} + c.$$

5) $I = \int \frac{x}{\sqrt[3]{5-3x^2}} dx$

Let $u = 5 - 3x^2 \Rightarrow du = -6x\ dx \Rightarrow x\,dx = -\frac{du}{6} \Rightarrow I = \int -\frac{du}{6\sqrt[3]{u}}$

$$\Rightarrow I = -\frac{1}{6} \int \frac{du}{u^{\frac{1}{3}}} = -\frac{1}{6} \int u^{\frac{-1}{3}} du = -\frac{1}{6} \times \frac{u^{\frac{-1}{3}+1}}{\frac{-1}{3}+1} + c = -\frac{1}{6} \times \frac{u^{\frac{2}{3}}}{\frac{2}{3}} + c$$

$$I = -\frac{1}{4}\sqrt[3]{u^2} + c = -\frac{1}{4}\sqrt[3]{(5-3x^2)^2} + c.$$

6) $I = \int \sec^8 x \tan x\ dx$

Let $u = \sec x$, then $du = \sec x \tan x\ dx \Rightarrow I = \int u^7 dx$

$$I = \frac{u^8}{8} + c = \frac{\sec^8 x}{8} + c.$$

Section 7.3

The Definite Integral

Definition 7.3-1

Let f be a continuous function on an interval I, F anti-derivative of f on I, a and b two points of I. We call integral of f from a to b, the real number F (b) – F (a). This number is denoted by $\int_a^b f(x)\,dx = F(x)]_a^b = F(b) - F(a)$, the variable x takes all values between a and b.

Remark 7.3-1

$\int_a^b f(x)\,dx$ is a real number independent of the variable x.

Example 7.3-1

1) $\int_0^{\frac{\pi}{2}} (1+\sin x)\, dx = [\, x - \cos x \,]_0^{\frac{\pi}{2}} = \frac{\pi}{2} + 1.$

2) $\int_1^3 (t+t^2)\, dt = \left[\frac{t^2}{2}+\frac{t^3}{3}\right]_1^3 = \frac{38}{3}.$

Property 7.3-1 **(Direct Properties Integration)**

Let f be a continuous function on an interval I, and let a, b $\in$ I, then:

1) $\int_a^a f(x)\, dx = 0$

2) $\int_b^a f(x)\, dx = -\int_a^b f(x)\, dx$

3) $\int_a^b dx = b-a$

Property 7.3-2 **(Chasles Relation)**

Let f be a continuous function on an interval I, and let a, b and c $\in$ I, then:

$$\int_a^c f(x)\, dx = \int_a^b f(x)\, dx + \int_b^c f(x)\, dx$$

Property 7.3- 3 (Integration by parts)

Let f and g be two differentiable functions on an interval I, a and b two elements of I. If the functions f' and g' are continuous on I, then:

$$\int_a^b f'(x)\, g(x)\, dx = [\, f(x) g(x)\,]_a^b - \int_a^b f(x) g'(x)\, dx.$$

Example 7.3-2

Calculate $\int_0^1 x e^x\, dx$.

Solution

Let u = x then du = dx; dv = e^x dx then v = e^x

$$\int_0^1 xe^x\,dx = uv\Big]_0^1 - \int_0^1 v\,du$$
$$= xe^x\Big]_0^1 - \int_0^1 e^x\,dx$$
$$= e - e^x\Big]_0^1$$
$$= e - e + 1$$
$$= 1.$$

Section 7.4

Graphical Interpretation of The Definite Integral

Let f be a positive continuous function on [a , b], $a \leq b$. The area of the region bounded by the curve of f, the x-axis and the two vertical lines x = a and x = b, is given by $S = \int_a^b f(x)\,dx$. (Figure 7.4 – 1).

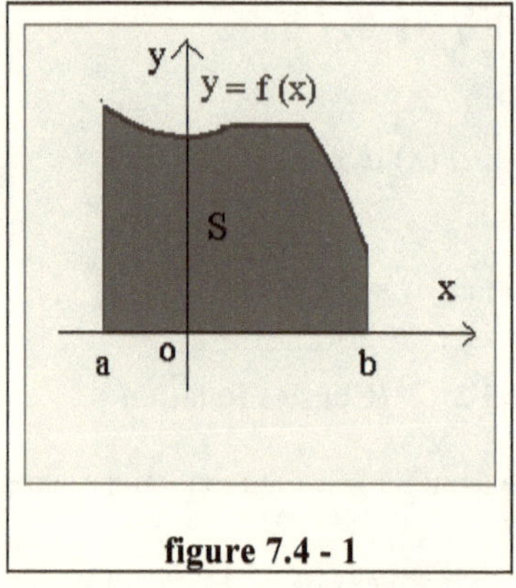

figure 7.4 - 1

Remark 7.4-1

1) For the case where f is negative it suffices to consider " - f " and consequently the area is given by $S = \int_a^b -f(x)\,dx$.

2) When f is continuous and has arbitrary sign on [a , b], the integral $\int_a^b f(x)\,dx$ is called the algebraic area bounded by the curve of f on [a , b].

3) To avoid getting a negative area , you can write the area in absolute value, that is $S = \left| \int_a^b f(x)\,dx \right|$.

4) The value of a given integral could be negative, that is $\int_a^b f(x)\,dx < 0$. However, if this integral is calculated as an area then it is impossible to be negative because the area is always positive; so in this case take its absolute value, that is $S = \left| \int_a^b f(x)\,dx \right|$.

Example 7.4-1

1) The area A of the region under the curve of the function $f(x) = \frac{1}{x^2}$ on the interval $[1, \alpha]$ where $\alpha > 1$ is

$$A = \int_1^{\alpha} \frac{1}{t^2}\, dt = \left[\frac{-1}{t}\right]_1^{\alpha} = 1 - \frac{1}{\alpha} \text{ square unit.}$$

2) Since the function $f(t) = -\frac{1}{2\sqrt{t}}$ is negative on $[1, 4]$, the area A of the region bounded by the curve of the function f and the x-axis on this interval is $A = \int_1^4 -f(t)\, dt = \int_1^4 -\frac{1}{2\sqrt{t}}\, dt = \sqrt{t}\,]_1^4 = 1$ square unit.

3) The area A of the region bounded by the curve of the function $f(x) = |x|$, the x-axis and the two straight lines $x = -2$ and $x = 1$ is

$$A = \int_{-2}^{0} -x\, dx + \int_0^1 x\, dx$$

$$= \left[\frac{-x^2}{2}\right]_{-2}^{0} + \left[\frac{x^2}{2}\right]_0^1$$

$$= \frac{5}{2} \text{ square unit. (Figure 7.4 – 2).}$$

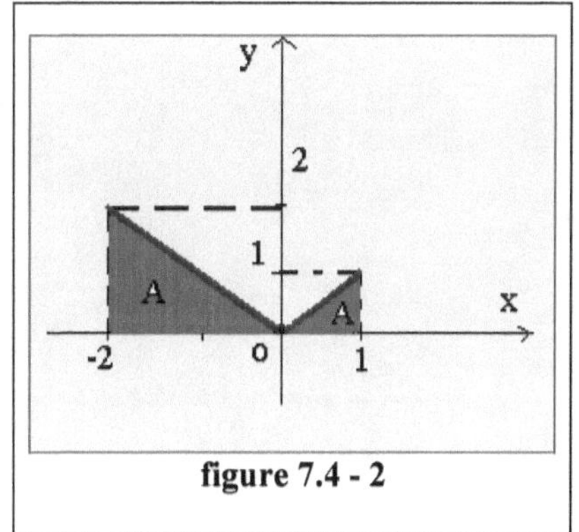

figure 7.4 - 2

4) The area of the region under the curve of the function $f(x) = \sin x$ on $\left[\frac{-\pi}{3}, \frac{\pi}{2}\right]$ is

$$A = \int_{\frac{-\pi}{3}}^{\frac{\pi}{2}} |\sin x|\, dx =$$

$$\int_{\frac{-\pi}{3}}^{0} -\sin x\, dx + \int_0^{\frac{\pi}{2}} \sin x\, dx =$$

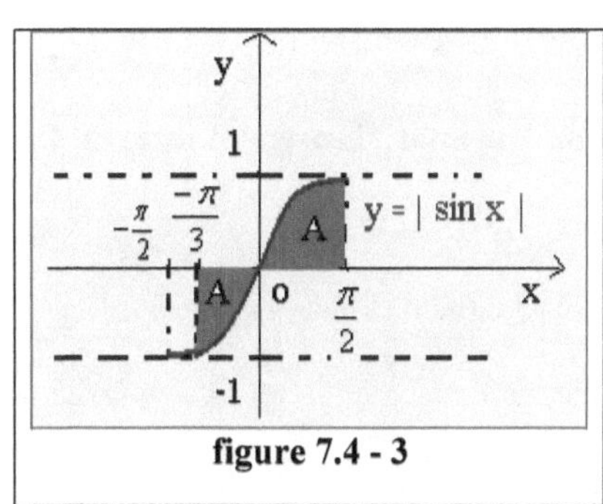

figure 7.4 - 3

$\frac{3}{2}$ square unit. (Figure 7.4 – 3).

Linearity of the Integral

Property 7.4-1

Let f and g be two continuous functions on an interval I, a and b are two real numbers of I. Then for every real numbers α and β we have:

$$\int_a^b (\alpha f(x) + \beta\, g(x))\, dx = \alpha \int_a^b f(x)\, dx + \beta \int_a^b g(x)\, dx .$$

Example 7.4-2

$$\int_1^3 \left(5x^2 + 12x\right) dx = 5 \int_1^3 x^2\, dx + 12 \int_1^3 x\, dx$$

$$= \left[5\frac{x^3}{3}\right]_1^3 + \left[12\frac{x^2}{2}\right]_1^3$$

$$= 5\,(\frac{27}{3}-\frac{1}{3}) + 12\left(\frac{9}{2}-\frac{1}{2}\right)$$

$$= 5\times\frac{26}{3} + 12\times 4$$

$$= \frac{130}{3} + 48$$

$$= \frac{274}{3}.$$

Section 7.5

The Fundamental Theorem of Integral Calculus

Property 7.5-1

Let f be a continuous function on an interval I and a real number of I.

The function defined on I by $\int_a^x f(t)\, dt$ is the antiderivative of f on I that vanishes at a.

Proof

Let F be an arbitrary antiderivative of f on I. Denote by φ the function defined on I by $\varphi(x) = \int_a^x f(t)\,dt$. Since $\varphi(x) = F(x) - f(a)$, then $\varphi(a) = 0$ and φ is differentiable on I with $\varphi' = F' = f$.

Example 7.5-1

The derivative of the function F defined by $F(x) = \int_0^x (u^2 + 2u + 1)\,du$ is given by $F'(x) = x^2 + 2x + 1$. We can verify it, because

$$F(x) = \left[\frac{u^3}{3} + u^2 + u\right]_0^x = \frac{x^3}{3} + x^2 + x.$$

Comparison Between Integrals

Property 7.5-1

Let f be a continuous function on an interval I, a and b are two real numbers of I. If $a \le b$ and if $f \ge 0$ on $[a, b]$ then $\int_a^b f(x)\,dx \ge 0$.

Indeed, if $f \ge 0$, and arbitrary antiderivative F of f is increasing on [a,b] (since $F' = f \ge 0$), hence $F(b) \ge F(a)$, therefore $\int_a^b f(x)\,dx \ge 0$.

Property 7.5-2

Let f and g be two continuous functions on an interval I, a and b are two real numbers of I such that $a < b$. If $f \le g$ on $[a, b]$, then $\int_a^b f(t)\,dt \le \int_a^b g(t)\,dt$.

Example 7.5-2

For every t we have $\cos t \le 1$, hence, integrating from 0 to x ($x \ge 0$), we get $\sin x \le x$.

Integration of Even And Odd Functions

The relation between area and integrals and the fact the symmetries and translations preserves areas, induce the following results:

Property 7.5-3 **(Even And Odd Functions)**

Let f be a function defined on an interval I whose center is O, and a an arbitrary real number of I.

1. If f is even then $\int_{-a}^{a} f(t)\ dt = 2\int_{0}^{a} f(t)\ dt$.

2. If f is odd then $\int_{-a}^{a} f(t)\ dt = 0.$

Example 7.5-3

1) Let $f(t) = t^2$

$$\int_{-1}^{1} t^2\ dt = 2\int_{0}^{1} t^2\ dt = \frac{2}{3}.$$

(Figure 7.5 – 1).

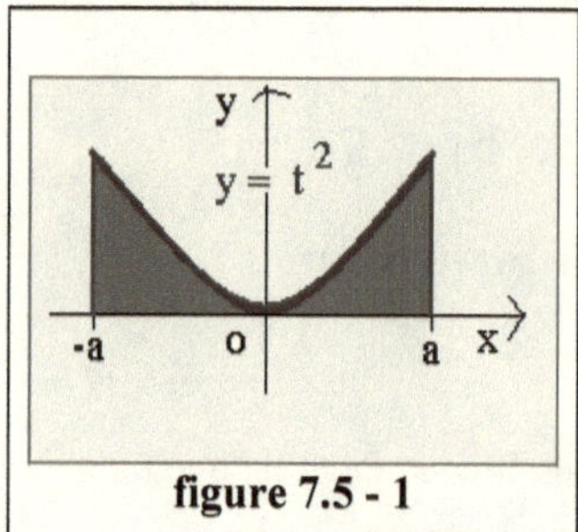

figure 7.5 - 1

2) Let $f(t) = t^3$

$\int_{-1}^{1} t^3\ dt = 0$. However, the area of the region bounded by $f(t) = t^3$, the x-axis, and the two straight lines x = -1 and x = 1 is $\int_{-1}^{1} f(t)\ dt =$

$$2\int_{-1}^{0} -t^3\ dt = 2\left[\frac{-t^4}{4}\right]_{-1}^{0} = \frac{1}{2} \text{ square unit.}$$

(Figure 7.5 – 2).

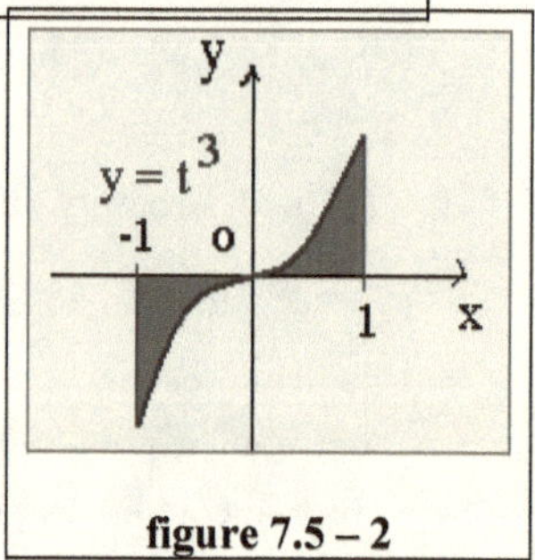

figure 7.5 – 2

Section 7.6

Area Between Two Curves

Property 7.6-1

Let f and g be two continuous functions on an interval [a , b]. If $f \leq g$ on [a , b], then the area A of the region bounded by the curves of f and g and the straight lines x = a and x = b is given by $A = \int_{a}^{b} [g(x) - f(x)]\ dx$.

Example 7.6-1

The two functions $f(x) = 1 + x^2$ and $g(x) = \sqrt{x}$ satisfy $0 \le g(x) \le f(x)$ for $x \in [0,1]$. Hence, the area A between the two curves and the two vertical lines x = 0 and x = 1 is given by $A = \int_0^1 (1 + x^2 - \sqrt{x})\ dx = \frac{2}{3}$. (Figure 7.6 – 1).

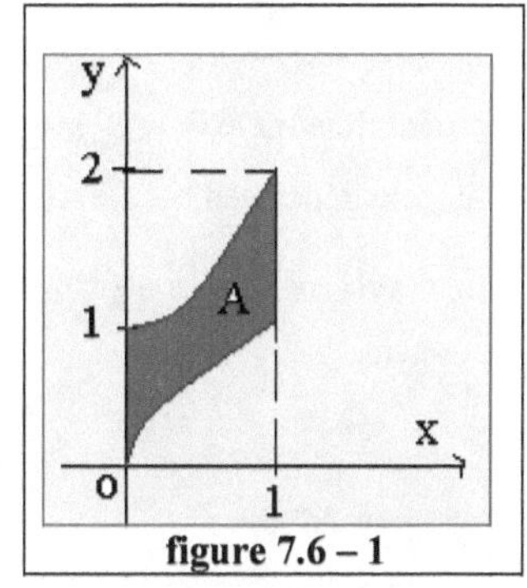

figure 7.6 – 1

Section 7.7

Volume of a Solid of Revolution

Property 7.7-1

The volume V of solid D of revolution about the x-axis, bounded by the two planes x = a and x = b is calculated by $V = \int_a^b S(x)\ dx$, where S(x) is the area of a cross section of D by a plane perpendicular to the x-axis.

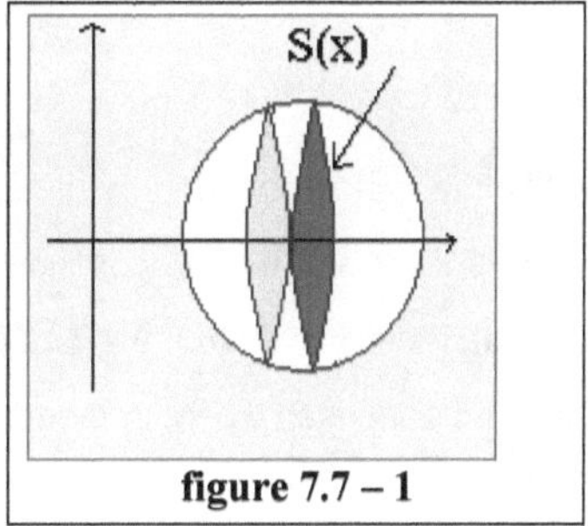

figure 7.7 – 1

Property 7.7-2

Let D be the plane region bounded by the curve of continuous positive function y= f(x), the x-axis and the two straight lines x = a and x = b. The volume of the solid of revolution, generated by the rotation of the region D around its revolution axis (o x), is given by $V = \int_a^b \pi y^2\ dx$.

Example 7.7-1

Let D be the region bounded by the curve of the function $f(x) = x^2$, the x-axis and the two vertical lines $x = 0$ and $x = 1$. The volume of the solid generated by the rotation of D about the x-axis is $V = \int_a^b \pi y^2 \, dx = \int_0^1 \pi (x^2)^2 \, dx = \int_0^1 \pi x^4 \, dx = \pi \frac{x^5}{5} \Big]_0^1 = \frac{\pi}{5}$. (Figure 7.7 – 2).

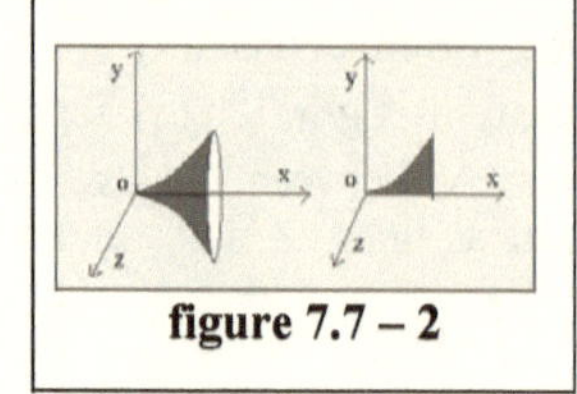
figure 7.7 – 2

Property 7.7-3

Let D be the plane region bounded by the curve of continuous positive function $y = f(x)$, the y-axis and the two straight lines $y = a$ and $y = b$. The volume of the solid of revolution, generated by the rotation of the region D around its revolution axis (o y), is given by $V = \int_a^b \pi x^2 \, dy$.

Example 7.7-2

Let D be the region bounded by the curve of the function $f(x) = x^2$, the y-axis and the two horizontal lines $y = 0$ and $y = 1$. The volume of the solid generated by the rotation of D about the y-axis is $V = \int_a^b \pi x^2 \, dy = \int_0^1 \pi (\sqrt{y})^2 \, dy = \int_0^1 \pi y \, dy = \pi \frac{y^2}{2} \Big]_0^1 = \frac{\pi}{2}$ cubic units.

(Figure 7.7 – 3).

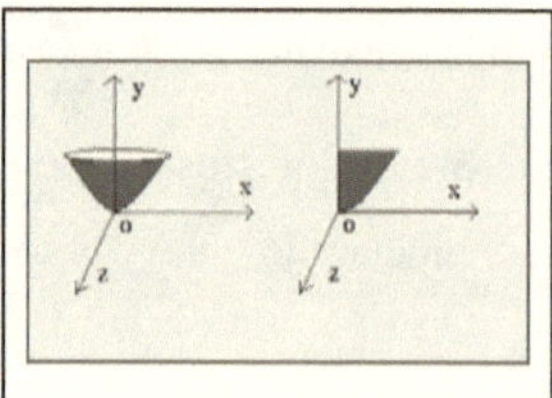
figure 7.7 – 3

Formulas

$u = \tan x$	$du = \sec^2 x \, dx$
$u = \sec x$	$du = \sec x \tan x \, dx$
$u = \cot x$	$du = - \csc^2 x \, dx$
$u = \csc x$	$du = - \csc x \cot x \, dx$

$\int 0 \, dx = c$

$\int a\ f(x)\ dx = a \int f(x)\ dx = a\ F(x) + c$
$\int [f(x) \pm g(x)]\ dx = \int f(x)\ dx \pm \int g(x)\ dx = F(x) \pm G(x) + c$
$\int a\ dx = a \int dx = a\,x + c$
$\int x^n\ dx = \frac{x^{n+1}}{n+1} + c$
$\int \cos x\ dx = \sin x + c$
$\int \sin x\ dx = -\cos x + c$
$\int \cos ax\ dx = \frac{\sin ax}{a} + c$
$\int \sin ax\ dx = \frac{-\cos ax}{a} + c$
$\int \sec^2 x\ dx = \int \frac{1}{\cos^2 x}\ dx = \int (1 + \tan^2 x)\ dx = \tan x + c$
$\int \csc^2 x\ dx = \int \frac{1}{\sin^2 x}\ dx = -\cot x + c$
$\int \sec x\ \tan x\ dx = \sec x + c$
$\int \csc x\ \cot x\ dx = -\csc x + c$
$\int \frac{1}{2\sqrt{x}}\ dx = \sqrt{x} + c$

Section 7.8

Finance

Deriving Totals (Revenue/Cost) from Marginals (Revenue/Cost)

MR is the derivative of the total revenue function TR. Therefore, the anti-derivative of the MR function gives the TR function.

Similarly, MC is the derivative of the total cost function TC. Therefore, the anti-derivative of the MC function gives the TC function.

Let $MC = 100 + x$

$$TC = \int MC\, dx = \int (100+x)\, dx$$

$$= 100x + \left(\frac{1}{2}\right)x^2 + c.$$

More information is needed to determine c in the TC function.

Thus, if

$$TC = 40\,000 \text{ at } x = 100,$$

then $40\,000 = 100(100) + \left(\frac{1}{2}\right)(100)^2 + c$

Therefore, $c = 40\,000 - 10\,000 - 5\,000$

$$= 25\,000$$

Hence: $TC = \left(\frac{1}{2}\right)x^2 + 100x + 25\,000.$

Capital Formation

Net investment I is defined as the rate of change in capital stock formation K over time t. If the process of capital formation is continuous over time,

$I(t) = \dfrac{dK(t)}{dt} = K'(t)$. From the rate of investment, the level of capital stock can be estimated. Capital stock is the integral with respect to time of net investment:

$$K_t = \int I(t)\, dt = K(t) + c = K(t) + K(0)$$

where $c =$ the initial capital stock K_0.

Example 7.8 -1

The rate of net investment is given by $I(t) = 140\, t^{\frac{3}{4}}$, and the initial capital stock at time $t = 0$ is 150. Find the time path $K(t)$.

Solution

$$K(t) = \int 140\, t^{\frac{3}{4}}\, dt = 140 \int t^{\frac{3}{4}}\, dt = 140\left(\frac{4}{7}t^{\frac{7}{4}}\right) + c = 80\, t^{\frac{7}{4}} + c$$

But $c = K_0 = 150$; therefore, $K(t) = 80\, t^{\frac{7}{4}} + 150$.

The value of Continuous Investment Flows Under Continuous Interest; Compounding

Remember that the general formula for solving the future value of an investment at the end of n years where interest is paid m times per year is

$$FV = PV\left(1 + \frac{i}{m}\right)^{mn}$$

In practice, interest is sometimes compounded continuously. That is, as the number of times per year m that interest's compounded approaches infinity, the term

$$\left(1 + \frac{i}{m}\right)^{mn}$$

approaches e^{in}, and the future value at the end of n years of an initial deposit PV_0 where interest is compounded continuously at the rate of i percent is

$$FV = PV_0\left(e^{in}\right)$$

By the same token, the formula for the present value of a single cash flow received at the end of year t, when interest is compounded continuously, is :

$$PV_0 = \frac{FV}{e^{in}}$$

The present value of a stream of future income (money to be received each year for t years) amounting to S dollars per year for n years, is found by the integral:

$$PV = \int_0^n S e^{-it}\, dt$$

Example 7.8 -2

Find the present value of $ 1 000 to be paid each year for three years when the interest rate is 5 % compounded continuously.

$$PV = \int_0^n S e^{-it}\, dt = S \int_0^n S e^{-it}\, dt = S\left[-\frac{1}{i}e^{-it}\right]_0^n = -\frac{S}{i}\left[e^{-it}\right]_0^n$$

$$= -\frac{S}{i}\left(e^{-in} - 1\right) = \frac{S}{i}\left(1 - e^{-in}\right)$$

$$= \frac{1000}{0.05}\left(1 - e^{-(0.05)(3)}\right) = 20\,000\left(1 - e^{-0.15}\right) = \$\,2786.$$

If the $ 1000 is to be received indefinitely for each year, then the present value of this income stream is given by:

$$PV = \int_0^\infty S e^{-it}\, dt = -\frac{S}{i}\left[e^{-it}\right]_0^\infty = -\frac{S}{i}\left(0 - 1\right) = \frac{S}{i} = \frac{1000}{0.05} = \$\,20\,000.$$

Summary

- F is said to be the antiderivative of f if:
  a) F is differentiable over I.
  b) $F'(x) = f(x)$ for all $x \in I$.
- If F is the antiderivative of f over the interval I , then the general signification f of F on I is: $f(x) = F(x) + c$ where c is any constant $\in \mathfrak{R}$.
- $\int f(x)dx = F(x)+c$, where $F'(x) = f(x)$ for all $x \in I$, and c is any constant $\in \mathfrak{R}$.
- $\int_a^a f(x)\,dx = 0$.
- $\int_b^a f(x)\,dx = -\int_a^b f(x)\,dx$.
- $\int_a^b dx = b-a$.
- $\int_a^c f(x)\,dx = \int_a^b f(x)\,dx + \int_b^c f(x)\,dx$.
- $\int_a^b f'(x)\,g(x)\,dx = [f(x)g(x)]_a^b - \int_a^b f(x)g'(x)\,dx$
- $\int_a^b (\alpha f(x) + \beta\, g(x))\,dx = \alpha \int_a^b f(x)\,dx + \beta \int_a^b g(x)\,dx$.
- If $a \le b$ and if $f \ge 0$ on $[a,b]$ then $\int_a^b f(x)\,dx \ge 0$.
- If $f \le g$ on $[a,b]$, then $\int_a^b f(t)\,dt \le \int_a^b g(t)\,dt$.
- If f is even then $\int_{-a}^a f(t)\,dt = 2\int_0^a f(t)\,dt$.
- If f is odd then $\int_{-a}^a f(t)\,dt = 0$.
- The area A of the region bounded by the curves of f and g and the straight lines $x = a$ and $x = b$ is given by $A = \int_a^b [g(x)-f(x)]\,dx$.
- The volume of the solid of revolution, generated by the rotation of the region D around its revolution axis (o x), is given by $V = \int_a^b \pi y^2\,dx$.
- The volume of the solid of revolution, generated by the rotation of the region D around its revolution axis (o y), is given by $V = \int_a^b \pi x^2\,dy$.

Exercises

7.1. Verify that:

a) $F(x) = x^2\sqrt{x} + 1$ is an antiderivative of $f(x) = \frac{5}{2}x\sqrt{x}$ *on* $[0,+\infty[$.

b) $F(x) = \frac{2}{3}\left(x\sqrt{x}-3\right)$ is an antiderivative of $f(x) = \sqrt{x}$ *on* $[0,+\infty[$.

In problems 7.2 – 7.23, find the given integrals :

7.2. $\int\left(x^4 - 3x^2 + 2x - 5\right) \, dx$

7.3. $\int\left(2x^2 + \frac{1}{x^2} - 4x + 12\right) \, dx$

7.4. $\int 2(x-1)^3 \, dx$

7.5. $\int \sin(3x) \, dx$

7.6. $\int \sin x \cos x \, dx$

7.7. $\int \frac{\sin 2x}{(1+\cos 2x)^3} \, dx$

7.8. $\int \sec 2x \tan 2x \, dx$

7.9. $\int \tan^2 x \sec^2 x \, dx$

7.10. $\int \frac{3x}{\sqrt[3]{3+6x^2}} \, dx$

7.11. $\int \left(x^3 - \csc x \cot x + 7\right) dx$

7.12. $\int\left(x + \frac{1}{\sqrt{x}} + \frac{1}{x^4}\right) \, dx$

7.13. $\int \frac{\arctan t}{1+t^2} \quad dt$

7.14. $\int x^2 \cos x \quad dx$

7.15. $\int x^5 \sin(x^6+2)\ dx$

7.16. $\int \frac{x+1}{x^2+2x+3}\ dx$

7.17. $\int \sec^2(x-\pi)\ dx$

7.18. $\int \frac{x}{\sin^2\left(\frac{x^2}{2}\right)}\ dx$

7.19. $\int \left(3x^2 - \sin x + 2\sec^2 x\right)\ dx$

7.20. $\int \frac{3x-1}{\left(3x^2-2x+1\right)^4}\ dx$

7.21. $\int \frac{(arc\cos t)^2}{\sqrt{1-t^2}} \quad dt$

7.22. $\int \frac{\ln x}{x^2} \quad dx$

7.23. $\int x\, e^{3x} \quad dx$

In problems 7.24 – 7.52, calculate the given integrals:

7.24. $\int_{-2}^{-1} \left(x + \frac{1}{x^2}\right)\ dx$

7.25. $\int_0^x \quad 2\sin t \cos t\ dt$

7.26. $\int_0^1 \quad x\sqrt{x^2+1}\ dx$

7.27. $\int_0^2 \frac{x}{\sqrt{2x^2+1}}\,dx$

7.28. $\int_{\frac{\pi}{6}}^{\frac{\pi}{8}} \tan^2\left(3x+\frac{\pi}{6}\right)dx$

7.29. $\int_{\frac{\pi}{8}}^{\frac{\pi}{6}} \frac{1}{\cos^2(2x)}\,dx$

7.30. $\int_0^{\frac{\pi}{2}} \frac{\sin x}{2-\cos x}\,dx$

7.31. $\int_{\frac{\pi}{10}}^{\frac{\pi}{4}} \cot 5x\ dx$

7.32. $\int_0^{\frac{\pi}{2}} \sin^2 x\cos x\ dx$

7.33. $\int_0^{\frac{\pi}{6}} \tan 2x\ dx$

7.34. $\int_0^{\frac{\pi}{12}} \frac{1}{\cot 3y}\,dx$

7.35. $\int_1^4 x|2-x|\ dx$

7.36. $\int_{-3}^5 |x+1|\ dx$

7.37. $\int_1^2 \sqrt{8x}\ dx$

7.38. $\int_0^\pi (\pi-|2x-\pi|\sin x)\ dx$

7.39. $\int_1^2 \frac{x-x^2+2x^3}{x}\,dx$

7.40. $\int_1^2 \left(-x+\frac{1}{2\sqrt{x}}+\frac{1}{x\sqrt{x}}\right)dx$

7.41. $\int_0^2 \frac{x^2}{\sqrt{x^3+1}}\,dx$

7.42. $\int_{0}^{\frac{\pi}{12}} \cos\left(2x+\frac{\pi}{3}\right) dx$

7.43. $\int_{\frac{\pi}{8}}^{\frac{\pi}{4}} \cot^2(2x)\ dx$

7.44. $\int_0^{\frac{\pi}{8}} \tan^2(2x)\ dx$

7.45. $\int_{\frac{\pi}{4}}^{\frac{\pi}{2}} \cos^3 x \sin x\ dx$

7.46. $\int_{\frac{\pi}{12}}^{\frac{\pi}{6}} \frac{1}{\sin^2(3x)}\ dx$

7.47. $\int_0^2 \left(x^2-|x-1|\right) dx$

7.48. $\int_0^4 |x-3|\ dx$

7.49. $\int_{-3}^{1} \left|x^2+x-2\right|\ dx$

7.50. $\int_0^{\pi} x\cos x\ dx$

7.51. $\int_0^2 (2x+1)\, e^{x}\ dx$

7.52. $\int_1^4 (x+1)\ln x\ dx$

In problems 7.53 – 7.56, using even and odd functions, give the values of the integrals:

7.53. $\int_{-1}^{1} \left(t\sqrt{1+t^4}\right)\ dt$

7.54. $\int_{\frac{-\pi}{2}}^{\frac{\pi}{2}} t^2\tan^3 t\ dt$

7.55. $\int_{-1}^{1} \left(3t^6+1\right)\ dt$

7.56. $\int_{\frac{-\pi}{3}}^{\frac{\pi}{3}} \sin^2 3x\ dx$

7.57. a. Justify that, for $\frac{\pi}{2} \le x \le \pi$, we have : $\frac{\sin x}{1+\pi^2} \le \frac{\sin x}{1+x^2} \le \frac{\sin x}{1+\frac{\pi^2}{4}}$.

b. Deduce a bound for the integral $\int_{\frac{\pi}{2}}^{\pi} \frac{\sin x}{1+x^2}\, dx$.

7.58. Find a, b, c verifying: $\frac{3x^2}{x^2(x+1)} = \frac{a}{x^2} + \frac{b}{x} + \frac{c}{x+1}$, and deduce the value

of $\int_1^2 \frac{3x^2+1}{x^2(x+1)}\, dx$.

7.59. Find a, b, c verifying: $\frac{3x^2+x+2}{(x+2)(x+1)^2} = \frac{a}{x+2} + \frac{b}{x+1} + \frac{c}{(x+1)^2}$,

and deduce the value of $\int_0^1 \frac{3x^2+x+2}{(x+2)(x+1)^2}\, dx$.

7.60. a. Find a, b, c verifying $f(x) = \frac{x^4 - x^3 + 3x^2 - 1}{x^2} = a x^2 + b x + c - \frac{1}{x^2}$.

b. Deduce an antiderivative F of f that verifies F (1) = 0.

7.61. Let F (x) = $\int \frac{x^2-4x}{(x-2)^2}\, dx$.

a. Show that F (x) = $\int \left(1 - \frac{k}{(x-2)^2}\right) dx$, where k is a real number to be determined.

b. Deduce the function F such that F (0) = 1.

In problems 7.62 – 7.66, sketch the curve of the indicated function f, and calculate the area of the region associated to f on [a , b]:

7.62. $a = -1$ $b = 2$ $f(x) = x^3 + 1$

7.63. $a = \frac{-\pi}{4}$ $b = \frac{\pi}{4}$ $f(x) = \sin x$

7.64. The x-axis , f (x) = x (x -1) (x – 2)

7.65. $a = 1$ $b = 3$ $f(x) = 4x - x^2$

7.66. $a = 1$ $b = 4$ $f(x) = -x^3$

In problems 7.67 – 7.73, find the intersection points of f and g then calculate the area bounded by the graphs of the given functions:

7.67. $f(x) = x^2$; $g(x) = \sqrt{x}$

7.68. $f(x) = 3x - x^2$; $g(x) = 3 - x$

7.69. $f(x) = \sin x$; $g(x) = \cos x$; $x = 0$; $x = \frac{\pi}{2}$

7.70. $f(x) = 2 - x^2$; $g(x) = -x$

7.71. $f(x) = x^2$; $g(x) = 4 - x^2$

7.72. $f(x) = |x|$; $g(x) = 1 - |x|$

7.73. $f(x) = x^4 - 2x^2$; $g(x) = 2x^2$

In problems 7.74 – 7.80, calculate the volume of the solid generated by the rotation about the x-axis of the region bounded by the graph of a function f, the x-axis and the two vertical lines x = a and x = b:

7.74. $f(x) = x + 1$; $a = 0$; $b = 1$

7.75. $f(x) = \cos x$; $a = 0$; $b = \frac{\pi}{2}$

7.76. $f(x) = 3 - x$; $a = 0$; $b = 3$

7.77. $f(x) = x - x^2$; $a = 0$; $b = 1$

7.78. $f(x) = 4 - x^2$; $a = -2$; $b = 2$

7.79. $f(x) = x^2$; $y = 0$; $y = 1$; y-axis

7.80. $x + y = 2$; $y = 0$; $x = 0$

7.81. Calculate the volume of the solid generated by the rotation about the x-axis of the region bounded by the two graphs of the functions:

a) $f(x) = 8x$ and $g(x) = x^2$.

b) $y^2 = x$ and $y = x^2$.

7.82. Let f be the function defined on the interval [0 , 11] and whose graph is given in figure 7.7 - 4:

The points M, N, A, B, D and C are defined by their coordinates as indicated on figure 7.7 - 4.

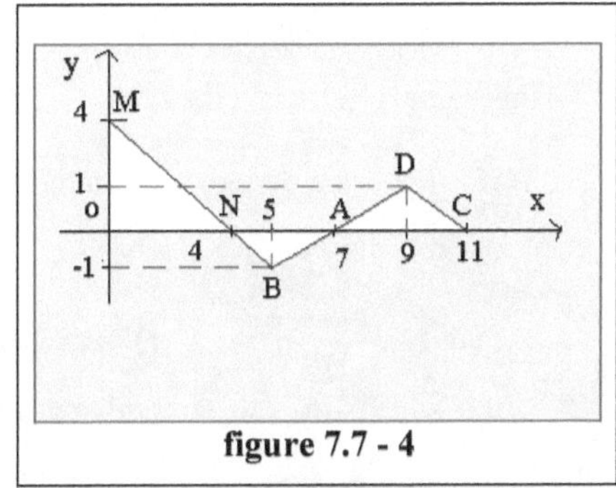

figure 7.7 - 4

a. Give the expression of f(x) on each of the following intervals [0 , 5], [5 , 9] and [9 , 11].

b. Calculate the values of the following integrals :

$$I = \int_{7}^{11} f(x)\, dx \quad ; \quad J = \int_{0}^{5} |f(x)|\, dx$$

7.83. The time equation of a moving particle M on an axis $(o,\vec{i})$ is x (t) , where t represents time , and the velocity is given by V (t) .

The average velocity v during t_1 to t_2 is given by $v = \dfrac{x(t_2)-x(t_1)}{t_2-t_1}$;

but $x(t_2) - x(t_1) = \int_{t_1}^{t_2} V(t)\, dt$ hence $v = \dfrac{1}{t_2-t_1} \int_{t_1}^{t_2} V(t)\, dt$.

(hence the name , average velocity of the function V on the interval $[t_1 , t_2]$).

a) Assume that $V(t) = \dfrac{t}{2} + 5,\ t \in \Re$. What is the average velocity of M during 0 to 5 ?

b) Same question when V is defined on [0 , 5] by :

$$V(t) = \begin{cases} 3t^2 & for \quad t \in [0,1] \\ 3 & for \quad t \in]1,4[\\ 3(t-5)^2 & for \quad t \in [4,5] \end{cases}$$

after graphical representation of V on the interval [0 , 5].

7.84. We call effective strength I of an alternating current i the strength of the direct current which passing through a dipoles during the same time of the alternating current ,makes the dipoles receive or dissipate the same electrical energy.

a) If T is the period of this alternating current and R the resistance of the dipoles, we have $\int_0^T Ri^2(t)\,dt = \int_0^T RI^2\,dt$. Show that $I^2 = \frac{1}{T}\int_0^T i^2(t)\,dt$, hence I^2 is defined as the average of the square of the strength of the current during a period.

b) Assume that, I_m and ω are two given real numbers (I_m the maximal strength and ω the pulse), $i(t) = I_m \cos \omega t$. Calculate I in terms of I_m.

7.85. A. Study the function f defined by f(x) = 1 + cos 2x on the interval $\left[\frac{-\pi}{2}, \frac{\pi}{2}\right]$. Draw its graph (C) in an orthonormal system x 'o x , y 'o y .

B. Calculate the area of the domain D limited by the curve (C), the x-axis and the two straight lines $x = -\frac{\pi}{2}$ *and* $x = \frac{\pi}{2}$.

C. a. Prove that : $(1+\cos 2x)^2 = \frac{1}{2}\cos 4x + 2\cos 2x + \frac{3}{2}$.

b. Calculate $\int_{\frac{-\pi}{2}}^{\frac{\pi}{2}} (1+\cos 2x)^2\,dx$.

c. Deduce the volume V of the solid of revolution generated by the rotation of D about the x-axis.

7.86. We want to calculate the volume V of a solid of a revolution that has the shape of a fishing plug. A model of this plug is obtained by rotation about the x-axis (o x) of the colorful plate as indicated in.

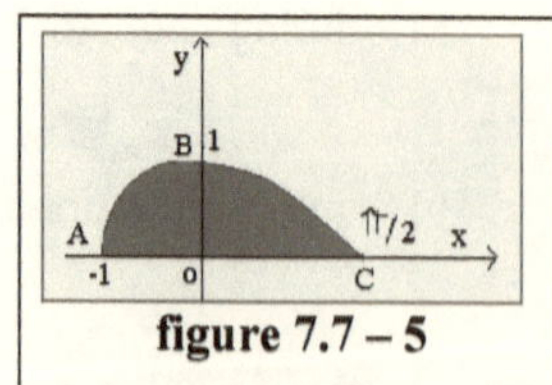

figure 7.7 – 5

- The arch A B is a quarter of circle of center O and of radius 1.

- The arch B C has for equation y = cos x, $x \in \left[0, \frac{\pi}{2}\right]$.

A. The section of the plug by a plane perpendicular to the axis is a disk.

a. Determine in terms of x the area A(x) of this disk for $x \in \left[0, \frac{\pi}{2}\right]$.

b. Calculate the volume V_1 of the part of the plug corresponding to interval $\left[0, \frac{\pi}{2}\right]$.

B. Calculate the volume V_2 of the part of the plug corresponding to interval $[-1,0]$, then deduce the volume V.

7.87. Can you find an integer n $(n \geq 2)$ for which the two regions, (in red and yellow) as shown in figure 7.7 - 6, has equal areas?

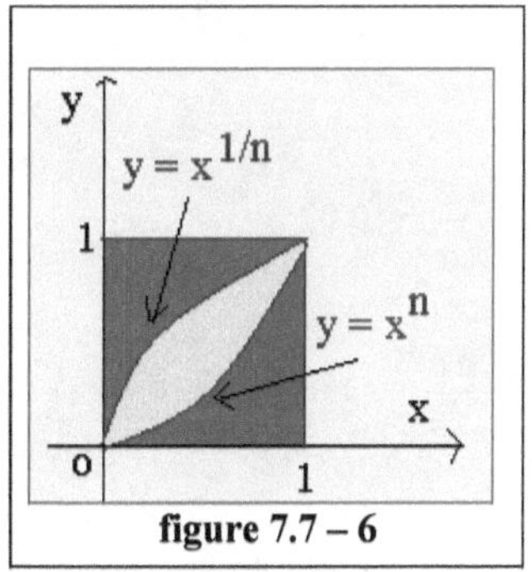

figure 7.7 – 6

7.88. Let (C) be the curve of equation $y = \frac{2}{\sqrt{x}}$. Consider the thin homogeneous plate (P), limited by the x-axis, the curve (C) and the two straight lines x = 1 and x = 4. (Figure 7.7 – 7).

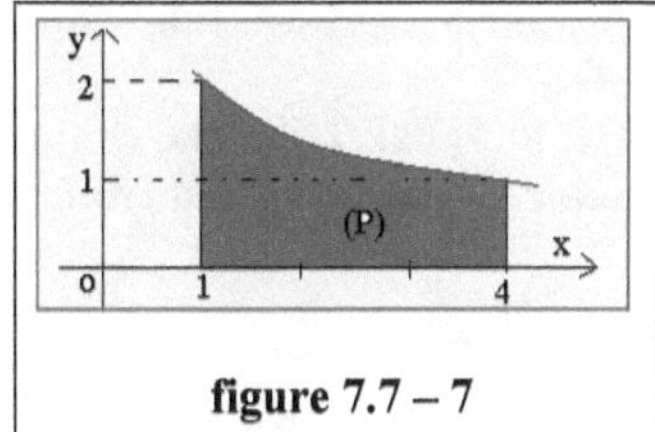

figure 7.7 – 7

a) Calculate the area of the plate (P).

b) Calculate the volume of the solid generated by rotation of the plate (P) about the x-axis.

In problems 7.89 – 7.91, given the values of the marginal – revenue functions, Find the demand function in each case (Hint : $p = \frac{r}{q}$).

7.89. $\frac{dr}{dq} = 0.6.$

7.90. $\frac{dr}{dq} = 4 - 0.1\,q.$

7.91. $\frac{dr}{dq} = 260 - q - 0.1\,q^2.$

In problems 7.92 – 7.96, given the values the marginal – cost functions and their corresponding fixed costs. Find the total cost for the indicated value of q:

7.92. $\frac{dc}{dq} = 2.34$; $q = 100$

7.93. $\frac{dc}{dq} = 2\,q + 65$; $q = 1000$

7.94. $\frac{dc}{dq} = 0.02\,q^2 - 0.2\,q + 2.3$; $q = 2000$

7.95. $\frac{dc}{dq} = 450 + 2\,q + 0.1\,q^2$; $q = 400$

7.96. $\frac{dc}{dq} = 260 - q - 0.1\,q^2$; $q = 300$

In problems 7.97 – 7.100, a manufacturer 's marginal - revenue function is $\frac{dr}{dq}$.

If r is in dollars, find the change in the manufacturer ' s total revenue if production is increased from a to b units:

7.97. $\frac{dr}{dq} = \frac{9000}{\sqrt{900\,q}}$; $a = 100$ and $b = 300$

7.98. $\frac{dr}{dq} = 200 + 80q + 2q^2$; $a = 10$ and $b = 20$

7.99. $\frac{dr}{dq} = 2000e^{-0.06q}$; $a = 0$ and $b = 5$

7.100. $\frac{dr}{dq} = 50 - 0.5q + 0.004q^2$; $a = 90$ and $b = 180$

7.101. The marginal cost of a production is given by $M(x) = 4x + 400$ where x is the number of units produced.

a. If the total cost of producing 20 units is $ 60 000, determine the total cost function.

b. Calculate the cost of producing 30 units.

7.102. The marginal profit in dollars of a product is given by $M(x) = -6x + 900$ where x is the number of units sold.

a. Find the profit function if $ 10 000 is the total profit of 100 units sold.

b. Calculate the total profit if 120 units are sold.

c. Determine x so that the profit is maximum.

7.103. The marginal revenue in dollars of a product is given by $M(x) = -0.02x + 10$ where x is the number of units sold.

a. Represent graphically this function.

b. Calculate the total revenue if 100 units are sold.

c. Calculate the supplementary revenue if the sales increase from 100 units to 200 units. What does this number represent graphically?

7.104. a. The rate of net investment is $I(t) = 250\,t^{\frac{2}{3}}$ and the initial capital stock is $ 110. Find the path of capital stock and determine the level of capital stock in 8 years.

b. Find the present value of an income stream of $ 5000 to be paid annually using a continuously compounded interest rate of 4 % when the income stream terminates in exactly 5 years and when it is received indefinitely.

7.105. a. The rate of net investment is $I(t) = 9\,t^{\frac{1}{2}}$ Find the level of capital formation in 8 years.

b. Find the present value of an income stream of $ 10 000 to be paid each year for 5 years when the continuously compounded interest rate is 5 %.

c. Marginal cost is given by the function $25 + 30q - 9q^2$. Fixed cost is 55. Find the total cost and average cost functions.

7.106. a. Given $Z = e^{(2x^2 - 12x - 2xy + y^2 - 4y)}$. Find the critical values and determine whether the function is maximized or minimized at these points.

b. The rate of net investment is $I(t) = 100\sqrt{t}$ and the initial capital stock is $ 120. Find the path of capital stock and its level after 20 years.

7.107. a. A savings and loan association pays 6 % per annum compounded continuously. If a person saves \$ 2000 per year for 5 years, will there be enough in the account to make a 20 % down payment on a \$ 100 000 house?

b. How much does a person need to save per year to accumulate the required down payment?

7.108. The marginal cost of producing a product is defined by:

$M(x) = x^2 - 100x + 3000$ where x is the quantity produced in units.

a. Knowing that the fixed cost is \$ 26244, express the total cost C(x) in terms of x.

b. 1) Express the average cost $\overline{C}(x)$ in terms of x.

2) Calculate $\overline{C}'(x)$.

3) Expand $(x - 81)(2x^2 + 12x + 972)$, then study the variations of $\overline{C}$ over $]0,+\infty[$.

4) determine the quantity produced for the average cost to be minimal, then calculate the corresponding cost.

7.109. An investor will receive an annuity of \$1000 in perpetuity. What is the present value of her income stream discounted at a continuously compounded interest rate of 6 %? What is the present value of the first 50 years of income?

7.110. The rate of consumption of a product is given by $g(t) = 6t^2 + 20t$, where t is the time in years starting from launching of the product and g(t) is expressed in thousands of units.

a. What is the total consumption of this product in the first 4 years of launching the product?

b. What is the total consumption of the product in the second five years?

7.111. The marginal cost M(x), expressed in dollars, of producing a certain article is defined by $M(x) = 6x + 300$ where x is the number of articles produced.

a. Knowing the total cost of production of 20 articles is \$ 8000, express the total cost C(x) in terms of x.

b. Calculate the total cost of production of the first 40 articles.

7.112. The annual rate cost of maintenance of an electronic device is given by $A(t) = 0.003\, t^2 + 5$ where t is the number of working hours and A(t) is expressed in dollars.

a. Determine the cost rate of maintenance after 30 working hours.

b. What is the cost of maintenance during the first 60 working hours?

c. What is the cost of maintenance during the second sixty working hours?

7.113. The marginal revenue of a certain product, expressed in dollars, is given by $M(p) = -0.01\, p + 5$.

a. Determine the total revenue by selling the first 100 units of this product.

b. Determine the total revenue resulting from increases the sells from 100 to 200 units of this product.

7.114. A manufacturer of smelting works produces x hundreds f units monthly where $0 \leq x \leq 8$. The marginal cost, in thousands of dollars for x hundreds of units is given by : $M(x) = x^2 + 18\, x + 53$.

a. Knowing that the fixed cost is \$ 45000, determine the total cost C(x) of x hundreds of units.

b. Calculate the total cost of production of the first 300 pieces.

7.115. The annual rate of cost of maintenance of some electronic devices of a company is represented by the function defined by $A(t) = 12\, t^2 + 100$ where t is the age of the device in years and A(t) is expressed in dollars per year.

a. What is the cost of maintenance of a device during the first 10 years?

b. What is the cost of maintenance of this device during its tenth year?

7.116. The reduction of the cost of consumption of fuel in a machine is given by $r(t) = 2000\, \dfrac{2t}{t^2+1}$, where t is the time expressed in years and r (t) is expressed in dollars. Calculate the economics on the facture of fuel during the first 4 years using the machine.

7.117. After the spread of a disease in a region, it was noted that the number of persons that have caught the disease x days after the manifestation of the first cases is given by $D(x) = 45\, x^2 - x^3$ for $0 \leq x \leq 25$. Calculate the average number of sick persons during the first eight days.

7.118. A company produces a quantity q of electronic devices (expressed in thousands of units) and whose marginal cost is given by $M(q) = q + \frac{8}{2q+1}$ with M(q) expressed in thousands of dollars and $\frac{1}{4} \le q \le 8$.

a. Determine the total cost function C(q) knowing that $C(0) = 0$.

b. 1) Verify that the average cost in thousands of devices is defined over $\left[\frac{1}{4}, 8\right]$ by $\overline{C}(q) = \frac{q}{4} + 4\frac{\ln(2q+2)}{q}$.

c. Show that $\overline{C}'(q) = \frac{f(q)}{2q^2}$, where $f(q) = q^2 + \frac{16q}{2q+1} - 8\ln(2q+1)$. .

d. Study the variations of f(q) .

e. Show that the equation f(q) = 0 admits a unique solution $\alpha \in]2.92, 2.93[$.

f. Study the variations of $\overline{C}$.

g. Assume $\alpha = 2.93$.

1) For what production , round to the nearest ten, is the average cost, in thousands of devices, minimum?

2) Verify that for this quantity obtained, the average cost and the marginal cost have the same value to the nearest \$ 5.

7.119. A print press has the capacity of producing 5000 books per day. A study showed that the marginal cost can be expressed by $M(n) = -2n + 5 + 3\ln(n+1)$ (in thousands of dollars) where n designates the number of books printed in thousands.

a. Show that $m(n) = -n^2 + 2n + 3(n+1)\ln(n+1)$ is an antiderivative of M(n).

b. Calculate, to the nearest thousands, the total cost of producing 5000 books knowing that the fixed cost is two thousands dollars.

c. The print press needs to carry out an order of 8000 textbooks in two days. There are two possible options:

Option I : 5000 books the first day the 3000 books the second day.

Option II : 4000 books each day.

Which option is better?

7.120. The function $M(q) = 0.8 + 4(1 - 2q)e^{-2q}$ represents the daily marginal cost of a factory for producing a liquid chemical product. q is being the quantity produced in thousands of liters. M (q) is expressed in thousands of dollars. Suppose that the fixed cost is \$ 1000.

a. Show that the total cost function C (q) is defined by:

$C(q) = 0.8q + 4qe^{-2q} + 1$.

b. The selling price of this liquid is \$ 18 per liter. The daily product is totally sold. Show that the daily revenue can be expressed as R(q) = 1.8 q and calculate the profit P(q).

c. The graph below represents the two functions R and C.

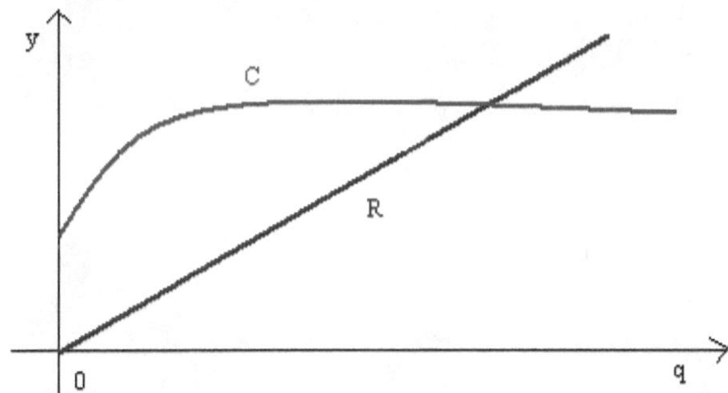

Show that the equations R(q) = C(q) admits a unique solution α such that $1.3 < \alpha < 1.4$. Give an approximate value of the minimum quantity produced for the factory to attain profit.

7.121. Suppose that the functions D and S are respectively the demand and supply functions of a company that transports goods. More precisely, for shipping one ton: $D(d) = 5(d + 2)e^{-d}$ is the price in dollars for 100 km of service accepted by the clients in terms of the distance d covered in hundred of kilometers. $S(d) = \frac{d+2}{5}e^{d}$ is the price in dollars for 100 km of service proposed by the company in terms of the distance d covered in hundreds kilometers.

a. What is the price p_1, in dollars for 100 km, a client is ready to pay and what price p_2, in dollars for 100 km, proposed by the company for a distance of 120 km covered?

b. Determine the price of equilibrium p_0 and find the corresponding distance d_0.

c. All the customers are ready to by the service at a price superior to the price of equilibrium which realizes a gain called the customer's surplus. We admit that this gain is measured by G = $\int_0^{d_0} D(d)\ d_d - p_0 \times d_0$.

Verify that F (d) = - 5 (d + 3) e^{-d} is an antiderivative of D(d) ,then calculate the exact value of G.

ANSWERS TO ODD – NUMBERED PROBLEMS

CHAPTER 1, pp. 29 - 34

1.1. 20 ; **1.3.** -32

1.5. $\frac{-4}{3}$; **1.7.** $\frac{1}{6}$

1.9. 4 ; **1.11.** 11

1.13. 15 ; **1.15.** 2

1.17. $4x - 6 + 3x - 3 = 6x - 7 \Rightarrow x = 2.$

1.19. Condition: $x \neq 0$. $\frac{5}{x} = \frac{5}{6} \Rightarrow x = 6$ acceptable.

1.21. Condition: $x \neq 0$. $6x^2 + 2x = 6x^2 - 12x \Rightarrow x = 0$ rejected.
Then the equation has no solution.

1.23. Condition: $x^2 - 3 \geq 0 \Rightarrow x \leq -\sqrt{3}$ *or* $x \geq \sqrt{3}$.
Square both sides : $x^2 - 3 = (3 - x)^2 \Rightarrow x = 2$ acceptable.
Then the equation has a single root $x = 2$.

1.25. Condition: $x^2 - 1 \geq 0 \Rightarrow x \leq -1$ *or* $x \geq 1$.
$\sqrt{x^2 - 1} = -(1 + x)$
Square both sides: $x^2 - 1 = 1 + 2x + x^2 \Rightarrow x = -1$ acceptable.
Then the equation has a single root $x = -1$.

1.27. $\Delta = b^2 - 4ac$
$= 25 - 4(-1)(3)$
$= 37 > 0$, then the equation has two distinct roots:

$$\begin{cases} x_1 = \dfrac{-b + \sqrt{\Delta}}{2a} \\ x_2 = \dfrac{-b - \sqrt{\Delta}}{2a} \end{cases} \Rightarrow \begin{cases} x_1 = \dfrac{-5 + \sqrt{37}}{-2} \\ x_2 = \dfrac{-5 - \sqrt{37}}{-2} \end{cases} \Rightarrow \begin{cases} x_1 = \dfrac{5 - \sqrt{37}}{2} \\ x_2 = \dfrac{5 + \sqrt{37}}{2} \end{cases}$$

1.29. $8x+7=\pm 15 \Rightarrow \begin{cases} x_1 = 1 \\ x_2 = \dfrac{-11}{4} \end{cases}$

1.31. $|-3x+2| = -3x+2 \quad for\ x < \dfrac{2}{3}$ and $|2x-3| = 2x-3 \quad for\ x > \dfrac{3}{2}$

x	- ∞ 2 / 3	3 / 2	+ ∞
$\|-3x+2\|$	- 3 x + 2	3 x - 2	3 x - 2
$\|2x-3\|$	- 2 x + 3	- 2 x + 3	2 x - 3
$\|-3x+2\|+\|2x-3\|=5$	-5 x + 5 = 5 ⇒ x = 0 acceptable	x + 1 = 5⇒ x = 4 rejected	5 x − 5 = 5 ⇒ x = 2 accepted

- For x = 2 / 3, the given equation gives $\dfrac{5}{3} = 5$ impossible.
- For x = 3 / 2, the given equation gives $\dfrac{5}{2} = 5$ impossible.

1.33. $S = (-\infty, 4)$

1.35. $S = \Re$

1.37. $-7 < 3x-2 < 7 \Rightarrow \dfrac{-5}{3} < x < 3$; $S = (\dfrac{-5}{3}, 3)$.

1.39. $\begin{cases} 10x-1 \le -4 \\ 10x-1 \ge 4 \end{cases} \Rightarrow \begin{cases} x \le -\dfrac{3}{10} \\ x \ge \dfrac{1}{2} \end{cases}$; $S = (-\infty, \dfrac{-3}{10}] \cup [\dfrac{1}{2}, +\infty)$.

1.41. $2(x+1)^2 - 4(x+1) = 2(x+1)(x-1)$, the roots are -1 and 1 .

x	- ∞ -1	1	+ ∞
x + 1	- 0	+	+
x − 1	-	- 0	+
2 (x + 1) (x − 1)	+ 0	- 0	+

$S = (-\infty, -1] \cup [1, +\infty)$

1.43. $\frac{4x^2-25-3(2x-5)}{x+4} = \frac{(2x-5)(2x+2)}{x+4}$, the roots are -1, $\frac{5}{2}$ and - 4.

x	$-\infty$ -4	-1	$\frac{5}{2}$	$+\infty$
2 x - 5	-	-	- 0	+
2 x + 2	-	- 0	+	+
x + 4	- 0	+	+	+
$\frac{4x^2-25-3(2x-5)}{x+4}$	- ||	+ 0	- 0	+

$S = (-4, -1) \cup (\frac{5}{2}, +\infty)$.

1.45. Let x be the first odd integer, then x + 2 and x + 4 are respectively the second and the third consecutive odd integers . So, x + x + 2 + x + 4 = 81 which implies that x = 25 . Hence the three consecutive odd integers are 25, 27, and 29 .

1.47. Let x be the price of one copybook before the reduction.
The price without the reduction in this case is ¢ 10 x.
The discount is at 10 %, then it is equal to ¢ $\frac{10x\times10}{100}$ = ¢ 10.
This customer paid ¢ 29250 , then 10 x - x = 29250 , that is x = 3250 .
Hence, the price of one copybook before the reduction is ¢ 3250.

1.49. Let x be the necessary number of years. Then 42 + x = 2 (15 + x) implies that x = 12. So after twelve years the age of the father will be double the age of his son .

1.51. $(a+3)^2 = a^2 + (a+1)^2 + (a+2)^2 \Rightarrow 2a^2 = 4 \Rightarrow a = \sqrt{2}$.

1.53. Let x be the velocity of the first, then x - 2 is the velocity of the second.
The time achieved by the first is $\frac{480}{x}$ and by the second is $\frac{480}{x-2}$.

So, $\frac{480}{x-2} - \frac{480}{x} = 1$ where x ≠ 0 and x ≠ 2. Solve the equation to get
$x^2 - 2x - 960 = 0$. The roots are x = - 30 rejected and x = 32 accepted.

Hence the velocity of the first is 32 km / hr.

1.55. Let $t = x^2$, then $t \geq 0$. Substitute to get $3t^2 + 2t - 5 = 0$. The roots are $t = \frac{-5}{3}$ rejected or $t = 1$ accepted. Hence, $x^2 = 1$ implies that $x = \pm 1$.

1.57. Let $t = x - 2$. Substitute to get $t^2 + 5t - 6 = 0$. The roots are $t = 1$ or $t = -6$. Hence, $x - 2 = 1$ or $x - 2 = -6$ implies that $x = 3$ or $x = -4$.

1.59. Condition: $x - 1 \neq 0$ implies that $x \neq 1$.

Let $t = \frac{2}{x-1}$. Substitute to get $t^2 + t - 6 = 0$. The roots are

$t = 2$ or $t = -3$. Hence, $\frac{2}{x-1} = 2$ or $\frac{2}{x-1} = -3$ implies that

$x = 2$ acceptable or $x = \frac{1}{3}$ acceptable.

1.61. $2x - 4 \geq x - 3 \Rightarrow x \geq 1$.

$S = [1, 2]$.

1.63. $S = [1, 2]$.

1.65. $|x+5| \geq 0 \quad \forall \quad x \in \Re$, then the equation has no solution.

1.67. $x^2 + 4 > 0 \quad \forall \quad x \in \Re$, then the equation has no solution.

1.69. $-x^2 - 4 < 0 \quad \forall \quad x \in \Re$, then the equation has no solution.

1.71. a. The cost of production of 100 machines is given by C (10) =18, then the cost of production of 100 machines is \$ 18000. The average cost of one machine is 18000 / 100 = \$180.

b. Revenue R(x) = Unit price × quantity = 225×10 x = \$ 2250 x.

Let R(x) = 2.250 x thousands of dollars.

The profit is $P(x) = R(x) - C(x) = 2.250x - 0.25x^2 + x - 3$. The roots of the equation P(x) = 0 are 12 and 1, so P(x) > 0 for $1 < x < 12$. Hence if the number of machines produced is between 10 and 120, the factory experiences profit.

1.73. a. The revenue resulting from selling x televisions is given by R(x) = \$ 1200 x.

b. The profit is given by:

$P(x) = R(x) - C(x) = 1200\,x - 20\,x^2 - 300\,x - 9000$. Then

$P(x) = -20\,x^2 + 900\,x - 9000$.

c. The profit is null when $-20\,x^2 + 900\,x - 9000 = 0$.Its roots are $x_1 = 30$ and $x_2 = 15$, Then for a production of 15 televisions or 30 televisions the profit is null.

d. $P(x) = -20\,x^2 + 900\,x - 9000$ is positive for the values of x between 15 and 30, then for a production of televisions between 15 and 30 there is profit and in the other cases there is loss.

1.75.

a. The simple interest is given by I = PV. r. t with r = 0.12, t = 3 and the future value is given by FV = PV (1 + r t), then

FV(3) = PV (1 + 3ri) = 5000(1+0.36) = \$ 6800 .

b. FV(t)= PV (1 + r t) = 5000(1+0.12t) = 5000 + 600t. .

c. FV(t) > 9500 gives 5000 + 600t > 9500 , then 600 t > 4500which implies t > 7.5 ; therefore, after 8 years the capital exceeds 9500 for the first time.

1.77.

a. $FV = PV\,(1 + i)^n = 40\,000\,000\,(1 + \frac{0.06}{12})^{6 \times 12} \approx \$\,57\,281\,771.$

The interest accumulated by this capital is then

57 281 771 – 40 000 000 = \$ 17 281 771

b. $FV = PV\,(1 + i)^n = 40\,000\,000\,(1 + \frac{0.06}{4})^{6 \times 4} \approx \$\,57\,180\,112.$

The interest accumulated by this capital is then

57 180 112 – 40 000 000 = \$ 17 180 112.

1.79. a. It is an annuity with 20 000 000 as the present value , k = 12 ,

$n = 5\,(12) = 60$ and $i = \frac{0.06}{12}$.

Then $A = R \times \dfrac{1-(1+i)^{-n}}{i}$ which gives $20\,000\,000 = R \times \dfrac{1-(1+\frac{0.06}{12})^{-60}}{\frac{0.06}{12}}$,

then $20\,000\,000 = R \times 51.725$. Consequently, $R = \dfrac{20\,000\,000}{51.725} \approx \$\ 386656$.

The value of the periodical payment is \$ 386656 .

b. Taan will pay $5 \times 12 = 60$ payments. The amount paid by Taan is

$60 \times R = 60 \times 388656 =$ \$ 23 319 360.

Interest accumulated is : 23 319 360 - 20 000 000 = \$ 3199360.

1.81. If Walid chooses the first offer, the sum accumulated after 10 years will be:

$FV = PV(1+i)^n = 10\,000\,000\,(1+\frac{0.08}{2})^{2\times10} = \$\ 21\,911\,231.43.$

If Walid chooses the second offer, the sum accumulated after 10 years will be:

$FV_1 = PV(1+i)^n = 10\,000\,000\,(1+\frac{0.08}{12})^{12\times10} = \$\ 22\,196\,402.34$.

Hence offer II is more advantageous for Walid.

CHAPTER 2, pp. 54 - 60

2.1. For n = 1, we have $12 + 2^2 + \ldots + n^2 = 1^2 = 1$ and

$\dfrac{n(n+1)(2n+1)}{6} = \dfrac{1(1+1)(3)}{6} = 1$. So the equality is true for n = 1.

Assume it is true for n = k – 1, that is $1^2 + 2^2 + \ldots + (k-1)^2 =$

$\dfrac{k(k-1)(2k-1)}{6}$

Prove it is true for n = k, that is we need to prove that

$$1^2 + 2^2 + 3^2 + \ldots\ldots + k^2 = \frac{k(k+1)(2k+1)}{6}$$

We have
$$\begin{aligned} 1^2 + 2^2 + 3^2 + \ldots\ldots + k^2 &= 1^2 + 2^2 + \ldots + (k-1)^2 + k^2 \\ &= \frac{k(k-1)(2k-1)}{6} + k^2 \\ &= \frac{k(k-1)(2k-1)+6k^2}{6} \\ &= \frac{k[2k^2+3k+1]}{6} = \frac{k(k+1)(2k+1)}{6} \text{ claim.} \end{aligned}$$

Therefore, $1^2 + 2^2 + 3^2 + \ldots\ldots + n^2 = \dfrac{n(n+1)(2n+1)}{6}$ for any natural integer n

2.3. For n = 1, we have $13 + 2^3 + 3^3 + \ldots\ldots + n^3 = 1^3 = 1$ and

$\frac{1}{4}n^2(n+1)^2 =$ 1. So the equality is true for n = 1.

Assume it is true for n = k – 1, that is $1^3 + 2^3 + \ldots + (k-1)^3 = \frac{1}{4}(k-1)^2(k^2)$

Prove it is true for n = k, that is we need to prove that

$$1^3 + 2^3 + 3^3 + \ldots\ldots + k^3 = \frac{1}{4}k^2(k+1)^2$$

We have $1^3 + 2^3 + 3^3 + \ldots\ldots + k^3 = 1^3 + 2^3 + \ldots + (k-1)^3 + k^3$

$$= \frac{1}{4}(k-1)^2(k^2) + k^3$$

$$= \frac{1}{4}k^2(k^2+2k+1)$$

$$= \frac{1}{4}k^2(k+1)^2 \text{ claim.}$$

Therefore, $1^3 + 2^3 + 3^3 + \ldots\ldots + n^3 = \frac{1}{4}n^2(n+1)^2$ for any natural integer n.

2.5. $u_{n+1} = 2^{n+1} - n - 1$. $u_{n-2} = 2^{n-2} - n + 2$. $u_{2n} = 2^{2n} - 2n$.

2.7. $n - 3 \geq 0 \Rightarrow n \geq 3$. So (u_n) is defined if $n \in N$ and $n \geq 3$.

The first term is $u_3 = \sqrt{3-3} = 0$, The second term is $u_4 = \sqrt{4-3} = 1$.

The third term is $u_5 = \sqrt{5-3} = \sqrt{2}$, and the fourth term is $u_6 = \sqrt{6-3} = \sqrt{3}$.

2.9. For n = 0, we have $u_0 = 3 - 2^0 = 3 - 1 = 2$. So the equality is true for n = 0.
Assume it is true for n = k, that is $u_k = 3 - 2^k$.
Prove it is true for n = k + 1, that is we need to prove that $u_{k+1} = 3 - 2^{k+1}$.
$u_{k+1} = 2u_k - 3 = 2(3 - 2^k) - 3 = 3 - 2^{k+1}$. Claim
Therefore, $u_n = 3 - 2^n$ for any natural integer n.

2.11.

a. $u_1 = 2u_0 + 1 = 4 + 1 = 5$.
$u_2 = 2u_1 + 1 = 11$ and $u_3 = 2u_2 + 1 = 23$.

b. For n = 0, we have $u_0 = 2 < u_1 = 5$. So the equality is true for n = 0.
Assume it is true for n = k -1, that is $u_{k-1} < u_k$.
Prove it is true for n = k + 1, that is we need to prove that $u_k < u_{k+1}$.
$u_{k-1} < u_k$ implies that $2u_{k-1} + 1 < 2u_k + 1$ which implies that
$u_k < u_{k+1}$. Claim.
Therefore, $u_n < u_{n+1}$ for any natural integer n.
Hence, the sequence (u_n) is strictly increasing for any natural integer n.

2.13. a, b and c are three consecutive terms of a geometric sequence , then

$$b = \frac{a+c}{2} \Rightarrow 2x+1 = \frac{x+2x+3}{2} \Rightarrow x = 1.$$

2.15. $u_{n+1} - u_n = \frac{n+1}{5} - \frac{1}{3} - (\frac{n}{5} - \frac{1}{3}) = \frac{1}{5}$. So (u_n) is an arithmetic sequence with common difference $d = \frac{1}{5}$ and with first term $u_0 = -\frac{1}{3}$.

2.17. $\begin{cases} u_1 + 10d = 20 \\ u_1 + 20d = 50 \end{cases}$, then $u_1 = -10$ and $d = 3$.

2.19. $u_3 = u_0 + 3d$ then $u_0 = 1 + 15 = 16$, and

$$S_3 = \frac{3+1}{2}[2\times 6 + 3(-5)] = 34.$$

2.21. $S_n = 329$ gives $\frac{n}{2}[10+(n-1)\times 14] = 329 \Rightarrow 14n^2 - 4n - 568 = 0 \Rightarrow$

$n = \frac{-47}{7}$ rejected or $n = 7$ accepted. $u_7 = u_1 + (7-1)(14) = 89$.

2.23. $q = \sqrt[5]{\frac{u_6}{u_1}} = \sqrt[5]{\frac{1}{243}} = \frac{1}{3}$ and $S_6 = u_1 \left[\frac{1-(\frac{1}{3})^6}{1-\frac{1}{3}} \right] = \frac{364}{81}$.

2.25. $u_6 = u_1 . q^5$ implies that $u_1 = \frac{u_6}{q^5} \Rightarrow u_1 = \frac{108}{(\frac{1}{3})^5} = 26244.$

$$S_n = u_1 \left(\frac{1-q^n}{1-q} \right) = 26244 \left(\frac{1-\frac{1}{729}}{1-\frac{1}{3}} \right) = 39312.$$

2.27.

a. $u_0 = 0$, $u_1 = 1$, $u_2 = 4$; $u_1 . u_0 = 1$; $u_2 . u_1 = 3$. Since $u_1 . u_0 \neq u_2 . u_1$, then the sequence (u_n) is not arithmetic .

b. i) $v_0 = 1, v_1 = \frac{1}{2}, v_2 = \frac{1}{3}$; $v_1 - v_0 \neq v_2 - v_1$, then the sequence (v_n) is not arithmetic .

ii) $\forall n \in N$, $w_{n+1} - w_n = \frac{1}{v_{n+1}} - \frac{1}{v_n} = \frac{1+v_n}{v_n} - \frac{1}{v_n} = 1$, then the sequence (w_n) is an arithmetic sequence with common difference $d = 1$.

2.29. Let $S_n = 1 + x^2 + x^4 + ... + x^{2n}$. S_n is the sum of (n + 1) first terms of a geometric sequence of first term 1 and with common ratio x^2.

- If x = 1 or x = -1, then $S_n = n + 1$.
- If $x \neq 1$ or $x \neq -1$, then $S_n = \dfrac{1-(x^2)^{n+1}}{1-x^2} = \dfrac{1-x^{2n+2}}{1-x^2}$.

2.31.

a. $u_0 = 2$, $u_1 = \frac{1}{2}\times 2 + 3 = 4$ and $u_2 = \frac{1}{2}\times 4 + 3 = 5$.

b. $u_1 - u_0 = 2$ and $u_2 - u_1 = 1$. Since $u_1 - u_0 \neq u_2 - u_1$, then the sequence (u_n) is not arithmetic.

$\frac{u_1}{u_0} = 2$ and $\frac{u_2}{u_1} = \frac{5}{4}$. Since $\frac{u_1}{u_0} = 2 \neq \frac{u_2}{u_1} = \frac{5}{4}$, then the sequence (u_n) is not geometric.

c. i) $v_{n+1} = u_{n+1} - 6 = \frac{1}{2} u_n + 3 - 6 = \frac{1}{2} u_n - 3 = \frac{1}{2}(u_n - 6) = \frac{1}{2} v_n$.

Then the sequence (v_n) is a geometric sequence of first term $v_0 = u_0 - 6 = 2 - 6 = -4$ and with common ratio $q = \frac{1}{2}$.

ii) $v_n = v_0 \cdot q^n = \frac{-4}{2^n}$ and $u_n = v_n + 6 = \frac{-4}{2^n} + 6$.

iii) $s = v_0 + v_1 + ... + v_n = v_0 \dfrac{1-q^{n+1}}{1-q} = -8\left(1 - \dfrac{1}{2^{n+1}}\right)$.

$$\begin{aligned} s' = u_0 + u_1 + ... + u_n &= (v_0 + 6) + (v_1 + 6) + ... + (v_n + 6) \\ &= (v_0 + v_1 + ... + v_n) + 6(n+1) \\ &= -8\left(1 - \frac{1}{2^{n+1}}\right) + 6(n+1). \end{aligned}$$

2.33. For n = 1, we have $u_1 = u_0 + 2 = -1 + 2 = 1 > 0$. So $u_n > 0$ for n = 1.
Assume it is true for n = k, that is $u_k > 0$.
Prove it is true for n = k + 1, that is we need to prove that $u_{k+1} > 0$.
$u_{k+1} = u_k + 2 > 0$. Claim.
Therefore, $u_n > 0$ for any natural integer n.
Moreover, $u_{n+1} - u_n = 2 > 0$, then the sequence (u_n) is strictly increasing.

2.35. Prove first $u_n > 0$ for any natural integer n.
Moreover, $u_{n+1} - u_n = u_n^2 - 3u_n + 4 - u_n = (u_n - 2)^2 \geq 0$, then the sequence (u_n) is increasing.

2.37. Prove first $u_n > 0$ for any natural integer n.
Moreover, $u_{n+1} - u_n = u_n - u_n - \frac{1}{n+1} = -\frac{1}{n+1} < 0$ for any natural integer n , then the sequence (u_n) is strictly decreasing.

2.39. $u_n = \dfrac{2n+3}{n+1} > 0$ for any natural integer n .

$u_{n+1} - u_n = \dfrac{2(n+1)+3}{n+1+1} - \dfrac{2n+3}{n+1} = \dfrac{-1}{(n+1)(n+2)} < 0$ for any natural integer n , then the sequence (u_n) is strictly decreasing .

2.41. $u_n = \dfrac{2^n}{n+2} > 0$ for any natural integer n .

$u_{n+1} - u_n = \dfrac{2^{n+1}}{n+1+2} - \dfrac{2^n}{n+2} = \dfrac{2^{n+1}(n+2)-2^n(n+3)}{(n+3)(n+2)}$

$= \dfrac{2^n(n+1)}{(n+3)(n+2)} > 0$ for any natural integer n, then the sequence (u_n) is strictly increasing .

2.43.

a. (u_n) is a geometric sequence with common ratio $0 < q = \dfrac{1}{5} < 1$, then

$$\lim_{n\to+\infty} u_n = 0 .$$

b. $S_n = \dfrac{u_0(1-q^{n+1})}{1-q} = \dfrac{-1(1-\frac{1}{5^{n+1}})}{1-\frac{1}{5}} = \dfrac{\frac{1}{5^{n+1}} - 1}{\frac{4}{5}} = \dfrac{1-5^{n+1}}{4\times 5^n}$.

c. (u_n) is a geometric sequence with common ratio $|q| < 1$, then

$$\lim_{n\to+\infty} S_n = \frac{u_0}{1-q} = \frac{-1}{1-\frac{1}{5}} = \frac{-5}{4} .$$

2.45. $u_n = \dfrac{n}{2n} - \dfrac{1}{2n} = \dfrac{1}{2} - \dfrac{1}{2n}$.

$\forall n \in \mathbb{N}$, if n = 1, $u_n = \dfrac{1}{2} - \dfrac{1}{2} = 0.$

$$\text{If } n = 2,\ u_n = \frac{1}{2} - \frac{1}{4} = \frac{1}{4}.$$

$$\text{as } n \to \infty,\ u_n \to \frac{1}{2}.$$

$$u_{n+1} - u_n = \frac{1}{2} - \frac{1}{2n+2} - \frac{1}{2} + \frac{1}{2n} = \frac{1}{n(2n+2)} > 0\ \forall n \in N,$$

then the sequence (u_n) is strictly increasing and bounded above by $\frac{1}{2}$.

2.47. $u_n = \dfrac{2n-1}{n+1} = 2 - \dfrac{3}{n+1}$.

$$u_{n+1} - u_n = 2 - \frac{3}{n+2} - 2 + \frac{3}{n+1} = \frac{3}{(n+1)(n+2)} > 0\ \forall n \in N,$$

then the sequence (u_n) is strictly increasing .
Moreover, If $n = 0$, $u_n = 2 - 3 = -1$ and if $n \to \infty$, $u_n \to 2$.
Hence, for all positive integer n the sequence (u_n) is bounded below by - 1 and bounded above by 2 .

2.49.

a. i) $P(60) = (60)^2 + 9(60) - 4140 = 0$.
ii) $P(x) = 0$ implies that $x^2 + 9x - 4140 = 0$; the roots are $x_1 = 60$ or $x_2 = -69$.
iii)

x	$-\infty$ -69	60	$+\infty$
P(x)	+ 0	- 0	+

If $x \in (-\infty, -69] \cup [60, +\infty)$, then $P(x) \geq 0$.
If $x \in [-69, 60]$, then $P(x) \leq 0$.

b. 1) The sequence (u_n) is an arithmetic sequence with common difference $d = 200$. $u_n = u_1 + (n-1)d = 1000 + 200(n-1) = 200n + 800$.

2) $S_n = \dfrac{n(u_1 + u_n)}{2} = \dfrac{n(1000 + 200n + 800)}{2} = 100n^2 + 900n$.

3) $200n + 800 = 41400$ implies that $n = 203$. Then the maximum depth that we can reach is 203 meters .

2.51.

a. 1) $a_1 = a_0 - \dfrac{2.4}{100} \times a_0 = (1 - 0.024)\,a_0 = \$\,14640$.

$a_2 = a_1 - \dfrac{2.4}{100} \times a_1 = (1 - 0.024)\,a_1 = 0.976\,(14640) = \$\,14288.64$.

2) $a_{n+1} = a_n - \dfrac{2.4}{100} \times a_n = 0.976\,a_n$, then (a_n) is a geometric sequence of common ratio 0.976 and its first term a_0 , then $a_n = a_0\,(0.976)^n$. So

$a_n = 15000\,(0.976)^n$.

3) $b_{n+1} = b_n - \frac{1.8}{100} \times b_n = (1 - 0.018)\, b_n = 0.982\, b_n$. Then (b_n) is a geometric sequence of common ratio 0.982 and of first term $b_0 = 12000$, then $b_n = b_0 (0.982)^n$. So $b_n = 12000\,(0.982)^n$.

b. $a_n < \frac{15000}{2}$ is equivalent to $15000\,(0.976)^n < 7500$, which gives $(0.976)^n < \frac{7500}{15000}$, consequently $n \ln 0.976 < \ln \frac{1}{2}$, and since $\ln 0.976 < 0$, we get $n > \frac{-\ln 2}{\ln 0.976}$, then $n > 28.5$. Then after 29 months the price of a car of type E will be less than half its original price.

c. $b_n > a_n$, gives $12000\,(0.982)^n > 15000\,(0.976)^n$.

Then $(\frac{0.982}{0.976})^n > \frac{15000}{12000}$, consequently $n \ln \frac{0.982}{0.976} > \ln 1.25$.

Since $\ln \frac{0.982}{0.976} > 0$, we get $n > \frac{\ln 1.25}{\ln \frac{0.982}{0.976}}$, then $n > 36.4$. So after 37 months the value of the vehicle of type D become more than that of E .

2.53. Part I

a. $a_1 = 0.6\, a_0 + 200 = 0.6\,(900) + 200 = 740$.
$a_2 = 0.6\, a_1 + 200 = 0.6\,(740) + 200 = 644$.

b. 1) $b_{n+1} = a_{n+1} - 500 = 0.6\, a_n + 200 - 500 = 0.6\, a_n - 300$
$= 0.6\,(a_n - 500) = 0.6\, b_n$. Then (b_n) is a geometric sequence of common ratio $q = 0.6$ and of first term $b_0 = a_0 - 500 = 400$.

2) $b_n = b_0\, q^n = 400\,(0.6)^n$ and since $b_n = a_n - 500$, we conclude that $a_n = b_n + 500 = 400\,(0.6)^n + 500$.

3) $\lim_{n \to \infty} a_n = \lim_{n \to \infty} 400\,(0.6)^n + 500 = 500$ since $\lim_{n \to \infty} (0.6)^n = 0$.

Part II

a. 1) In 2006, company D has 900 clients and company E has 100 clients. In 2007, company D will lose 20 % of its 900 clients, then 0.2(900)= 180 and gain 20 % of company's E clients, then 0.2(100) = 20.So, in 2007, company D will have 900 – 180 + 20 = 740 clients. Notice that company E will have 1000- 740 = 260 clients. In 2008, company D will lose 20 % of its 740 clients, then 0.2 (740) = 148 and gain 20% of the 260 clients of company E , so 0.2(260) = 52. Then in 2008, company D will have 740 – 148 + 52 = 644 clients.

2) If c_n is the number of client of company D in the year 2006 + n then $1000 - c_n$ is the number of clients of company E in the year 2006 + n.

Then, $c_{n+1} = c_n - \frac{20}{100} c_n + \frac{20}{100}(1000 - c_n) = 0.6\, c_n + 200$.

b. Notice that the sequence (c_n) is the sequence (a_n) studied in part I .

Then $\lim_{n \to \infty} c_n = 500$, this means after many years, the number of clients of company D will get near to 500 and that of E will get near to

500, then the two companies tend to have the same number of clients.

2.55.

a.1) $r_{n+1} = r_n + \frac{5}{100} r_n = 1.05\, r_n$, then (r_n) is a geometric sequence of common ratio $q = 1.05$ and of first term $r_1 = 4200$.
$r_n = r_1 \cdot q^{n-1} = 4200\,(1.05)^{n-1}$.

2) $r_7 = r_1 \cdot q^6 = 4200\,(1.05)^6 \approx \$\,5628$.

3) $S = r_1 + r_2 + \ldots\ldots. + r_7 = r_1 \frac{1-q^7}{1-q} = 4200 \frac{1-(1.05)^7}{1-1.05} \approx \$\,34196$.

b. 1) (u_n) is an arithmetic sequence since $u_{n+1} = u_n + 340$, then
$u_n = u_1 + (n-1)d = 4200 + (n-1)340 = 340n + 13860$.

2) $u_7 = u_1 + 6d = 4200 + 6(340) = \$\,6240$.

3) $S = \frac{7}{2}(u_1 + u_7) = 3.5\,(4200 + 6240) = \$\,36540$.

c. The more advantageous contract is that of first contract.

2.57.

a. We have $s_0 = 10000$ and $s_{n+1} = s_n + 500$, then (s_n) is arithmetic sequence of common difference $d = 500$.In the first year, the member pays $s_1 = s_0 + 500 = 10000 + 500 = \$\,10500$. In the second year, the member pays $s_2 = s_1 + 500 = 10500 + 500 = \$\,11000$.During n years, the member grants the club \$ 10000 to which he adds \$ 500 n , then $s_n = 10000 + 500n$.

b. 1) $A_{n+1} = A_n + \frac{10}{100} A_n - 50 = 1.1\,A_n - 50$.

2) $E_{n+1} = A_{n+1} + a = 1.1\,A_n - 50 + a$. For the sequence (E_n) to be geometric of common ratio 1.1, $E_{n+1} = 1.1\,E_n$, which gives $1.1\,A_n - 50 + a = 1.1\,(A_n + a)$. Then, $-50 + a = 1.1\,a$ and consequently $a = -500$. The first term of this sequence is $E_1 = A_1 - 500 = 500$.

c. 1) $E_n = E_1 \cdot q^{n-1} = 500\,(1.1)^{n-1}$.
$A_n = E_n - a = E_n + 500 = 500\,(1.1)^{n-1} + 500$.

2) $t_n = A_1 + A_2 + \ldots\ldots\ldots.. + A_n$
$= (E_1 + 500) + (E_2 + 500) + \ldots\ldots\ldots + (E_n + 500)$
$= (E_1 + E_2 + \ldots\ldots\ldots + E_n) + (500n)$

But $(E_1 + E_2 + \ldots\ldots\ldots + E_n)$ is the sum of n terms of a geometric sequence, then :

$$E_1 + E_2 + \ldots\ldots\ldots + E_n = E_1 \frac{1-q^n}{1-q} = 500\frac{1-(1.1)^n}{1-1.1}.$$

Then $E_1 + E_2 + \ldots\ldots\ldots + E_n = 5000\,[(1.1)^n - 1]$,
consequently: $t_n = 5000\,[(1.1)^n - 1] + 500n$.

3) $s_n < t_n$, then $10000 + 500n < 5000\,[(1.1)^n - 1] + 500n$, then $10000 < 5000\,(1.1)^n - 5000$, so $5000\,(1.1)^n > 15000$. Hence, $(1.1)^n > 3$ and consequently $n \ln(1.1) > \ln 3$, which gives

$n > \frac{\ln 3}{\ln 1.1}$. Thus, $n > 11.52$. So, they should at least subscribe for 12 years for category I to be more advantageous to that of II.

CHAPTER 3 , pp. 109 - 121

3.1. $D_f = \Re$; **3.3.** $D_f = \Re^*$; 3.5. $D_f = \Re - \{2,-3\}$;
3.7. $D_f = \Re - \left\{1,-\frac{4}{3}\right\}$; **3.9.** $D_f = \Re$; **3.11.** $D_f = [-2, +\infty)$;
3.13. $D_f = (-\infty, -3] \cup [3, +\infty)$; **3.15.** $D_f = (-\infty, -1) \cup (1, +\infty)$;
3.17. $D_f = (-\infty, 0)$; 3.19. $D_f = (-2, 2)$; **3.21.** $D_f = \Re^*$;
3.23. $D_f = (-\infty, -3)$; **3.25.** $D_f = \Re - \left\{ \pm\frac{\pi}{2} + 2 + 2k\pi \ ; k \in Z\right\}$;
3.27. $D_f = (2, +\infty)$; **3.29.** $D_f = \Re$; **3.31.** $D_f = [-5, 4]$;
3.33. $D_f = (0, +\infty)$; **3.35.** $D_f = \Re - \{ 2+2k\pi, \pi+2+2k\pi \ ; k \in Z\}$.

3.37.

a. $P(-1) = (-1)^3 + 2(-1)^2 - 8(-1) - 9 = 0$, then P(x) is divisible by $(x+1)$

b. $P(\frac{1}{2}) = 0$, then P(x) is divisible by $(2x - 1)$.

3.39.

a. $P(-2) = 0$ implies that $-16 + 48 - 2a - 84 = 0$, then $a = -26$.

b. The long division of P (x) by (x + 2) gives $q(x) = 2x^2 + 8x - 42$.

c. $q(3) = 0$, then $(x-3)$ is a factor of $q(x)$. The long division of $P(x)$ by $(x+2)$ gives $q(x) = (x-3)(2x+14)$.
Hence $P(x) = (x+2)(x-3)(2x+14)$.

3.41.

a. $P(x) = (x+1)(2x^2 - 2x + 3)$.

b. $P(x) = (x+3)(x^2 - 3x + 9)$.

3.43. $\dfrac{-x^2 - x - 8}{(x-1)(x^2+4)} = \dfrac{(a+c)x^2 + (b-a)x + 4c - b}{(x-1)(x^2+4)}$. By identification,
$a + c = -1$, $b - a = -1$ and $-b + 4c = -8$. Solve this system to get $c = -2$, $a = 1$ and $b = 0$.

3.45. By long division we get, $a = 1$, $b = -1$ and $c = -1$.

3.47. $\dfrac{3x^2 - 4x + 2}{x-1} = \dfrac{ax^2 + (b-a)x + c - b}{x-1}$. By identification,
$a = 3$, $b - a = -4$ implies that $b = -1$ and $c - b = 2$ implies that $c = 1$.

3.49.

a. $C(10) = \dfrac{98\times106}{400-100^{2}} = \$\ 34.63.$

$$\overline{C}(q) = \frac{98\times106}{400q-q^{3}}\ .$$

b.

x	0	1	2	10	16	17
y	25.97	26.04	26.23	34.6	72.14	93.59

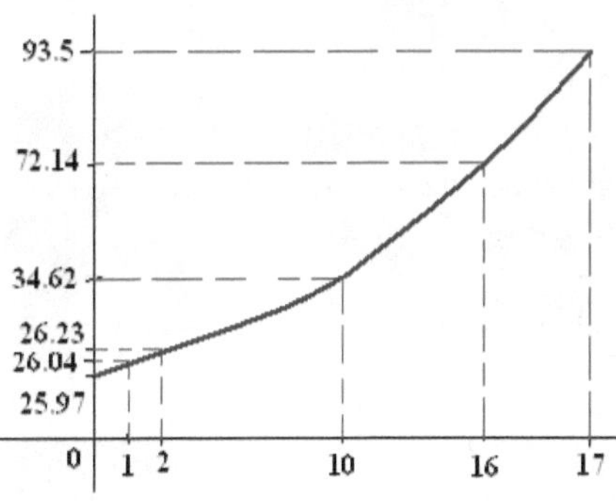

3.51.

a. $D(t) = \dfrac{C-S}{n}t$ for $0 \le t \le n$. Then

$D(t) = \dfrac{1000-200}{4}t = 400\text{ t}$ for $0 \le t \le 2$.

b. $V(t) = C - D(t)$ for $0 \le t \le n$

$V(t) = 1000 - 400\text{ t}$ for $0 \le t \le 2$.

c. $V(1) = 1000 - 400 = 600$ and $V(2) = 1000 - 800 = 200$.

d. $V(0) = 1000$

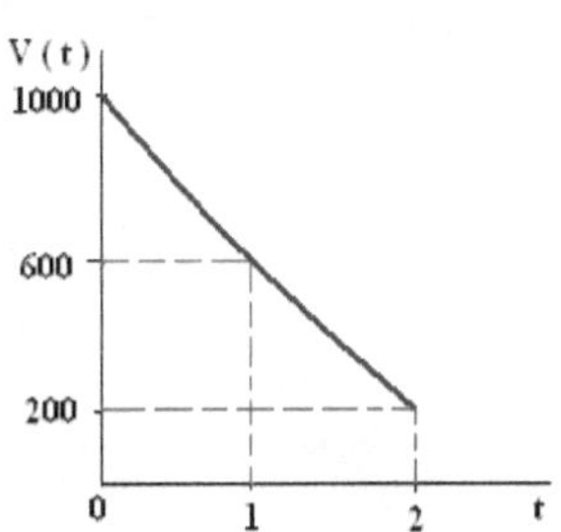

3.53. Total cost = Fixed costs + variable costs

$$= 300 + \frac{5}{x}.$$

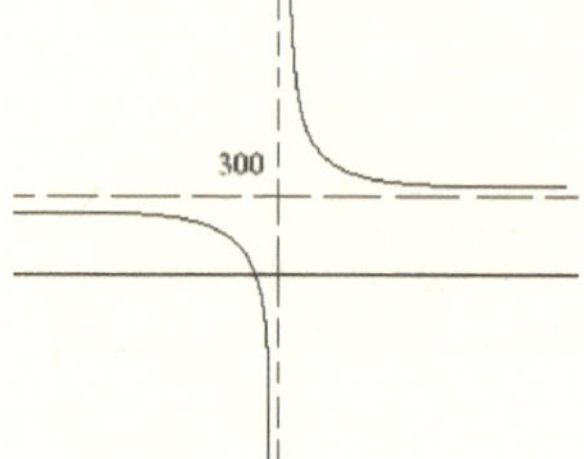

3.55.

a. Range of f = [0, 5].

b. $f(x) = (x^2 + 2x + 1) - 4 = (x+1)^2 - 4$.

c. i) $a < b \Rightarrow a + 1 < b + 1 \Rightarrow$ a and b belong to $[-1, +\infty[$, then $(a+1)^2 < (b+1)^2 \Rightarrow (a+1)^2 - 4 < (b+1)^2 - 4 \Rightarrow$ $f(a) < f(b)$.

ii) $a < b \Rightarrow a + 1 < b + 1 \Rightarrow$ a and b belong to $]-\infty, -1]$, then $(a+1)^2 > (b+1)^2 \Rightarrow (a+1)^2 - 4 > (b+1)^2 - 4 \Rightarrow$ $f(a) > f(b)$.

d. If $x \in [-1, +\infty[$, then f is strictly increasing.
If $x \in]-\infty, -1]$, then f is strictly decreasing.

3.57.

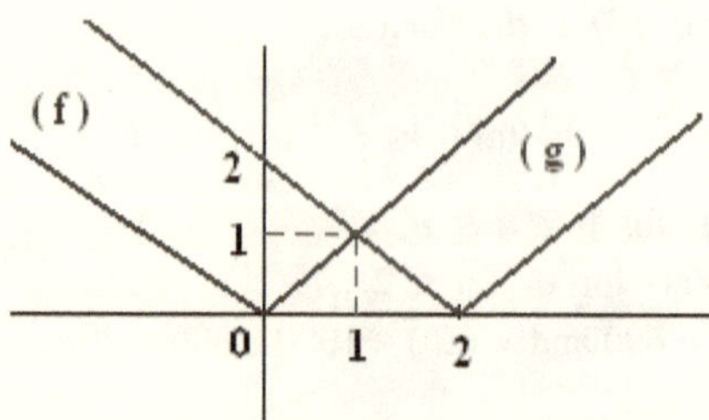

The graph of g is deduced from the graph of f by shifting it 2 units to the right.

3.59. The vertex of the given parabola is $S\left(\frac{-b}{2a}, \frac{-b^2+4ac}{4a}\right)$. Then $S(3, -4)$.

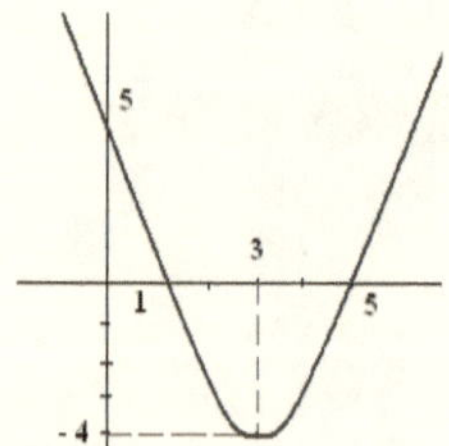

3.61. $D_f = (-\infty, 0)$.

x	0^-	-1	-2
y	$-\infty$	0	0.69

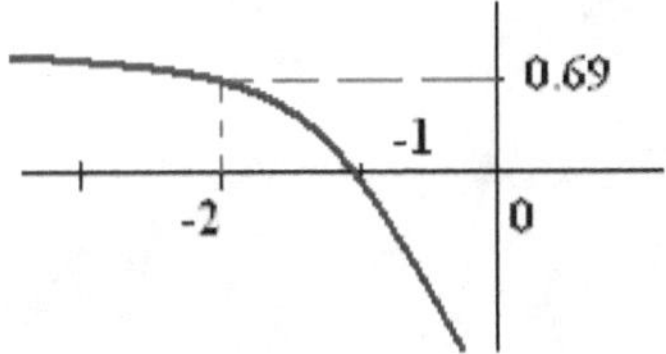

3.63. $D_f = \Re$.

x	-2	-1	0	1
y	0.02	0.14	1	7.39

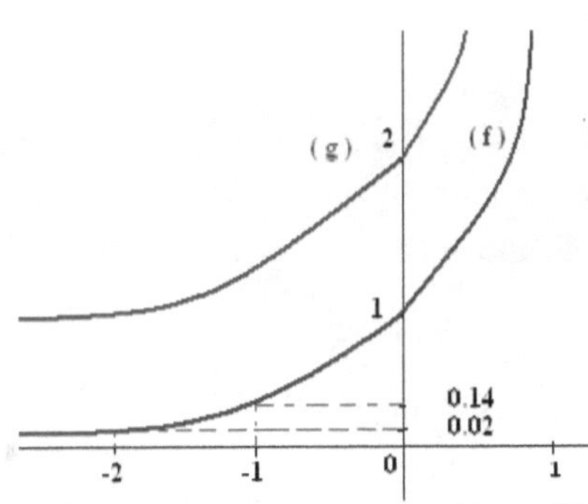

The graph of g is deduced from the graph of f by shifting it one unit upwards .

3.65. This function is defined for every x in $\Re - \{k\pi, k \in Z\}$. It is periodic with period $k\pi$, $k \in Z$, then we can reduce the interval of discussion to $\left[\frac{-\pi}{2}, 0\right[\cup \left]0, \frac{\pi}{2}\right]$. Moreover, $f(-x) = \cot(-x) = -\cot(x) = -f(x)$; the function is odd and the origin O is center of symmetry . So the interval of discussion is reduced to $\left]0, \frac{\pi}{2}\right]$. Finally the graph continue periodically with period $k\pi$.

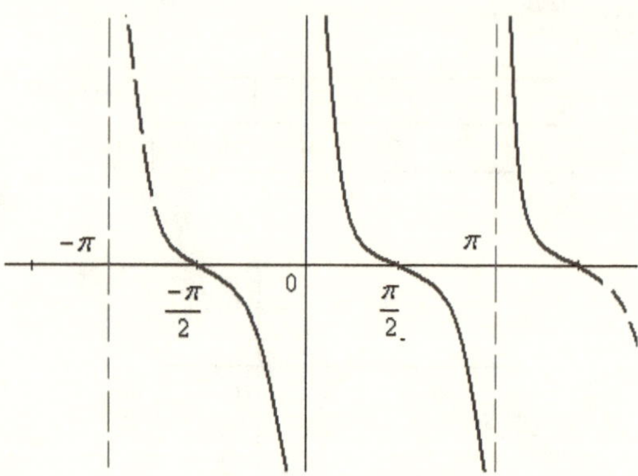

The graph of g^{-1} is deduced from the graph of g with respect to the first bisector.

3.67.

a. $O(\frac{1}{2}, \frac{1}{2})$

b. $r = OA = \sqrt{\frac{1}{4}+\frac{1}{4}} = \frac{\sqrt{2}}{2}$; $(\vec{i}, \overrightarrow{AO}) = \frac{\pi}{4} + 2k\pi$, $k \in Z$

Then the polar coordinates of O: $O(\frac{\sqrt{2}}{2}, \frac{\pi}{4})$.

3.69.

a. $r = \sqrt{1+1} = \sqrt{2}$. $x = r\cos\theta$ implies that $\cos\theta = \frac{x}{r} = \frac{\sqrt{2}}{2}$, similarly $\sin\theta = \frac{y}{r} = \frac{\sqrt{2}}{2}$, then $(\vec{i}, \overrightarrow{OB}) = \frac{\pi}{4} + 2k\pi$, $k \in Z$. So the polar coordinates of B : $B(\sqrt{2}, \frac{\pi}{4})$.

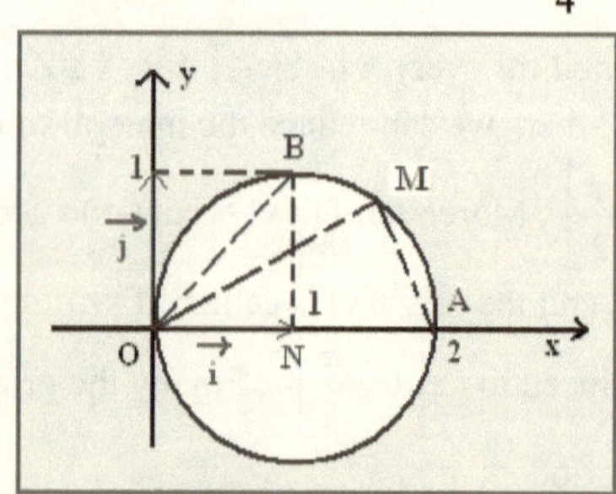

b. In triangle O A M, $\cos\theta = \frac{OM}{2} \Rightarrow OM = 2\cos\theta$.

If $B \equiv M$, then in triangle O B N, $(OB)^2 = 1^2 + 1^2 = 2 \Rightarrow OB = \sqrt{2}$

$\forall$ M, and O B = 2 cosθ $\Rightarrow$ cos θ $= \dfrac{OB}{2} = \dfrac{\sqrt{2}}{2} \Rightarrow \theta = \dfrac{\pi}{4} + 2k\pi \ , k \in Z$.

So B ($\sqrt{2}$, $\dfrac{\pi}{4}$) .

3.71.

a. $(\cos x + \sin x)^2 + (\cos x - \sin x)^2 = \cos^2 x + \sin^2 x + 2\cos x \sin x + \cos^2 x + \sin^2 x - 2\cos x \sin x$
$= 2(\cos^2 x + \sin^2 x)$
$= 2 \times 1 = 2$. Claim.

b. $\cos^4 x + \sin^4 x = \cos^2 x + \sin^2 x + 2\cos^2 x \sin^2 x - 2\cos^2 x \sin^2 x$

$= (\cos^2 x + \sin^2 x)^2 - 2\cos^2 x \sin^2 x$

$= 1^2 - 2\left(\dfrac{\sin 2x}{2}\right)^2$

$= 1 - \dfrac{1}{2}\sin^2(2x)$. Claim.

3.73. $x = \pm\dfrac{\pi}{4} + 2k\pi \ , k \in Z$.

3.75. $x = \dfrac{\pi}{3} + k\pi \ , k \in Z$.

3.77. $x = -\dfrac{\pi}{3} + 2k\pi \ , k \in Z$. or $x = \dfrac{4\pi}{3} + 2k\pi \ , k \in Z$.

3.79. $x = \pm\dfrac{\pi}{4} + 2k\pi \ , k \in Z$, $x = \pm\dfrac{3\pi}{4} + 2k\pi \ , k \in Z$

$x = \dfrac{-\pi}{4} + 2k\pi \ , k \in Z$, or $x = \dfrac{5\pi}{4} + 2k\pi \ , k \in Z$.

3.81. $x = \dfrac{k\pi}{5} \ , k \in Z$

3.83. $x = \pm\dfrac{\pi}{4} + 2k\pi \ , k \in Z$, $x = \pm\dfrac{3\pi}{4} + 2k\pi$, $x = 2k\pi$,
or $x = \pi + 2k\pi \ , k \in Z$

3.85. $x = \dfrac{-\pi}{4} + k\pi \ , k \in Z$ or $x = \dfrac{\pi}{4} + k\pi \ , k \in Z$.

3.87. $x = \pm\dfrac{\pi}{3} + 2k\pi \ , k \in Z$ or $x = 2k\pi$.

3.89. $x = \dfrac{\pi}{4} + k\pi \ , k \in Z$, or $x = \arctan\dfrac{1}{2} + k\pi$.

3.91. $D_f = (-\dfrac{1}{2}, +\infty)$, $x = -0.495$.

3.93. $D_f = (1, +\infty)$, $x = 2$.

3.95. $D_f = (0, \frac{\pi}{2})$, $x = 0$.

3.97. $D_f = (-7, -1) \cup (3, +\infty)$, $x = 5$.

3.99. $D_f = \Re$, $x = 0$.

3.101. $D_f = \Re^*$, $x = 0.49$.

3.103. $D_f = \Re$, $x = 0$.

3.105. $D_f = [0, \frac{2}{3}]$. $\text{Arc}\cos(3x - 1) + \text{Arc}\sin x = \frac{\pi}{2} \Leftrightarrow$

$\text{Arc}\cos(3x - 1) = \frac{\pi}{2} - \text{Arc}\sin x \Leftrightarrow \text{Arc}\cos(3x - 1) = \text{Arc}\cos x$

$\Leftrightarrow 3x - 1 = x \Leftrightarrow x = \frac{1}{2}$ acceptable.

3.107. $\begin{cases} x \neq 0 \\ -\frac{\pi}{2} < 2\,Arc\tan x < \frac{\pi}{2} \end{cases} \Leftrightarrow \begin{cases} x \neq 0 \\ -\frac{\pi}{4} < Arc\tan x < \frac{\pi}{4} \end{cases} \Leftrightarrow$

$\begin{cases} x \neq 0 \\ -1 < x < 1 \end{cases} \Leftrightarrow x \in (-1, 0) \cup (0, 1)$.

Let $\text{Arc}\tan x = \alpha$, then $\tan\alpha = x$ and $\beta = Arc\tan\frac{2}{x} \Leftrightarrow \tan\beta = \frac{2}{x}$,

then $2\alpha = \beta \Leftrightarrow \tan 2\alpha = \tan\beta \Leftrightarrow \frac{2\tan\alpha}{1-\tan^2\alpha} = \tan\beta \Leftrightarrow$

$\frac{2x}{1-x^2} = \frac{2}{x} \Leftrightarrow x = \pm\frac{\sqrt{2}}{2}$ acceptable.

3.109. $f[g(x)] = -2x + 5$.

3.111. $f[g(x)] = 6x + 6$.

3.113. $f \circ g(x) = f[g(x)] = x^2 + 1$ and $g \circ f(x) = g[f(x)] = (x+1)^2$.

3.115. $f \circ g(x) = f[g(x)] = \frac{2}{x}$ and $g \circ f(x) = g[f(x)] = \frac{1}{2x}$.

3.117. Range of $f = (0, 16)$ and $f^{-1}(x) = -\sqrt{x} - 1$.

3.119. Range of $f = (-\infty, +\infty)$ and $f^{-1}(x) = \frac{1}{x+1}$.

3.121. $f(x_1) = f(x_2) \Leftrightarrow x_1^2 = x_2^2 \Leftrightarrow x_1 = x_2$ (since $x > 0$).
Then f is one – to – one function.

3.123. Let $x_1 = 1$ and $x_2 = 0$, then $x_1 \neq x_2$ but $f(x_1) = 1 = f(x_2)$.
So f is not one – to – one function.

3.125. $f^{-1}(x) = x^2 - 1$.

3.127. $f^{-1}(x) = \frac{\sqrt{x}}{2} + 2$.

3.129. $f^{-1}(x) = \text{Arc}\sin x$.

3.131. a) The annual depreciation of the machine is : $\frac{15000 - 5000}{10} = \$1000$ .Then TD (t) = 1000 t.

b) The value of the machine after t years is DV (t) = 15000-1000 t. Then, after three years it is : DV(3) = 15000 – 1000(3) = \$ 12000.

3.133. The value of this equipment is: 8000 + 100 = \$ 8100. The total depreciation is TD(t) = $\frac{8100-500}{10}t = 760\,t$ .The value of this equipment after t years is DV (t) = 8100 – 760 t.

3.135. a) TD(t) = $\frac{12500-4000}{10}t$.

b) DV (7) = 1250 - $\frac{12500-4000}{10} \times 7$.

3.135. $FV = PV(1+i)^n = 80\,000\,000\,(1 + \frac{0.1}{2})^n$.

If the capital is doubled, then $80\,000\,000\,(1 + \frac{0.1}{2})^n = 160\,000\,000$, then

$(1+\frac{0.1}{2})^n = 2$ which implies that n ln 1.05 = ln, then $n \approx 14.2$. Since the interest is compounded semiannually the $\frac{14.2}{2} = 7.1$. Hence after 7 years six months the capital is doubled.

CHAPTER 4 , pp. 141 - 149

4.1. $-\infty$; **4.3.** 4; **4.5.** 0 ; **4.7.** 1 / 2 ; **4.9.** 0; **4.11.** 1 or -1

4.13. $-\infty$; **4.15.** 0; **4.17.** 2 ; **4.19.** 0 ; **4.21.** – 1 / 3; 4.23. 2

4.25. 2 / 3; **4.27.** 4 ; 4.29. 0 ; **4.31.** 0

4.33. Continuous

4.35. Continuous

4.37. $\lim_{x\to 0} f(x) = \lim_{x\to 0} \frac{\sin x}{x} = 1 = f(0)$.

4.39.

a. $\frac{\infty}{\infty}$

b. * $\frac{\sqrt{1+x}-1}{x} = \frac{\sqrt{1+x}-1}{x} \times \frac{\sqrt{1+x}+1}{\sqrt{1+x}+1} = \frac{x}{x(\sqrt{1+x}+1)} = \frac{1}{\sqrt{1+x}+1}$

* $\lim_{x\to +\infty} f(x) = \lim_{x\to +\infty} \frac{1}{\sqrt{1+x}+1} = 0$.

4.41.

a. $D_f =]-1\ ,\ 1[$

b. $\lim_{x\to 1^-} f(x) = 0$ and $\lim_{x\to 1^+} f(x)$ doesn't exist because f is undefined at 1^+.

4.43. $\lim_{x\to \infty} f(x) = 0$.

4.44. $\lim_{x\to 1} g(x) = \pm\infty$.

4.45. $\lim_{x\to -2} P(x) = \pm\infty$ and $\lim_{x\to \infty} P(x) = 2$.

4.47. $\lim_{x \to 2^+} f(x) = 1$, $\lim_{x \to 2^-} f(x) = -1 = f(2)$

Then f is continuous to left of 2 but not to the right. Therefore, it is not continuous at 2 .

4.49. $\lim_{x \to 1^+} g(x) = \lim_{x \to 1^+} (x^2 - 1) = 0 = g(0)$

$\lim_{x \to 1^-} g(x) = \lim_{x \to 1^-} (x^2 + 1) = 2$

Therefore, g is continuous to the right of 1 but not to the left. Therefore, it is not continuous at 1 .

4.51. $\lim_{x \to 3^+} g(x) = \lim_{x \to 3^+} \frac{1}{\sqrt{1+x}+2} = \frac{1}{4}$

$\lim_{x \to 3^-} g(x) = \lim_{x \to 3^-} \frac{1(3)}{4} - \frac{1}{2} = \frac{1}{4} = g(3)$

Then g is continuous at 3.

$\lim_{x \to -1^+} g(x) = \lim_{x \to -1^+} \frac{1(-1)}{4} - \frac{1}{2} = \frac{-3}{4} = g(-1)$

Then g is continuous over $[-1, +\infty[$

4.53. $\lim_{x \to 1^-} h(x) = h(1)$ then $3 + a + 1 = 2/3$ which implies that $a = -10/3$.

4.55. f is continuous over [1 , 2] .

$f(1) = 1 - \sqrt{2}$ and $f(2) = 4 - \sqrt{3}$. Then $f(1) \neq f(2)$.
So if $d \in [1, 2]$ then , there exists at least one number $c \in (1, 2)$ such that $f(1) = d$.

4.57. $f(x) = x^3 - x^2 + x$
$f(2) = 6$ and $f(3) = 33$, then $f(2) \neq f(3)$.
Moreover , $6 = f(2) < f(3) = 33$. Hence, there exists at least one number $c \in (2, 3)$ such that $f(c) = 0$.

4.59.

a. $\overline{C}\,(x) = \dfrac{C(x)}{x} = \dfrac{1000 + 10x}{x}$ for x < 0.

b. 10.

c. y = 10 is the horizontal asymptote .

d. x = 0 is the vertical asymptote .

4.61.

a. $\lim_{x \to 0} C(x) = +\infty$, then the y-axis is a vertical asymptote to f.

b. $\lim_{x \to \infty} C(x) = +\infty$ and $\lim_{x \to \infty} [C(x) - y_d] = 0$, then (d) is an oblique asymptote to f.

c.

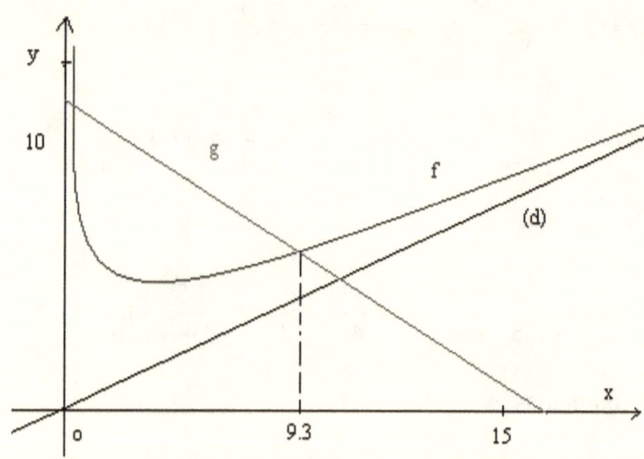

d. 1) g is a semi-straight line passing through points (0, 13) and (16.25, 0).

2) The factory realizes profit when R(x) > C(x). Then, $-0.8x + 13 > 0.5x + \frac{8}{x}$ which is equivalent to $-1.3x + 13 - \frac{8}{x} > 0$. It gives $-1.3x^2 + 13x - 8 > 0$ since x is positive, the roots of this equation are about 0.65 and 9.34. $-1.3x^2 + 13x - 8 > 0$ for $0.65 < x < 9.34$.The factory attains profit when producing from 0.65 to 9.34 hectoliters.

CHAPTER 5 , pp. 173 - 179

5.1. $\lim_{h \to 0} \dfrac{f(2+h)-f(2)}{h} = 6$

5.3. $\lim_{h \to 0} \dfrac{f(1+h)-f(1)}{h} = \dfrac{1}{2\sqrt{2}}$

5.5. $\lim_{h \to 0} \dfrac{f(2+h)-f(2)}{h} = -2$

5.7. $f'(x) = 8x^3 - 15x^2 + 2x$

5.9. $p'(x) = \dfrac{2(x+1)-2x}{(x+1)^2} = \dfrac{2}{(x+1)^2}$

5.11. $A'(x) = \cos\left(\dfrac{\pi}{6} + x\right)$

5.13. $C'(x) = 2\cos 2x \tan 3x + \dfrac{3\sin 2x}{\cos^2 3x}$

5.15. $z'(x) = -\sin x[-\sin(\cos x)] = \sin x[\sin(\cos x)]$

5.17. $f'(x) = 2x\ln x + x$

5.19. $f'(x) = -\tan x$

5.21. $f'(x) = \dfrac{e^{\frac{x}{x+1}}}{(x+1)^2}$

5.23. $f'(x) = e^{\frac{1}{x}} - \dfrac{e^{\frac{1}{x}}}{x}$

5.25. $f'(x) = \dfrac{3x^3\ln x + 2^x x\ln 2\ln x - 1}{x\ln^2 x}$

5.27. $f'(x) = \dfrac{-\ln 2}{2^x}$

5.29. $f'(x) = \cos x\, \pi^{\sin x} \ln \pi$

5.31. $y = -2x - 2$

5.33. $y = 1$

5.35. $A'(x) = \dfrac{1}{2\sqrt{1+x}}$ and $A''(x) = \dfrac{-1}{4(1+x)\sqrt{1+x}}$

5.37. $C'(x) = 2\cos x + 3$ and $C''(x) = -2\sin x$

5.39. $E'(x) = \dfrac{5}{\sqrt{10x+1}}$ and $E''(x) = \dfrac{-25}{(10x+1)\sqrt{10x+1}}$

5.41. $G'(x) = \cos x - \sin x$ and $G''(x) = -\sin x - \cos x$

5.43. $I'(x) = 3x^2 - 2\sqrt{x} - \dfrac{x}{\sqrt{x}}$ and $I''(x) = 6x - \dfrac{3}{2\sqrt{x}}$

5.45.

a. $f'(x) = 0$, then $3x^2 - 6x = 0$. So $x = 0$ or $x = 2$

b. $f'(x) = 1$, then $3x^2 - 6x - 1 = 0$. So $x \approx 2.15$ or $x \approx -0.15$

5.47. -864 ; **5.49.** $\dfrac{-\sqrt{2}}{2\sqrt{3}}$; **5.51.** 0 ; **5.53.** 6 ; **5.55.** 2

5.57. $V(t) = x'(t)\big|_{t=2} = \dfrac{-16}{(4t+3)^2}\Big|_{t=2} = \dfrac{-16}{121}$

5.59. $V(t) = x'(t)\big|_{t=3} = \dfrac{4}{(t+2)^2}\Big|_{t=3} = \dfrac{4}{25}$

5.61. $V(t) = x'(t)\big|_{t=1} = \dfrac{1}{2\sqrt{t+2}}\Big|_{t=1} = \dfrac{1}{2\sqrt{3}}$

5.63. $\lim\limits_{x \to 2^+} f(x) = f(2)$, then $2b = 4a + 1$.

$f'_{2^+} = f'_{2^-}$, then $b = 4a$. Solve the system formed by $2b = 4a + 1$ and $b = 4a$ to get $a = 1/4$ and $b = 1/2$.

5.65. $V(t) = x'(t) = 0$, then $4t^3 - 6t^2 + 2t = 0$. So $t = 0$, $t = 1$ or $t = 1/2$.

5.67. $\frac{dy}{dx} = \frac{dy}{du} \times \frac{du}{dx} = (3u^2 - 2)(2x) = 6x^5 - 4x.$

5.69. $\frac{dy}{dx} = \frac{dy}{du} \times \frac{du}{dx} = (3u^2)(\frac{1}{2\sqrt{x}}) = \frac{3(\sqrt{x}+1)^2}{2\sqrt{x}}$

5.71. 49 ; **5.73.** 6

5.75.

a. $2e^x - 4k(x^{k-1}) = 0$, then $2e - 4k = 0$. So $k = e/2$.

b. $e^x - k(x^{k-1}) = 0$, then $e - k = 0$. So $k = e$.

5.77. $y' = \frac{\sec x \tan x(1+\tan x) - (1+\tan^2 x)\sec x}{(1+\tan x)^2} = \frac{\sec x(\tan x - 1)}{(1+\tan x)^2}$

5.79. $y' = 3\sec^2(3x^3 - 2)(9x^2)\sec(3x^3 - 2)\tan(3x^3 - 2)$
$= 27x^2\sec^3(3x^3 - 2)\tan(3x^3 - 2).$

5.81. $y' = \frac{-\csc^2 x(1+\csc x) + \csc x \cot x(\cot x)}{(1+\csc x)^2}$
$= \frac{\csc x(\cot^2 x - \csc x - \csc^2 x)}{(1+\csc x)^2}.$

5.83. $y' = \frac{6\tan 3x}{\cos^2(3x)}$

5.85. Differentiate both sides : $2x + 2yy' = 0$, then $\frac{dy}{dx} = \frac{-x}{y}$.

5.87. Differentiate both sides : $y'\ln(x+1) + \frac{1}{x+1}y = e^y + xy'e^y$,

then $\frac{dy}{dx} = \frac{(x+1)e^y - y}{(x+1)\left[\ln(x+1) - xe^y\right]}$.

5.89. Differentiate both sides : $1 + y + xy' + 2yy' = 0$ then $\frac{dy}{dx} = \frac{-1-y}{x+2y}$.

5.91.

a. $M(q) = 6q + 6$.

b. $AC(q) = \frac{C(q)}{q} = 3q + 6$.

c. $M(25) = 6(25) + 6 = 156$.

$AC(25) = 3(25) + 6 = 81$.

d. $M(q) = AC(q) \Leftrightarrow 6q + 6 = 3q + 6 \Leftrightarrow q = 0$.

5.93.

a. $E(p) = -p\,\dfrac{D'(p)}{D(p)} = -p\dfrac{-\dfrac{1}{2\sqrt{p+1}}}{10-\sqrt{p+1}} = \dfrac{p}{2(10-\sqrt{p+1})\sqrt{p+1}}$.

b. $E(80) = \dfrac{80}{2(10-\sqrt{81})\sqrt{81}} \approx 4.4$.

$E(80) > 1$, then the demand is elastic for p = 80.If the price increases by 1 % starting from a price of \$ 80 , the demand decreases by 4.4 % .

5.95.

a. D(25) = 200 – 2 (25) = 150 units.

b.1) $E(p) = \dfrac{p}{100-p}$.

2) $E(25) \approx 0.3$.

c. $\dfrac{p}{100-p} = 1$ gives p = 50.

d. $E(p) > 1$ gives $\dfrac{p}{100-p} > 1$, which is equivalent to $\dfrac{2p-100}{100-p} > 0$, which implies $50 < p < 100$.

5.97.

a. The coordinates of the equilibrium points are the solution of the point of intersection of D and S . D(p) = S(p) gives $4\ln\left(\dfrac{6}{p}\right) = 4\ln(p-1)$, then $\dfrac{6}{p} = p - 1$ which gives the quadratic equation $p^2 - p - 6 = 0$ that has as a solution $p_1 = 3$ and $p_2 = -2$. But $-2 \notin [2, 6]$ then p = 3 is the accepted solution and consequently the equilibrium point is the point A (3 , 4 ln 2) and the price of equilibrium is p = \$ 3.

b. For p = 3, the quantity produced S(3) is equal to the quantity bought D(3). The total revenue R is given by $R = 3 \times S(3) = 3\times 4\ln 2 = 12\ln 2$ thousands of dollars, which is \$ 8318.

c. $E(p) = -p\dfrac{D'(p)}{D(p)} = -p\dfrac{\dfrac{-4}{p}}{4\ln\left(\dfrac{6}{p}\right)} = \dfrac{1}{\ln\left(\dfrac{6}{p}\right)}$.

d. $E(2) = \dfrac{1}{\ln 3} \approx 0.9 < 1$. The demand is inelastic for the price p = 2.If the price increases 1 % starting with a price of \$ 3 , the demand decreases 0.9 % .

5.99.

a. D(p) = S(p) gives f(p) = 0 , then p = 4.15. Therefore, the case of equilibrium the price is \$ 4.15 .

b. $E(p) = -p\dfrac{D'(p)}{D(p)} = -p\dfrac{(-0.5p-0.25)e^{-0..5p+1}}{(2.5+p)e^{-0..5p+1}} = p\dfrac{(0.5p+0.25)}{(2.5+p)}$.

c. $E(4) = 4\dfrac{(0.54+0.25)}{(2.5+4)} \approx 1.38 > 1$ then for the price p = 4 , the demand is elastic. If the price increases 1 % starting from the price p = 4, the demand decreases 1.38 % .

CHAPTER 6 , pp. 218 - 228

6.1. f (x) is a polynomial function and it is continuous over [-1 , 2]. $f'(x) = 3x^2 - 4x$; $f'(x) = 0$ if $3x^2 - 4x = 0$, then x = 0 and x = 4 / 3 . The critical numbers are 0 and 4 / 3 . Now , f (- 1) = - 3 and f (2) = 0 ; f (0) = 0 and f (4 / 3) = - 32 / 27 . So the maximum value is the highest which is 0 and it occurs at 0 and 2 , and the minimum value is - 3 ; it occurs at -1 , so the slope of the function admits a maximum value on the points (2 , 0) and (0 , 0) , and the slope of the function admits a minimum value on the point (-1 , -3) .

6.3. f (x) is a polynomial function and it is continuous over [-1 , 2]. $f'(x) = \dfrac{2}{3}x^{-\frac{1}{3}} = \dfrac{2}{3\sqrt[3]{x}}$,then f ' (x) is not defined at 0 , so x = 0 is the critical number. Now , f (- 1) = 1 , f (2) = $\sqrt[3]{4}$ and f (0) = 0 ; the maximum value is the highest which is $\sqrt[3]{4}$ and it occurs at 2 , and the minimum value is 0 ; it occurs at 0 . Hence, the slope of the function admits a maximum value on the point (2 , $\sqrt[3]{4}$) , and the slope of the function admits a minimum value on the point (0 , 0) .

6.5. f (x) is a polynomial function and it is continuous over [0 , 2 π]. f ' (x) = 2 cos x + 2 sin 2 x = 2 cos x + 4 sin x cos x = 2 cos x (1 + 2 sin x) ; f ' (x) = 0 if cos x = 0 or 1 + 2 sin x = 0 , that is cos x = 0 or sin x = $\dfrac{-1}{2}$, then ($x = \dfrac{\pi}{2}$, $x = \dfrac{3\pi}{2}$) or ($x = \dfrac{7\pi}{6}$, $x = \dfrac{11\pi}{6}$) .The critical numbers are $\dfrac{\pi}{2}$, $\dfrac{3\pi}{2}$, $\dfrac{7\pi}{6}$,and $\dfrac{11\pi}{6}$. Now , $f(\dfrac{\pi}{2}) = 3$, $f(\dfrac{3\pi}{2}) = -1$, $f(\dfrac{7\pi}{6}) = \dfrac{-3}{2}$, $f(\dfrac{11\pi}{6}) = \dfrac{-3}{2}$ and f (0) = f (2 π) = - 1. So the maximum value is the highest which is 3 and it occurs at $\dfrac{\pi}{2}$, and the minimum value is $\dfrac{-3}{2}$; it occurs at $\dfrac{7\pi}{6}$ and $\dfrac{11\pi}{6}$, so the slope of the function admits a maximum value on the point ($\dfrac{\pi}{2}$, 3) , and the slope of the function admits a minimum value on the points

$(\frac{7\pi}{6},\frac{-3}{2})$ and $(\frac{11\pi}{6},\frac{-3}{2})$.

6.7. f is a polynomial function , then :
1) f is continuous over [0 , 4]
2) f is differentiable over (0 , 4)
3) f (4) = 1 = f(0)

Then Rolle's theorem is satisfied . So there exists a number $c \in (0,4)$ such that $f'(c) = 0$. $f'(x) = 2x - 4$; $f'(c) = 0$ then $2c - 4 = 0$, then $c = 2 \in (0,4)$

6.9. f is a polynomial function over [0 , 1] and over [1 , 2], then :
1) f is continuous over [0 , 1] and over [1 , 2]
2) f is differentiable over (0 , 1) and over (1 , 2)
3) f (2) = 0 = f(0)

Now we study the behavior of the function at x = 1 , whether it is continuous and differentiable :

f(1) =2-1 =1

$$\lim_{x\to 1^+} f(x) = \lim_{x\to 1^+} (2-x) = 2-1 = 1$$

$$\lim_{x\to 1^-} f(x) = \lim_{x\to 1^-} x^2 = 1$$

So f is continuous at x = 1 and then continuous over [0 , 2] .

$$f'_+(1) = \frac{d}{dx}(2-x)\Big|_{x=1} = -1 \text{ and } f'_-(1) = \frac{d}{dx}(x^2)\Big|_{x=1} = 2$$

$f'_+(1) \neq f'_-(1)$; $f'(1)$ does not exist and then f is not differentiable over the interval (0 , 2) . Hence , Rolle's theorem is not satisfied over the interval [0 , 2] .

6.11. g(x) = cos x is continuous and differentiable over $\left[\frac{\pi}{4},\frac{7\pi}{4}\right]$, then it satisfies the mean value theorem . Now we find the numbers that satisfy this theorem . Since g satisfies the mean value theorem , then there exists at least one value $c \in \left[\frac{\pi}{4},\frac{7\pi}{4}\right]$ such that : $f'(c) = \frac{f(b)-f(a)}{b-a}$.

$f'(x) = -\sin x \Rightarrow f'(c) = -\sin c$

$$f(a) = f(\frac{\pi}{4}) = \cos\frac{\pi}{4} = \frac{\sqrt{2}}{2}$$

$$f(b) = f(\frac{7\pi}{4}) = \cos\frac{7\pi}{4} = \frac{\sqrt{2}}{2}$$

Then $f'(c) = \dfrac{\dfrac{\sqrt{2}}{2} - \dfrac{\sqrt{2}}{2}}{\dfrac{7\pi}{4} - \dfrac{\pi}{4}} = 0$. So $-\sin c = 0 \Rightarrow c = \pi$.

Hence $c = \pi \in \left(\dfrac{\pi}{4}, \dfrac{7\pi}{4} \right)$ is the number that satisfies the mean value theorem .

6.13. f is a polynomial function , then it is continuous over $\Re$. We need to find the critical values . $f'(x) = 0$ implies that $2x - 2 = 0 \Rightarrow x = 1$. The critical value is 1 . Now we study the sign of $f'(x)$ over $\Re$.

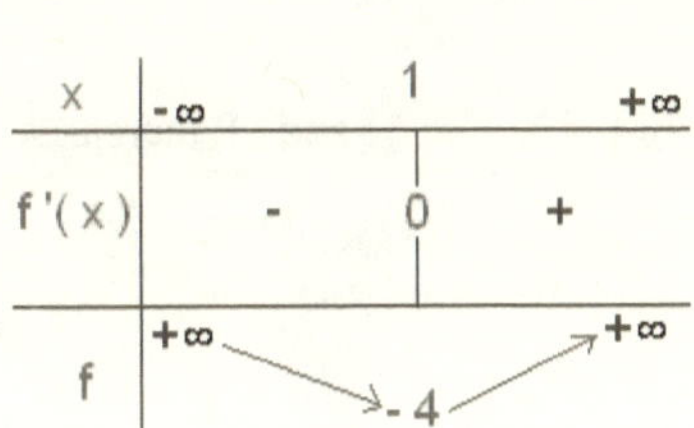

x	$-\infty$		1		$+\infty$
f'(x)		-	0	+	
f	$+\infty$	↘	-4	↗	$+\infty$

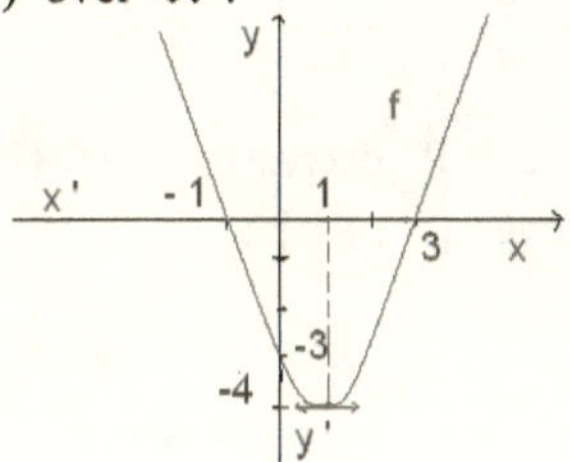

So , f decreases over $(-\infty, 1]$ and f increases over $[1, +\infty)$.

6.15. h is a polynomial function , then it is continuous over $\Re$. We need to find the critical values . $h'(x) = 0$ implies that $4x^2(-x+3) = 0 \Rightarrow x = 0$ or $x = 3$. The critical values are 0 and 3 . Now we study the sign of $h'(x)$ over $\Re$.

x	$-\infty$		0		3		$+\infty$
x^2		+	0	+		+	
$-x+3$		+		+	0	-	
h'(x)		+	0	+	0	-	
h	$-\infty$	↗	0	↗	27	↘	$-\infty$

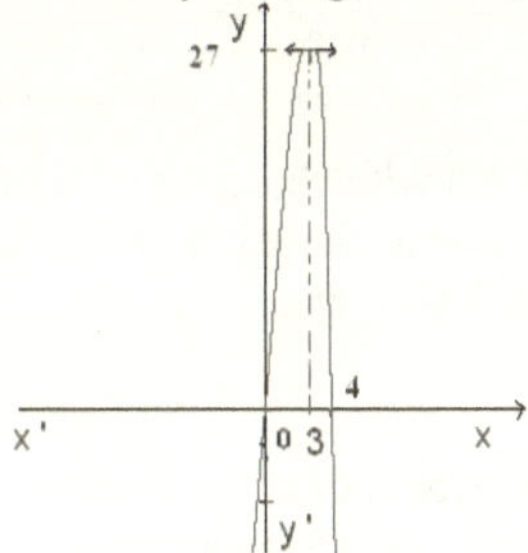

So , h decreases over $[3, +\infty)$ and h increases over $(-\infty, 3]$.

6.17. g is continuous over $[0, \pi]$. We need to find the critical values . $g'(x) = 0$ implies that $-\sin x = 0 \Rightarrow x = 0$ or π . The critical values are 0 and π . Now we study the sign of $g'(x)$ over $[0, \pi]$.

x	0		π
g'(x)	0	-	0
g	1	↘	- 1

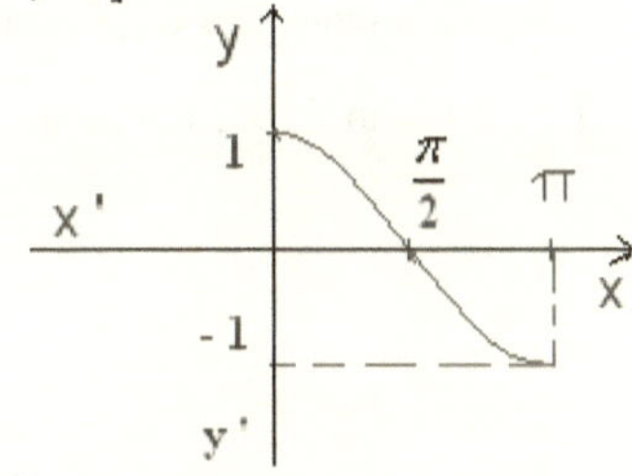

So , g decreases over $[0, \pi]$.

6.19. P(x) is continuous over $\Re$. We need to find the critical values .

$P'(x)=\dfrac{-x^2+1}{(x^2+1)^2}$. $P'(x)$ admits the sign of $-x^2+1$.$P'(x)=0$ implies that $-x^2+1=0 \Rightarrow x=-1$ or $x=1$. The critical values are -1 and 1 . Now we study the sign of $P'(x)$ over $\Re$.

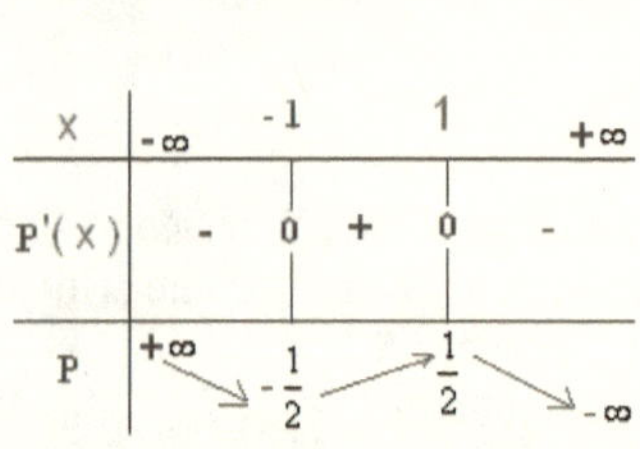

x	$-\infty$		-1		1		$+\infty$
P'(x)		-	0	+	0	-	
P	$+\infty$	↘	$-\frac{1}{2}$	↗	$\frac{1}{2}$	↘	$-\infty$

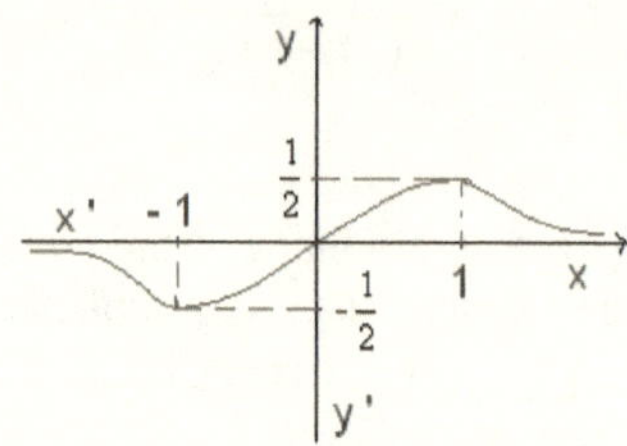

So , P decreases over $(-\infty, -1] \cup [1, +\infty)$ and P increases over $[-1, 1]$.

6.21. $g'(x)=1+\cos x$ and $g''(x)=-\sin x$.

$g''(x)=0$ if $-\sin x=0$ implies that $x=0$, $x=\pi$, $x=-\pi$, -2π or 2π .

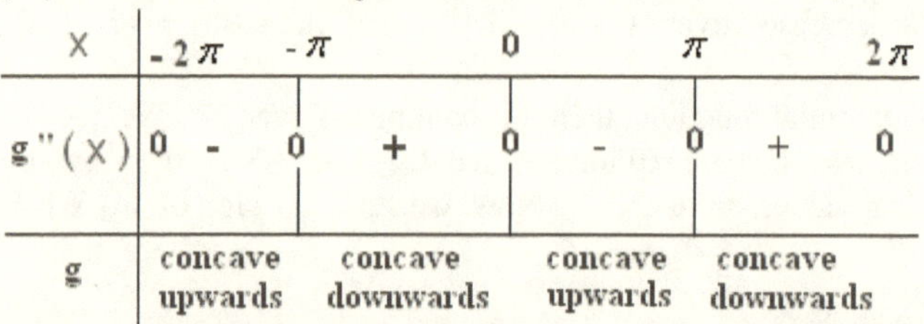

x	-2π		$-\pi$		0		π		2π
g''(x)	0	-	0	+	0	-	0	+	0
g		concave upwards		concave downwards		concave upwards		concave downwards	

$g''(-\pi)=0$ with changing of sign , then $(-\pi, -\pi)$ is an inflection point .
$g''(0)=0$ with changing of sign , then $(0, 0)$ is an inflection point .
$g''(\pi)=0$ with changing of sign , then (π, π) is an inflection point .

6.23. Concave upwards over the interval $(0, 3)$ and $(3, +\infty)$.
Concave downwards over the interval $(-\infty, 0)$.
No inflection points .

6.25. Concave upwards over the interval $(-\frac{\pi}{2}, 0)$.

Concave downwards over the interval $(0, \frac{\pi}{2})$.

$(0, 0)$ is an inflection point .

6.27. f is continuous over $\Re$. $f'(x) = -4x^3 + 4x$; $f'(x) = 0$, then $x = 0$ or $x = \pm 1$

So $f(0) = 12$ and $f(\pm 1) = 13$.

x	$-\infty$		-1		0		1		$+\infty$
4x		-	|	-	0	+	|	+	
$-x^2+1$		-	0	+	|	+	0	-	
f'(x)		+	0	-	0	+	0	-	
f	$-\infty$	↗	13	↘	12	↗	13	↘	$-\infty$

Interpretation :

* $f'(\pm 1) = 0$ with changing of sign from positive to negative , then (- 1 , 13) and (1 , 13) are local maximum .
* $f'(0) = 0$ with changing of sign from negative to positive , then (0 , 12) is a local minimum .

6.29. (0 , 3) is a local maximum .
(2 , $\sqrt{5}$) is a local minimum .

6.31. $(-1, \frac{1}{4})$ and $(1, \frac{1}{4})$ are local maximum .
(0 ,0) is a local minimum .

6.33. $\left(\frac{\pi}{4}, 2\sqrt{2}+1\right)$ is a local maximum .
$\left(\frac{-\pi}{6}, \frac{\sqrt{3}}{3}\right)$ is a local minimum .

6.35. g is defined , continuous and differentiable over $\Re$. $g'(x) = 4x(x^2-1)$.
$g'(x) = 0$ implies that $4x(x^2-1) = 0 \Rightarrow x = 0$ or $x = \pm 1$.
$g(0) = 1$ and $g(\pm 1) = 0$.

X	$-\infty$		-1		0		1		$+\infty$
4x		-	|	-	0	+	|	+	
x^2-1		+	0	-	|	-	0	+	
g'(x)		-	0	+	0	-	0	+	
g	$+\infty$	↘	0	↗	1	↘	0	↗	$+\infty$

Concavity : $g''(x) = 12x^2 - 4$. $g''(x) = 0$ if $12x^2 - 4 = 0$ implies that $x = \pm\dfrac{\sqrt{3}}{3}$. $g(\pm\dfrac{\sqrt{3}}{3}) = \dfrac{4}{9}$.

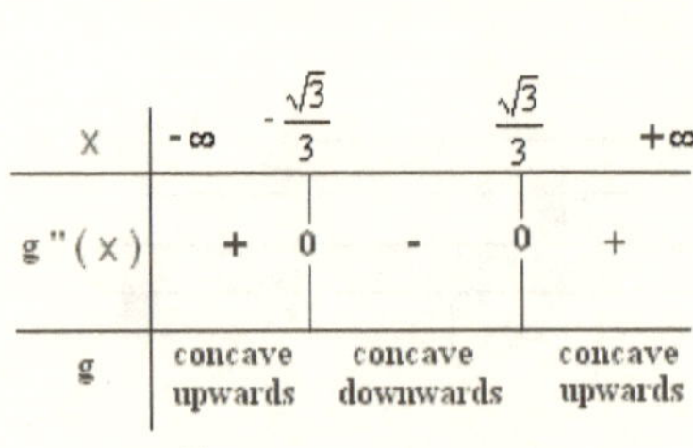

x	$-\infty$		$-\dfrac{\sqrt{3}}{3}$		$\dfrac{\sqrt{3}}{3}$		$+\infty$
g"(x)		+	0	-	0	+	
g		concave upwards		concave downwards		concave upwards	

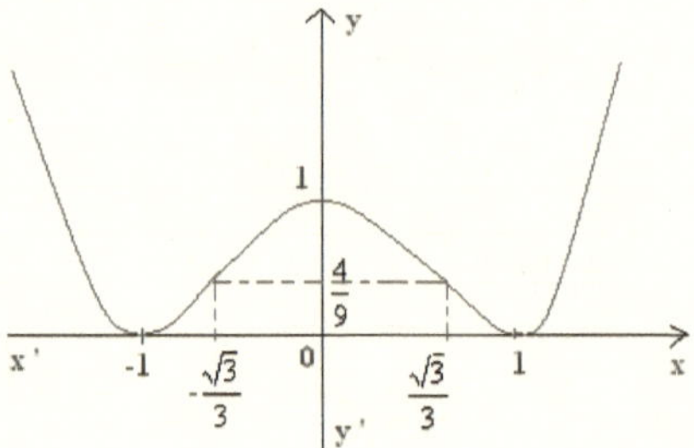

$(\pm\dfrac{\sqrt{3}}{3}, \dfrac{4}{9})$ are inflection points .

6.37. f is a polynomial function , then it is continuous over $\Re$.$f'(x) = 6(x-1)^2$.
$f'(x) = 0$ implies that $x - 1 = 0 \Rightarrow x = 1$. $f(1) = 0$.

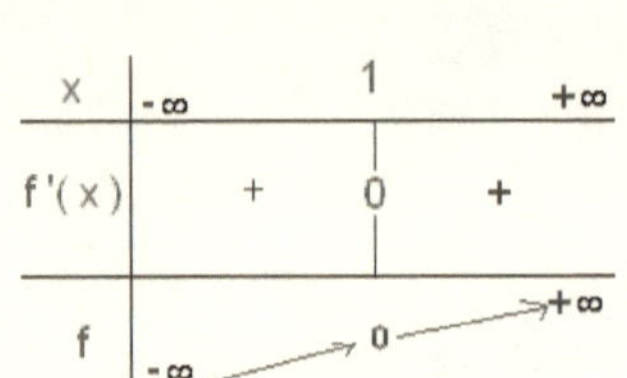

x	$-\infty$		1		$+\infty$
f'(x)		+	0	+	
f	$-\infty$	↗	0	↗	$+\infty$

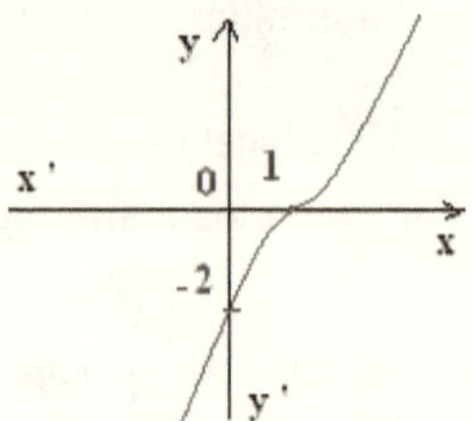

Concavity : $f''(x) = 12x - 12$. $f''(x) = 0$ if $12x - 12 = 0$ implies that $x = 1$. $f(1) = 0$.

x	$-\infty$		1		$+\infty$
f"(x)		-	0	+	
f		concave upwards		concave downwards	

(1 , 0) is an inflection point .

6.41. This function is defined , continuous and differentiable over $(0, +\infty)$.

$\lim\limits_{x \to 0^+} f(x) = +\infty$ and the y-axis is a vertical asymptote .

$\lim\limits_{x \to +\infty} f(x) = \lim\limits_{x \to +\infty} x(1 - \dfrac{\ln x}{x}) = +\infty$. Possibility of oblique asymptote $y = ax + b$, where

$a = \lim\limits_{x \to +\infty} \dfrac{f}{x} = \lim\limits_{x \to +\infty} \dfrac{x}{x} - \dfrac{\ln x}{x} = 1$. and

$b = \lim\limits_{x \to +\infty} f - ax = \lim\limits_{x \to +\infty} x - \ln x - x = -\infty$. Then f admits an asymptotic direction parallel to the first bisector $y = x$.

$f'(x) = 1 - \frac{1}{x} = \frac{x-1}{x}$. f ' admits the same sign as x – 1 over (0 , + ∞).

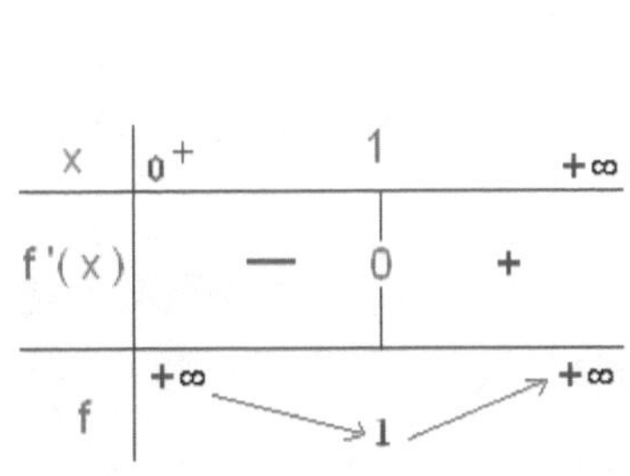

x	0^+		1		$+\infty$
f '(x)		—	0	+	
f	$+\infty$	↘	1	↗	$+\infty$

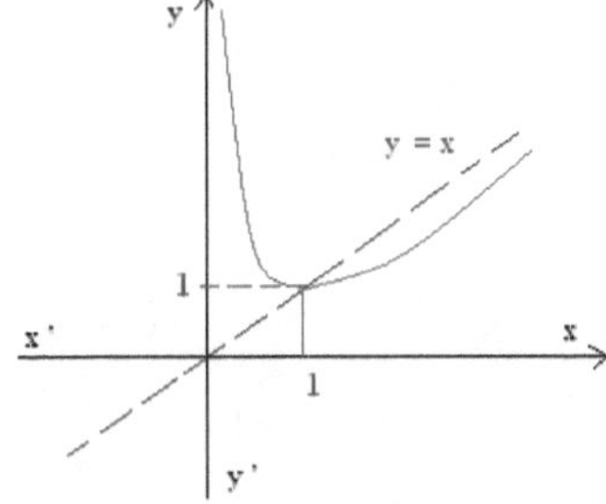

6.43. This function is defined , continuous and differentiable over $\Re$.

$\lim_{x \to -\infty} f(x) = -\infty$ and $\lim_{x \to +\infty} f(x) = +\infty$. Possibility of oblique asymptote $y = ax + b$, where

$a = \lim_{x \to -\infty} \frac{f}{x} = \lim_{x \to -\infty} \frac{e^x}{x} - \frac{e^{-x}}{x} = 0.$ and

$b = \lim_{x \to -\infty} f - ax = \lim_{x \to -\infty} e^x - e^{-x} = -\infty$. Then f admits an asymptotic direction parallel to the x-axis .Similarly , when $x \to +\infty$, f admits an asymptotic direction parallel to the x-axis .

$f'(x) = e^x + e^{-x} > 0$ for every $x \in \Re$ and . f is strictly increasing .

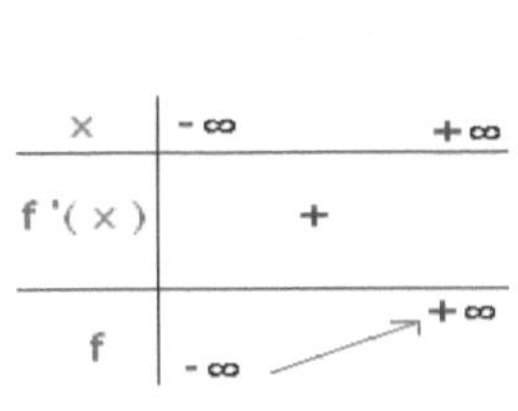

x	$-\infty$		$+\infty$
f '(x)		+	
f	$-\infty$	↗	$+\infty$

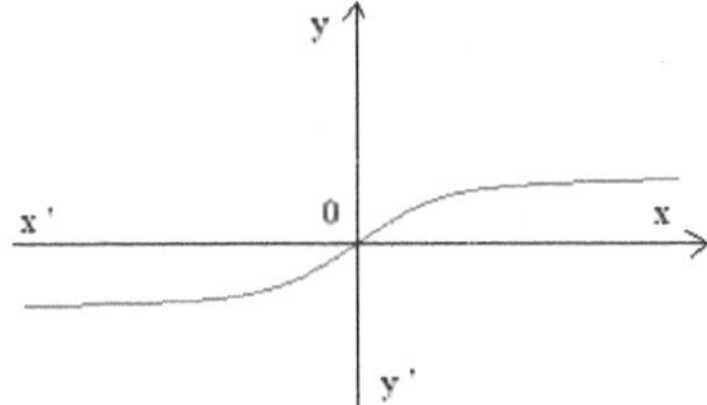

6.45. $-\infty$; **6.47.** 1 (hint : let $X = \frac{1}{x}$) ; **6.49.** -1 ; **6.51.** $\frac{1}{e}$.

6.53. 1 ; **6.55.** 0 ; **6.57.** ln 10 .

6.59.

6.59.

- for every x in the interval $[0, 1)$, $f(x) = x^2$ is a polynomial function continuous over $[0, 1)$ and differentiable over $(0, 1)$.
- for every x in the interval $[1, 2]$, $f(x) = 2 - x$ is a polynomial function continuous over $[1, 2]$ and differentiable over $(1, 2)$.
- $f(0) = 0 = f(2)$.
- We need to study now the behavior of f at $x = 1$ whether it is continuous and differentiable. $f(1) = 2 - 1 = 1$.

$$\lim_{x \to 1^+} f(x) = \lim_{x \to 1^+} 2 - x = 1 \text{ , and}$$

$$\lim_{x \to 1^-} f(x) = \lim_{x \to 1^-} x^2 = 1 \text{ , then}$$

$$\lim_{x \to 1^+} f(x) = \lim_{x \to 1^-} f(x) = f(1)$$, so it is continuous at 1.

Now, $f'_{1^+}(1) = \lim_{x \to 1^+} \dfrac{f(x) - f(1)}{x - 1} = \lim_{x \to 1^+} \dfrac{2 - x - 1}{x - 1} = -1$

and

$$f'_{1^-}(1) = \lim_{x \to 1^-} \frac{f(x) - f(1)}{x - 1} = \lim_{x \to 1^-} \frac{x^2 - 1}{x - 1} = 2 \text{ , then}$$

$f'_{1^+}(1) \neq f'_{1^-}(1)$. So $f'(1)$ does not exist, and f is not differentiable at 1 and then not differentiable over $(0, 2)$. Hence f does not satisfy Rolle's theorem over the interval $[0, 2]$.

6.61. f is defined and continuous over $[0, \pi]$.
$f'(x) = 8 \cos 2x$ and $f''(x) = -16 \sin 2x$.
$f''(x) = 0$ if $-16 \sin 2x = 0 \Rightarrow \sin 2x = 0 \Rightarrow \sin 2x = \sin 0 \Rightarrow$

$$\begin{cases} x = k\pi \\ \text{or} \\ x = \dfrac{\pi}{2} + k\pi \end{cases} \Rightarrow \begin{cases} x = 0 \text{ or } \pi \text{ in } [0, \pi] \\ \text{or} \\ x = \dfrac{\pi}{2} \text{ in } [0, \pi] \end{cases}$$

x	0	$\frac{\pi}{2}$	π
f"(x)	—		+
f	∩		∪

* $x \in \left[0, \dfrac{\pi}{2}\right] \Rightarrow 2x \in [0, \pi] \Rightarrow \sin 2x > 0$ and $f''(x) < 0$.

* $x \in \left[\dfrac{\pi}{2}, \pi\right] \Rightarrow 2x \in [\pi, 2\pi] \Rightarrow \sin 2x < 0$ and $f''(x) > 0$.

* $f''\left(\dfrac{\pi}{2}\right) = 0$ with changing of sign from negative to positive, then f

admits an inflection point $I(\frac{\pi}{2}, f(\frac{\pi}{2}))$ i.e $I(\frac{\pi}{2}, 0)$. Therefore , there exists a tangent line that cut the curve at I of abscissa $x = \frac{\pi}{2}$.

6.63. $f'(x) = 3ax^2 + 2bx + c \Rightarrow f''(x) = 6ax + 2b$.

$f''(\frac{1}{2}) = 0 \Rightarrow 3a + 2b = 0$, and $f'(-1) = 0 \Rightarrow 3a - 2b + c = 0$.

$f(1) = 13 \Rightarrow a + b + c = 13$. Then we get a system of three equations

with three unknowns : $\begin{cases} 3a + 2b = 0 \\ 3a - 2b + c = 0 \\ a + b + c = 13 \end{cases} \Rightarrow a = -2 , \; b = 3 , \; c = 12$.

6.65. Rate of change : $\frac{dp}{dq}\Big|_{q=10} = (30q^2 + 100)\Big|_{q=10} = 30(100) + 100 = 3100$, so the Rate of change of price p with respect to quantity q = 10 is \$ 3100 .

6.67. Rate of change : $\frac{dp}{dq}\Big|_{q=100} = (\ln q + 1)\Big|_{q=100} = \ln(100) + 1 = 5.605$, so the Rate of change of price p with respect to quantity q = 100 is \$ 5.605 .

6.69. Marginal - cost function : $\frac{dc}{dq} = 10$.

The Marginal - cost function when q = 100 : $\frac{dc}{dq}\Big|_{q=100} = 10$.

So the value of the Marginal - cost function when q = 100 is \$ 10 .

6.71. Marginal - cost function : $\frac{dc}{dq} = \frac{24}{q+2}$.

The Marginal - cost function when q = 6 : $\frac{dc}{dq}\Big|_{q=6} = (\frac{24}{q+2})\Big|_{q=6} = \frac{24}{8} = 3$.

So the value of the Marginal - cost function when q = 6 is \$ 3 .

6.73. Total cost function : $c = q\bar{c} \Rightarrow c = 0.02q^2 + 3q + 400$.

Marginal - cost function : $\frac{dc}{dq} = 0.04q + 3$.

Marginal - cost function when q = 50 : $\frac{dc}{dq}\Big|_{q=50} = (0.04q + 3)\Big|_{q=50} = 5$.

So the value of the Marginal - cost function when q = 50 is \$ 5 .

6.75. Total cost function : $c = q\bar{c} \Rightarrow c = \frac{240q}{\ln(q+2)}$.

Marginal - cost function : $\frac{dc}{dq} = \frac{240\,(q+2)\ln(q+2)\ +240q}{(q+2)\ln^2(q+2)}$.

Marginal - cost function when q = 30 :

$\left.\frac{dc}{dq}\right|_{q=30} = \left.\left(\frac{240(q+2)\ln(q+2)+240q}{(q+2)\ln^2(q+2)}\right)\right|_{q=30} =$

$\frac{240\,(30+2)\ln(30+2)\ +240\times 30}{(30+2)\ln^2(30+2)} = 87.98168279$.

So the value of the Marginal - cost function when q = 30 is \$ 87.98168279

6.77. Total cost function : $c = q\ \bar{c} \Rightarrow c = 0.01\,q^3 - 0.03\,q^2 + 3\,q + 200$.

Marginal - cost function : $\frac{dc}{dq} = 0.03\,q^2 - 0.06q + 3$.

Marginal - cost function when q = 20 :

$\left.\frac{dc}{dq}\right|_{q=20} = \left.(0.03q^2 - 0.06q + 3)\right|_{q=20} = 13.8$.

So the value of the Marginal - cost function when q = 20 is \$ 13.8 .

6.79. Marginal – Revenue Function : $\frac{dr}{dq} = 48\,q + 230 - 3\,q^2$;

Marginal – Revenue Function when q = 25 :

$\left.\frac{dr}{dq}\right|_{q=25} = \left.(48q + 230 - 3\,q^2)\right|_{q=25} = -445$.

6.81. Marginal – Revenue Function : $\frac{dr}{dq} = \frac{2q+1}{2\sqrt{q^2+q-1}}$;

Marginal – Revenue Function when q = 10 :

$\left.\frac{dr}{dq}\right|_{q=10} = \left.\left(\frac{2q+1}{2\sqrt{q^2+q-1}}\right)\right|_{q=10} = 1.00572$.

6.83. Marginal – Revenue Function : $\frac{dr}{dq} = 2\,e^{2q} + e^q + q^2 e^q$;

Marginal – Revenue Function when q = 15 :

$\left.\frac{dr}{dq}\right|_{q=15} = \left.(2e^{2q} + e^q + q^2 e^q)\right|_{q=15} = 2.1374 \times 10^{13}$.

6.85. Revenue is $r = p\,.\,q = \frac{q(12)}{\ln(q+1)}$.

Marginal – Revenue Function : $\dfrac{dr}{dq} = \dfrac{12\,(q+1)\ln(q+1)\ -12\,q}{(q+1)\ln^{2}(q+1)}$.

6.87. Revenue is $r = p\,.\,q = (q^{2}+2q)\ln q$.

Marginal – Revenue Function : $\dfrac{dr}{dq} = (2q+2)\ln q + q + 2$.

6.89. 1. Rate of change of p with respect to q is $\dfrac{dp}{dq} = 1$.

2. Relative Rate of Change is : $\dfrac{p'}{p} = \dfrac{1}{q+4}$. When q = 5 ,

$RRC = \dfrac{p'}{p}\Big|_{q=5} = \dfrac{1}{9}$.

3. Percentage rate of change when q = 5 is : $\dfrac{p'}{p}\times 100\Big|_{q=5} = \dfrac{1}{9}\times 100 = 11.1$

So the percentage rate of change of the demand equation when q = 5 is about 11 % .

6.91. 1. Rate of change of p with respect to q is $\dfrac{dp}{dq} = \dfrac{q}{\sqrt{q^{2}+20}}$.

2. Relative Rate of Change is : $\dfrac{p'}{p} = \dfrac{q}{100\sqrt{q^{2}+20}+q^{2}+20}$.

When q = 2 , $RRC = \dfrac{p'}{p}\Big|_{q=2} = 0.0039$.

3. Percentage rate of change when q = 2 is :
$\dfrac{p'}{p}\times 100\Big|_{q=2} = 0.0039\ \times 100 = 0.39$.

So the percentage rate of change of the demand equation when q = 2 is about 0.39 % .

6.93. 1. Rate of change of p with respect to q is $\dfrac{dp}{dq} = 4q+5$.

2. Relative Rate of Change is : $\dfrac{p'}{p} = \dfrac{4q+5}{(2q+3)(q+1)}$. When q = 4 ,

$RRC = \dfrac{p'}{p}\Big|_{q=4} = 0.38$.

3. Percentage rate of change when q = 4 is :

$$\frac{p'}{p} \times 100 \Big|_{q=4} = 0.38 \times 100 = 38 \ .$$

So the percentage rate of change of the demand equation when q = 4 is about 38 % .

6.95. a. R (x) = (Price per unit) × (Quantity demand) = 2000 x .
b. Profit = R (x) – C (x) = $100\,x^2 - 700\,x + 1000$
Profit = 0 ⇔ $100\,x^2 - 700\,x + 1000 = 0$ ⇔ x = 5 or x = 2 .
The production should be 5000 or 2000 units .
c. P ' (x) = 200 x – 700 . P ' (x) = 0 if 200 x – 700 = 0 implies that x = 3.5 . $P''(x) = 200 > 0$ for every x . Then P admits a local minimum , that is the production must be of 3500 units to get a minimum profit .
d. Profit = P (3500) = $100\,(3500)^2 - 700\,(3500) + 1000 = 1222551000$.

6.97. a. $C'(x) = \dfrac{3}{4} > 0$, and $C''(x) = 0$, then the cost is strictly increasing .
Then for x = 2 , the cost is minimized .That is the cost is minimized for a quantity of 2 .
b. $R'(x) = \dfrac{-22}{54}\,x + \dfrac{235}{108}$. R' (x) = 0 if $\dfrac{-22}{54}\,x + \dfrac{235}{108} = 0$ which implies that $x \approx 5.341 \notin [\,2\,,5\,]$. So the revenue could not be maximized .

c. Profit : P (x) = R (x) – C (x) = $\dfrac{-11}{54}\,x^2 + (\dfrac{235}{108} - \dfrac{3}{4})x - \dfrac{28}{27} - 1$

= $-0.20\,x^2 + 1.43\,x - 2.04$.
$P'(x) = -0.40\,x + 1.43 > 0 \Leftrightarrow x = 3.575$.
$P''(x) = -0.40 < 0$, then the profit is maximized for a quantity of 3.575 .

6.99.

a. R (x) = x D (x) = $\dfrac{x^3}{3} - \dfrac{25x^2}{2} + 150x$ .Then , by solving R ' (x) = 0 for x , we obtain x = 15 and x = 10 . Since R "(15) = 5 > 0 and R "(10) = - 5 < 0 then for x = 10 the revenue is maximum .
b. The maximum value of revenue is R (10) = $ 583.33 .

6.101.

a. $D'(P) = -8\,p^{-3} = \dfrac{-8}{p^3}$ for all p > 0 .

b. $E(p) = -p\dfrac{D'(p)}{D(p)} = -p\dfrac{-8p^{-3}}{4p^{-2}} = 2$.

c. $E(p) = 2 > 1$, then D is elastic for all values of $p > 0$.

d. The result in (c) implies that small changes in price (up or down from this level) will cause a relatively large change in demand . That is , an elasticity of 2 means that a price change of 1 % will cause about a 2 % change in demand .

6.103.

a. $U'(W) = \dfrac{\alpha}{W+1}$; $\alpha > 0$ and $W + 1 > 1$, then $\ln(W+1) > 0$ and $U'(W) > 0$.

b. $U''(W) = \dfrac{-\alpha}{(W+1)^2} < 0$, then U is risk – averse .

c. $RRA = R(W) = \dfrac{-WU''(W)}{U'(W)} = \dfrac{W}{W+1}$.

6.105.

a. $\mu_1 = 0.15$, $\mu_2 = 0.10$, $\sigma_1 = 0.20$, $\sigma_2 = 0.12$, and $\rho = -0.2$.

$\min \sigma_p^2 = w'\Sigma_x w = w_1^2\sigma_1^2 + w_2^2\sigma_2^2 + 2w_1w_2\rho\sigma_1\sigma_2 \quad (*)$

such that $w_1 + w_2 = 1 \Rightarrow w_2 = 1 - w_1$. Substitute in (*) to get :

$\sigma_p^2 = w_1^2\sigma_1^2 + (1-w_1)^2\sigma_2^2 + 2w_1(1-w_1)\rho\sigma_1\sigma_2$

$$\frac{d\sigma_p^2}{dw_1} = 2w_1\sigma_1^2 - 2(1-w_1)\sigma_2^2 + 2\rho\sigma_1\sigma_2 - 4w_1\rho\sigma_1\sigma_2 = 0$$

then $w_1 = \dfrac{2\sigma_2^2 - 2\rho\sigma_1\sigma_2}{2\sigma_1^2 + 2\sigma_2^2 - 4\rho\sigma_1\sigma_2} = 0.3$ and then $w_2 = 1 - 0.3 = 0.7$.

b. The second derivative test :

$\dfrac{d^2\sigma_p^2}{dw_1^2} = 2\sigma_1^2 + 2\sigma_2^2 - 4\rho\sigma_1\sigma_2 > 0$ (since $\rho < 0$).

Therefore , the minimum variance is attained and is equal to :

$\sigma^2_{P,\min} = 0.864\%$.

$E(R_p) = w'E(R) = w_1\mu_1 + w_2\mu_2 = 11.50\%$.

6.107.

a. $P = 12x - 3x^2 + 20y - 3y^2 - 2xy$.

FOC : $\dfrac{dp}{dx} = 12 - 6x - 2y = 0$ and $\dfrac{dp}{dy} = 20 - 6y - 2x = 0$. Solve this system to get $x = 1$ and $y = 3$. Hence $(x, y) = (1, 3)$.

SOC : $\dfrac{d^2p}{dx^2} = -6 < 0$; $\dfrac{d^2p}{dy^2} = -6 < 0$ and $\dfrac{d^2p}{dx\,dy} = -2$.

$(-6)(-6) > 4$ so maximum .

b. Profit maximizing prices : $p = 12 - 2 = 10$; $q = 20 - 3 = 17$.
$P_{max} = 12 - 3 + 60 - 27 - 6 = 36$.

6.109.

a. $U'(W) = a - 2bWe^{cW} - bcW^2e^{cW} = a - bWe^{cW}(2 + cW)$

$$U''(W) = -2be^{cW} - 2bcWe^{cW} - 2bcWe^{cW} - bc^2W^2e^{cW}$$
$$= -be^{cW}(c^2W^2 + 4cW + 2) < 0$$, hence risk-averse .

b. $$RRA = \frac{bWe^{cW}(c^2W^2 + 4cW + 2)}{e^{cW}(ae^{-cW} - 2bW - bcW^2)} = \frac{bW(c^2W^2 + 4cW + 2)}{(ae^{-cW} - 2bW - bcW^2)} .$$

6.111. $f(x) = \ln x$, then $f(1) = 0$; $f'(x) = \frac{1}{x}$, then $f'(1) = 1$;

$f''(x) = \frac{-1}{x^2}$, then $f''(1) = -1$; $f'''(x) = \frac{2}{x^3}$, then $f'''(1) = 2$;

Hence , $P_0(x) = 0$; $P_1(x) = (x-1)$; $P_2(x) = (x-1) - \frac{1}{2}(x-1)^2$;

$P_3(x) = (x-1) - \frac{1}{2}(x-1)^2 + \frac{1}{3}(x-1)^3$.

6.113. $f(x) = \cos x$, then $f(\frac{\pi}{4}) = \frac{1}{\sqrt{2}}$; $f'(x) = -\sin x$, then $f'(\frac{\pi}{4}) = -\frac{1}{\sqrt{2}}$;

$f''(x) = -\cos x$, then $f''(\frac{\pi}{4}) = -\frac{1}{\sqrt{2}}$; $f'''(x) = \sin x$, then $f'''(\frac{\pi}{4}) = \frac{1}{\sqrt{2}}$;

Hence , $P_0(x) = \frac{1}{\sqrt{2}}$; $P_1(x) = \frac{1}{\sqrt{2}} - \frac{1}{\sqrt{2}}(x - \frac{\pi}{4})$;

$P_2(x) = \frac{1}{\sqrt{2}} - \frac{1}{\sqrt{2}}(x - \frac{\pi}{4}) - \frac{1}{2\sqrt{2}}(x - \frac{\pi}{4})^2$;

$P_3(x) = \frac{1}{\sqrt{2}} - \frac{1}{\sqrt{2}}(x - \frac{\pi}{4}) - \frac{1}{2\sqrt{2}}(x - \frac{\pi}{4})^2 + \frac{1}{6\sqrt{2}}(x - \frac{\pi}{4})^3$.

6.115. $f(x) = \sin x$, then $f(0) = 0$; $f'(x) = \cos x$, then $f'(0) = 1$;
$f''(x) = -\sin x$, then $f''(0) = -1$; $f'''(x) = -\cos x$, then $f'''(0) = -1$;

•　　　　　•

•　　　　　•

•　　　　　•

$f^{(2k)}(x) = (-1)^k \sin x$; $f^{(2k)}(0) = 0$; $f^{(2k+1)}(x) = (-1)^k \cos x$; $f^{(2k+1)}(0) = (-1)^k$.

So , $\sin x = x - \frac{x^3}{3!} + \frac{x^5}{5!} + \dots\dots\dots\dots + \frac{(-1)^k x^{2k+1}}{(2k+1)!}$ + Remainder .

6.117. $f(x) = x^4 - 2x^3 - 5x + 4$, then $f(0) = 4$;

$f'(x) = 4x^3 - 6x^2 - 5$, then $f'(0) = -5$;

$f''(x) = 12x^2 - 12x$, then $f''(0) = 0$;

$f'''(x) = 24x - 12$, then $f'''(0) = -12$;

$f^{(4)}(x) = 24$, then $f^{(4)}(0) = 24$;

$f^{(n)}(x) = 0$ if $n \geq 5$, then $f^{(n)}(0) = 0$ if $n \geq 5$.

Hence , $x^4 - 2x^3 - 5x + 4 = 4 - 5x - \frac{12x^3}{3!} + \frac{24x^4}{4!}$.

6.119.

a. The fixed cost is given by $C_F = C(0) = 2590 + 2\sqrt{900+0} = \$\ 2650$.

b. The total cost of production of 4000 units is :
$C(4000) = 2590 + 2\sqrt{900+4000} = \$\ 2730$.
The average cost of production of one of these 4000 units is
2730/4000 = $ 0.6825 .

c. $C'(x) = 2 \times \frac{1}{2\sqrt{900+x}} > 0 \ \forall x \in \Re$. The table of variations of C is as follows :

x	0	$+\infty$
C'(x)	+	
C(x)	2650 ↗	$+\infty$

d. If $x > 2700$ then $\sqrt{900+x} > \sqrt{900+2700}$ and consequently
$2590 + 2\sqrt{900+x} > 2590 + 2\sqrt{900+2700}$, but
$2590 + 2\sqrt{900+2700} = 2710$, then $C(2700) > 2710$.

e. $C(x) \leq 3000$, then $2590 + 2\sqrt{900+x} \leq 3000$ and $\sqrt{900+x} \leq 205$ which gives $900 + x \leq 42025$, then $x \leq 41125$. So the maximum quantity to be produced to get total cost less than or equal to $ 3000 is 41125. The average cost of one unit is then 3000/ 41125 $\approx$ $ 0.072 .

6.121.

a. C'(q) = 2q – 20 . The table of variations of C is as follows :

q	0		10		55
C'(q)		—	0	+	
C	200	↘	100	↗	2125

b. 1) P'(q) = - 2q + 54. The table of variations of P is as follows :

q	0		27		55
P'(q)		+	0	—	
P	- 200	↗	529	↘	- 255

2) The revenue is given by R(q) = 34 q, then the profit attains from selling q articles is given by : F(q) = R(q) – C (q) = 34 q – q^2 + 20q -200.Then F(q) = - q^2 + 54 q – 200 = P(q). Hence, when the company produces and sells 27 articles it attains a maximum profit equal to \$ 529.

3) P(q) = 0 implies that - q^2 + 54 q – 200 =0 with roots $q_1 = 50$ and $q_2 = 4$.So the trinomial P(q) = - q^2 + 54 q – 200 takes the nonzero positive values for $4 \le q \le 50$, then if the number of articles is between 4 and 50 or equal to 4 or 50 , it attains a positive profit or zero.

6.123. Part A

a. $M(x) = C'(x) = \frac{1}{10}x^2 - 30x + 2500$.

b. $M'(x) = \frac{1}{5}x - 30 = \frac{x-150}{5}$, then the table of variations of M is :

x	0		150		300
C '(x)		—	0	+	
C	2500	↘	250	↗	2500

c. The minimum of M is 250, then M (x) > 0 for every x which gives C ' (x) > 0, consequently C is strictly increasing.

Part B

a. $R(q) = P(q) \times q = \dfrac{-45}{8}q^2 + 2750q.$

b. The marginal revenue is $R'(q) = \dfrac{-45}{4}q + 2750.$ The marginal revenue is equal to the marginal cost, then is $R'(q) = C'(q)$ which gives

$\dfrac{-45}{4}q + 2750 = \dfrac{1}{10}q^2 - 30q + 2500$, then $\dfrac{1}{10}q^2 - \dfrac{75}{4}q - 250 = 0$ with roots $q' = -12.5$ is rejected and $q'' = 200$ acceptable.

c. $P(q) = R(q) - C(q) = -\dfrac{1}{30}q^3 + \dfrac{75}{8}q^2 + 250q.$

d. $P'(q) = -\dfrac{1}{10}q^2 + \dfrac{75}{4}q + 250.$ It has two roots $q_1 = -12.5$ and $q_2 = 200.$

The table of variations of P is as follows:

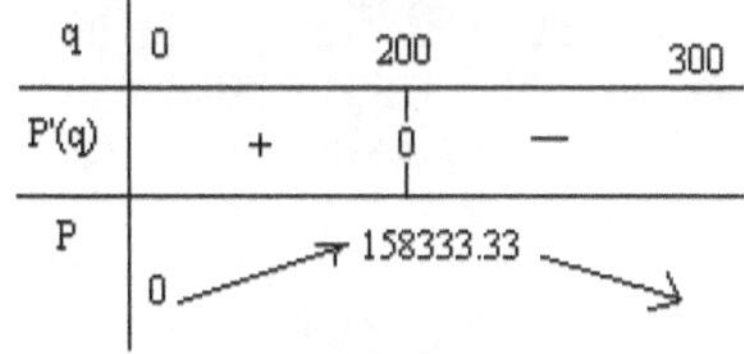

q	0		200		300
P'(q)		+	0	−	
P	0	↗	158333.33	↘	

Then $P(q)$ is maximum when $q = 200$.
The profit is maximum when $P'(q) = 0$, then $R'(q) = C'(q)$.
Then the marginal revenue is equal to the marginal cost.

6.125.

a. When the price is \$ 3, the number of bikes demanded is $D(3) \times 1000 \approx 610$. The revenue is $R = 3\,(610) = \$\,1830$.

b. $R(p) = p \times D(p) \times 1000 = \dfrac{5000 \ln p}{p}.$

c. $R'(p) = 5000\,\dfrac{1 - \ln p}{p^2}$. $R'(p) \geq 0$ when $1 - \ln p \geq 0$, which gives $p \leq e$.

Then the table of variations of R is :

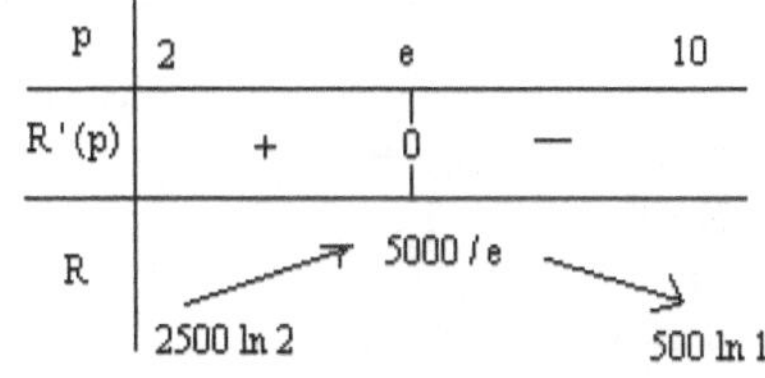

p	2		e		10
R'(p)		+	0	−	
R	2500 ln 2	↗	5000 / e	↘	500 ln 10

d. The revenue is maximum when the price is \$ e which is \$ 2.72 .For this rent price, the revenue raises to $\dfrac{5000}{e} \approx \$\,1839$.

6.127.

6.127.

a. $\overline{C}(q)=\dfrac{C(q)}{q}=\dfrac{q}{4}+\dfrac{9}{2}\dfrac{\ln(q+1)}{q}$, then $\overline{C}'(q)=\dfrac{1}{4}+\dfrac{9}{2}\dfrac{\dfrac{q}{q+1}-\ln(q+1)}{q^2}$, then

$$\overline{C}'(q)=\frac{1}{4}+\frac{9}{2}\left[\frac{q}{q^2(q+1)}-\frac{\ln(q+1)}{q^2}\right]=\frac{1}{4}+\frac{9}{2q(q+1)}-\frac{9\ln(q+1)}{2q^2}.$$

$$\frac{f(q)}{2q^2}=\frac{\dfrac{q^2}{2}+\dfrac{9q}{q+1}-9\ln(q+1)}{2q^2}=\frac{q^2}{4q^2}+\frac{9q}{2q^2(q+1)}-\frac{9\ln(q+1)}{2q^2}$$

$\dfrac{f(q)}{2q^2}=\dfrac{1}{4}+\dfrac{9}{2q(q+1)}-\dfrac{9\ln(q+1)}{2q^2}$, then $\overline{C}'(q)=\dfrac{f(q)}{2q^2}$.

b. $f'(q)=\dfrac{q(q-2)(q+4)}{(q+1)^2}$, Over the interval $[0,5]$, $q\geq 0$ and $q+4>0$ then $f'(q)>0$ for $q>2$, $f'(q)<0$ for $0<q<2$, ad $f'(q)=0$ for $q=2$. Moreover, $f(0)=0$, $f(2)=8\text{-}9\ln 3<0$ and $f(5)=20\text{-}9\ln 6>0$, then the graph of f cut the x-axis at a unique point. So the equation $f(q)=0$ admits a unique root α .But $f(3.6)\approx -0.2<0$ and $f(3.7)\approx 0.002>0$, then $f(3.6)\times f(3.7)<0$ and consequently $3.6<\alpha<3.7$. Hence, $f(q)>0$ for $\alpha<q<5$ and $f(q)<0$ for $0<q<\alpha$.

c. $\overline{C}'(q)$ admits the sign of $f(q)$.Then $\overline{C}'(q)>0$ when $f(q)>0$ which is true for $\alpha<q<5$ and consequently $\overline{C}$ is strictly increasing over $]\alpha,5[$. $\overline{C}'(q)<0$ when $f(q)<0$ which is true for $0<q<\alpha$ and consequently $\overline{C}$ is strictly decreasing over $]0,\alpha[$.

d. The function $\overline{C}$ is strictly increasing over $]\alpha,5[$ and strictly decreasing over $]0,\alpha[$, then it admits a minimum at α .The minimum average cost is $\overline{C}(\alpha)=\overline{C}(3.7)\approx 2.807$ million dollars per thousands of tons, which is $\dfrac{2.807\times 10^6}{10^3}\approx \$2807$ per ton.

6.129.

a. The average cost is defined by $\overline{C}(q)=\dfrac{C(q)}{q}=3+\dfrac{4}{q}+\dfrac{(q-3)e^q}{q}$.

b. 1) On selling q hundreds of articles, the revenue is : $R(q)=100\,q\times 30$, then $R(q)=3000\,q$ dollars, so $R(q)=3\cdot q$ in thousands of dollars. The profit is $P(q)=R(q)-C(q)=3q-3q-4-(q-3)e^q=(3-q)e^q-4$.

2) $P'(q)=-e^q+(3-q)e^q=(2-q)e^q$. $P'(q)\geq 0$ for $0<q\leq 2$, then the table of variations of P is :

q	0		2		3
P'(q)		+	0	−	
P	-1	↗	e^2-4	↘	-4

3) The profit is maximum when q = 2, which is true when producing 200 objects.

c. If the factory produces 200 objects, which is 2 hundreds, the corresponding profit will be $P(2) = e^2 - 4 \approx 3.38$, so the factory attains profit. $P(0.4) = (3 - 0.4)e^{0.4} - 4 \approx -0.12$, so the factory does not attain profit.

CHAPTER 7 , pp. 249 - 260

7.1.

a. $F'(x) = 2x\sqrt{x} + \dfrac{x^2}{2\sqrt{x}}$

$= \dfrac{5}{2}x\sqrt{x}$

$= f(x)$

b. $F'(x) = \dfrac{2}{3}(\sqrt{x} + \dfrac{x}{2\sqrt{x}})$

$= \sqrt{x}$

$= f(x)$

7.3. $\dfrac{2x^3}{3} - \dfrac{1}{x} - 2x^2 + 12x + c$; **7.5.** $\dfrac{-\cos 3x}{3} + c$; **7.7.** $\dfrac{1}{4(1+\cos 2x)^2} + c$

7.9. $\dfrac{\tan^3 x}{3} + c$ (*Hint: take* $u = \tan x$) ; **7.11.** $\dfrac{x^4}{4} + \csc x + 7x + c$

7.13. $\dfrac{(\arctan t)^2}{2} + c$; **7.15.** $\dfrac{-\cos(x^6+2)}{6} + c$ (*Hint: take* $u = x^6 + 2$)

7.17. $\tan(x - \pi) + c = \tan x + c$ (*Hint: take* $u = x - \pi$)

7.19. $x^3 + \cos x + 2\tan x + c$; **7.21.** $\dfrac{-(\arccos t)^3}{3} + c$; **7.23.** $\dfrac{xe^{3x}}{3} - \dfrac{e^{3x}}{9}$.

7.24. -1 **7.25.** $\sin^2 x$ **7.26.** $\dfrac{\sqrt{2}-1}{3}$ **7.27.** 1 **7.28.** 0.1286

7.29. $\dfrac{\sqrt{3}-1}{2}$ **7.30.** ln 2 **7.31.** $\dfrac{1}{5}\ln(\dfrac{\sqrt{2}}{2})$ **7.33.** $\dfrac{\ln 2}{2}$ **7.35.** $\dfrac{22}{3}$

7.37. $\dfrac{-3}{64}$ **7.39.** $\dfrac{25}{6}$ **7.41.** $\dfrac{4}{3}$ **7.43.** $\dfrac{1}{2}$ **7.45.** $\dfrac{1}{16}$ **7.47.** $\dfrac{1}{6}$

7.49. $\dfrac{35}{3}$ **7.51.** $3e^2 + 1$ **7.53.** 0 **7.55.** $\dfrac{20}{7}$

7.57.

7.57.

a. For every $\frac{\pi}{2} \le x \le \pi$, we have $\begin{cases} \sin x \ge 0 \\ and \\ \frac{\pi^2}{4} \le x^2 \le \pi^2 \end{cases} \Rightarrow$

$$\begin{cases} \sin x \ge 0 \\ and \\ 1+\frac{\pi^2}{4} \le 1+x^2 \le 1+\pi^2 \end{cases} \Rightarrow \frac{\sin x}{1+\pi^2} \le \frac{\sin x}{1+x^2} \le \frac{\sin x}{1+\frac{\pi^2}{4}}$$

b. $\frac{\sin x}{1+\pi^2} \le \frac{\sin x}{1+x^2} \le \frac{\sin x}{1+\frac{\pi^2}{4}} \Rightarrow$

$$\int_{\frac{\pi}{2}}^{\pi} \frac{\sin x}{1+\pi^2}\, dx \le \int_{\frac{\pi}{2}}^{\pi} \frac{\sin x}{1+x^2}\, dx \le \int_{\frac{\pi}{2}}^{\pi} \frac{\sin x}{1+\frac{\pi^2}{4}}\, dx$$

$$\frac{1}{1+\pi^2} \le \int_{\frac{\pi}{2}}^{\pi} \frac{\sin x}{1+x^2}\, dx \le \frac{1}{1+\frac{\pi^2}{4}}$$

7.59.

$$\frac{3x^2+x+2}{(x+2)(x+1)^2} = \frac{a(x+1)+bx(x+1)+cx^2}{x^2(x+1)}$$

$$= \frac{(b+c)x^2+(b+a)x+a}{x^2(x+1)}$$

By identification , $\begin{cases} b+c=3 \\ a+b=1 \\ a=2 \end{cases} \Rightarrow \begin{cases} c = 4 \\ b=-1 \end{cases}$

Deduction :

$$\int_0^1 \frac{3x^2+x+2}{(x+2)(x+1)^2}\, dx = \int_0^1 \left(\frac{2}{x+2} - \frac{1}{x+1} + \frac{4}{(x+1)^2} \right) dx$$

$$= \left[2\ln|x+2| - \ln|x+1| - \frac{4}{x+1} \right]_0^1$$

$$= 2\ \ln 3 - 3\ln 2 + 2\ .$$

7.61.

a. $\frac{x^2-4x}{(x-2)^2} = 1 - \frac{k}{(x-2)^2}$

$= \frac{x^2 - 4x + 4 - k}{(x-2)^2}$

By identification , 4 - k = 0 ⇒ k = 4

b. $F(x) = \int \left(1 - \frac{4}{(x-2)^2} \right) dx$

$= x + \frac{4}{x-2} + c$

F (0) = -1 ⇒ 0 - 2 + c = - 1 ⇒ c = 1

Hence , $F(x) = x + \frac{4}{x-2} + 1$.

7.63. $\sqrt{2} - 1$ **7.65.** $\frac{22}{3}$

7.67. $x^2 = \sqrt{x}$ then x = 0 or x = 1.Then $f \cap g = \{A , B\}$ where $x_A = 0$ and $x_B = 1$.

Area = $\int_0^1 [g(x) - f(x)]\, dx = \int_0^1 [\sqrt{x} - x^2]\, dx = \frac{1}{3}$ square units .

7.69. cos x = sin x ⇒ $x = \frac{\pi}{4} - k\pi$ *where* $k \in Z$. $f \cap g$ at the points of abscissas $\frac{\pi}{4} - k\pi$ *where* $k \in Z$.

Area = $\int_0^{\frac{\pi}{2}} [g(x) - f(x)]\, dx = 2\int_0^{\frac{\pi}{4}} [\cos x - \sin x]\, dx = 2\sqrt{2} - 2$ square units .

7.71. $x^2 = 4 - x^2 \Rightarrow x = \pm\sqrt{2}$ .Then $f \cap g = \{A , B\}$ where $x_A = \sqrt{2}$ and $x_B = \sqrt{2}$.

Area = $\int_{-\sqrt{2}}^{\sqrt{2}} [g(x) - f(x)]\, dx = 2\int_0^{\sqrt{2}} [g(x) - f(x)]\, dx =$

$2\int_0^{\sqrt{2}} [4-x^2-x^2]\,dx = 2\int_0^{\sqrt{2}} [4-2x^2]\,dx = \frac{16\sqrt{2}}{3}$ square units .

7.72. $|x| = 1 - |x| \Rightarrow |x| = \frac{1}{2} \Rightarrow x = \pm\frac{1}{2}$. Then $f \cap g = \{A, B\}$ where $x_A = \frac{1}{2}$ and $x_B = -\frac{1}{2}$.

$$\text{Area} = \int_{-\frac{1}{2}}^{\frac{1}{2}} [g(x)-f(x)]\,dx = 2\int_0^{\frac{1}{2}} [g(x)-f(x)]\,dx =$$

$$2\int_0^{\frac{1}{2}} [1-|x|-|x|]\,dx = 2\int_0^{\frac{1}{2}} [1-2|x|]\,dx = \frac{1}{2} \text{ square units .}$$

7.73. $x^4-2x^2=2x^2 \Rightarrow x^4-4x^2=0 \Rightarrow \begin{cases} x=0 \\ x=2 \\ x=-2 \end{cases}$. Then $f \cap g = \{A, B, C\}$ where $x_A = 0$, $x_B = 2$ and $x_c = -2$.

$$\text{Area} = 2\int_0^2 [g(x)-f(x)]\,dx = 2\int_0^2 [2x^2-x^4+2x^2]\,dx = \frac{128}{15}$$

square units .

7.75. $\frac{\pi}{4}$ **7.77.** $\frac{\pi}{30}$ **7.79.** $\frac{\pi}{2}$

7.81.

a. $x^2-8x=0 \Rightarrow x=0$ or $x=8$. Then the volume

$$V = \int_0^8 \pi\,[f^2(x)-g^2(x)]\,dx = \int_0^8 \pi\,[64x^2-x^4]\,dx =$$

$4369.07\,\pi$ cubic units .

b. $y^2 = x$ and $y = x^2 \Rightarrow y^2 = x$ and $y^2 = x^4 \Rightarrow x^4 = x \Rightarrow x^4 - x = 0$

$\Rightarrow x=0$ or $x=1$ .Then the volume $V = \int_0^1 \pi\,[x-x^4]\,dx = \frac{3\pi}{10}$ cubic units .

7.83.

a. $x_{(5)} - x_{(0)} = \int_0^5 V(t)\,dt = \int_0^5 (\frac{t}{2}+5)\,dt = [\frac{t^2}{4}+5t]_0^5 = \frac{125}{4}$.

$$\text{Average velocity } \; v = \frac{\frac{125}{4}}{5-0} = \frac{25}{4} .$$

$$\text{b. } V(t) = \begin{cases} 3t^2 & for \quad t \in [0,1] \\ 3 & for \quad t \in]1,4[\\ 3(t-5)^2 & for \quad t \in [4,5] \end{cases} \Rightarrow$$

$$\begin{cases} x(t_2) - x(t_1) = \int_0^1 3t^2 \, dt = t^3 \,]_0^1 = 1 \\ x(t_2) - x(t_1) = \int_1^4 3 \, dt = 3t \,]_1^4 = 9 \\ x(t_2) - x(t_1) = \int_4^5 3(t-5)^2 \, dt = [(t-5)^3]_4^5 = 1 \end{cases}$$

$$\text{Hence , the average velocity } \; v = \frac{1}{1} + \frac{9}{4-1} + \frac{1}{5-4} = 5 .$$

7.85.

A . This function is defined and continuous for every $x \in \Re$. It is periodic with period 2π , then the interval of discussion could be reduced to $[0 , 2\pi]$ Moreover , $f(-x) = f(x)$, so the f is even and the y – axis is axis of symmetry . Therefore , the interval of discussion could be reduced to $[0 , \pi]$. Finally , f is drawn periodically with period $2k\pi$ where $k \in Z$.

$f(0) = 1 + \cos 0 = 2$, $f(\pi) = 1 + \cos 2\pi = 2$

$$f'(x) = -2 \sin 2x \; . \; f'(x) = 0 \text{ if } -2 \sin 2x = 0 \Rightarrow \begin{cases} x = 0 \\ or \\ x = \frac{\pi}{2} \end{cases} \text{ on } [0 , \pi]$$

$f(\frac{\pi}{2}) = 1 + \cos\pi = 1 - 1 = 0$. So the two points $(0 , 2)$ and $(\frac{\pi}{2} , 0)$ are vertices .

Sign of $f'(x)$:

If $x \in \left[0, \frac{\pi}{2}\right], 2x \in [0, \pi]$, then $\sin 2x \succ 0$ and $f'(x) \prec 0$. Consequently, f is strictly decreasing.

If $x \in \left[\frac{\pi}{2}, \pi\right], 2x \in [\pi, 2\pi]$, then $\sin 2x \prec 0$ and $f'(x) \succ 0$. Consequently, f is strictly increasing.

Table of Variation :

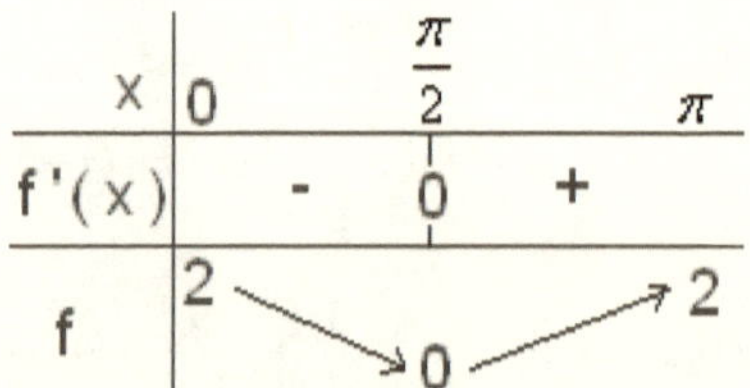

x	0		$\frac{\pi}{2}$		π
f'(x)		-	0	+	
f	2	↘	0	↗	2

Graph :

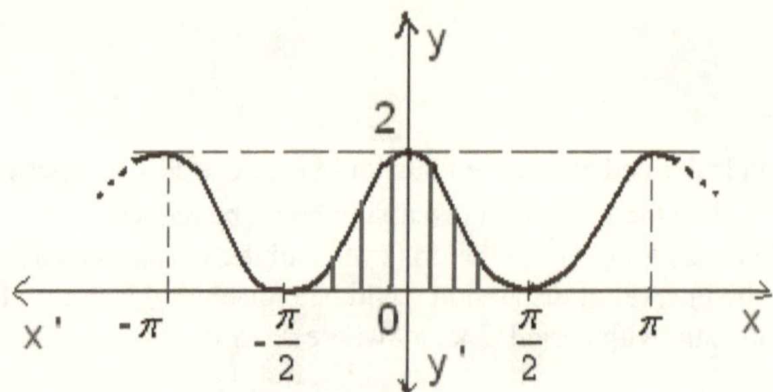

B) Area $= 2\int_0^{\frac{\pi}{2}} f(x)\ dx = 2\int_0^{\frac{\pi}{2}} (1+\cos 2x)\ dx = 2\left[x + \frac{\sin 2x}{2}\right]_0^{\frac{\pi}{2}}$

$= \pi$ square units .

C) a. $(1+\cos 2x)^2 = 1 + \cos^2 4x + 2\cos 2x$

$$= 1 + \frac{1+\cos 4x}{2} + 2\cos 2x$$

$$= 1 + \frac{1}{2} + \frac{\cos 4x}{2} + 2\cos 2x$$

$$= \frac{\cos 4x}{2} + 2\cos 2x + \frac{3}{2}$$

b. $\int_{-\frac{\pi}{2}}^{\frac{\pi}{2}} (1+\cos 2x)^2\ dx = \int_{-\frac{\pi}{2}}^{\frac{\pi}{2}} \left(\frac{1}{2}\cos 4x + 2\cos 2x + \frac{3}{2}\right)\ dx$

$$= \left[\frac{\sin 4x}{8} + \sin 2x + \frac{3}{2}x \right]_{-\frac{\pi}{2}}^{\frac{\pi}{2}}$$

$$= \frac{3\pi}{2} .$$

c. Deduction : $V = \int_{-\frac{\pi}{2}}^{\frac{\pi}{2}} \pi y^2 \, dx = \int_{-\frac{\pi}{2}}^{\frac{\pi}{2}} \pi (1 + \cos 2x)^2 \, dx$

$$= \int_{\frac{-\pi}{2}}^{\frac{\pi}{2}} \pi (1 + \cos 2x)^2 \, dx$$

$$= \pi \int_{\frac{-\pi}{2}}^{\frac{\pi}{2}} (1 + \cos 2x)^2 \, dx$$

$$= \pi \left(\frac{3\pi}{2}\right)$$

$$= \frac{3\pi^2}{2} \text{ cubic units .}$$

7.87. The yellow area = $A_1 = \int_0^1 (x^{\frac{1}{n}} - x^n) \, dx = \left[\frac{x^{\frac{1}{n}+1}}{\frac{1}{n}+1} - \frac{x^{n+1}}{n+1} \right]_0^1$

$$= \frac{n-1}{n+1} \textit{ square units .}$$

The red area = $A_2 = 1 - A_1$.

The two areas are equal implies that $A_1 = 1 - A_1$, then $A_1 = \frac{1}{2}$, So

$\frac{n-1}{n+1} = \frac{1}{2} \Rightarrow n = 3$ an integer ≥ 2 then it is acceptable . Thus, when n = 3 the two regions A_2 and A_1 are equal .

7.89. $r = \int \frac{dr}{dq} \, dq$, then $r = \int 0.6 \, dq \Rightarrow r = 0.6 q$.

So the demand function is $p = \frac{0.6 q}{q} = 0.6$.

7.91. $r = \int \frac{dr}{dq} \, dq$, then $r = \int (260 - q - 0.1 q^2) \, dq \Rightarrow$

$r = 260q - \frac{q^2}{2} - \frac{0.1q^3}{3}$. So the demand function is

$p = 260 - \frac{1}{2}q - \frac{1}{30}q^2.$

7.93. Total cost $C = \int \frac{dc}{dq}\, dq\Big|_{q=1000} = \int (2q+65)\,dq\Big|_{q=1000} =$

$(q^2 + 65q)\Big|_{q=1000} = (1000)^2 + 65(1000) = 1065000.$

7.95. Total cost $C = \int \frac{dc}{dq}\, dq\Big|_{q=400} = \int (450+2q+0.1q^2)\,dq\Big|_{q=400} =$

$(450q + q^2 + \frac{0.1q^3}{3})\Big|_{q=400} = 2473333.333.$

7.97. $r = \int_{100}^{300} \frac{9000}{\sqrt{900q}}\, dq \Rightarrow r = [\,300\, \frac{q^{\frac{1}{2}}}{\frac{1}{2}}\,]_{100}^{300} \Rightarrow r = 4392.30.$

7.98. $r = \int_{10}^{20} (200 + 8q + 2q^2)\, dq \Rightarrow r = [\,200q + 4q^2 + 2\frac{q^3}{3}\,]_{10}^{20} \Rightarrow$

$r = 7866.67.$

7.99. $r = \int_{0}^{5} (2000\, e^{-0.06q})\, dq \Rightarrow r = [\,2000\, \frac{e^{-0.06q}}{-0.06}\,]_{0}^{5} \Rightarrow$

$r = 8639.39$.

7.100. $r = \int_{90}^{180} (50 - 0.5q + 0.004q^2)\, dq \Rightarrow$

$r = [\,50q - 0.5\frac{q^2}{2} + 0.004\frac{q^3}{3}\,]_{90}^{180} \Rightarrow r = 5229$.

7.101.

a. Total – cost function $= \int M(x)\, dx = \int (4x+400)\, dx = 2x^2 + 400x + c$.
$60000 = 2(20)^2 + 400(20) + c \Rightarrow c = 51200$. Then the Total – cost function $= 2x^2 + 400x + 51200$.

b. The cost of producing 30 units $= 2(30)^2 + 400(30) + 51200 = 65000$.

7.103.

a.

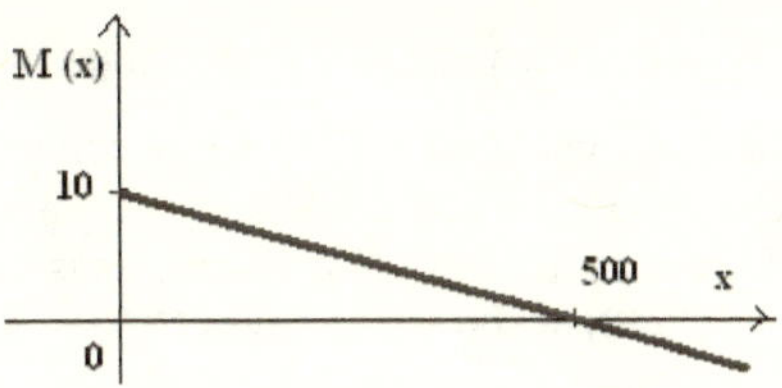

b. Total revenue = $\int \frac{dM}{dx} \; dx\Big|_{x=100} = \int (-0.02x+10)\; dx\Big|_{X=100} = 900$.

c. r = $\int_{100}^{200} (-0.02x+10)\; dx = (-0.01\,x^2 + 10\,x)\;\Big]_{100}^{200} = 700$.

7.105.

a. $K(t) = \int I(t)\; dt\Big|_{t=8} = \int 9t^{\frac{1}{2}}\; dt\Big|_{t=8} = 6\,t^{\frac{3}{2}}\Big|_{t=8} = 6 \times 8^{\frac{3}{2}}$.

b. $PV = \int_0^u Se^{-it}\; dt = \int_0^5 10000e^{-0.03t}\; dt = \$\,46430.67$.

c. $MC = 25 + 30q - 9q^2$. $\quad FC = 55$.

$$TC = \int MC\; dq = \int (25 + 30q - 9q^2)\, dq$$
$$= 25q + 15q^2 - 3q^3 + c$$

$q = 0$ gives $TC = FC = 55$, then $c = 55$.

$$TC = 55 + 25q + 15q^2 - 3q^3$$

$$AC = \frac{TC}{q} = \frac{55}{q} + 25 + 15q - 3q^2 \; .$$

7.107.

a. $FV = \int_0^5 2000\, e^{rt}\; dt = 2000\left[\frac{e^{rt}}{r}\right]_0^5 = \frac{2000}{0.06}(e^{0.06\times 5} - 1) = \$\,11662$.

b. $20\,000 = \int_0^5 Se^{rt}\; dt = S\left[\frac{e^{rt}}{r}\right]_0^5 = \frac{S}{0.06}(e^{0.06\times 5} - 1) \Rightarrow S = \$\,3430$.

7.109. $PV = \int_0^{\infty} 1000\, e^{-0.06t}\; dt = \$\,16667$.

$PV = \int_0^{50} 1000\, e^{-0.06t}\; dt = \$\,15837$.

7.111. a. $C(x) = \int M(x)\,dx = \int (6x+300)\,dx = 3x^2 + 300x + k$.

Since C(20) = 8000, we get 8000 = 7200 + k which implies k = 800.
Consequently, $C(x) = 3x^2 + 300x + 800$.

b. The total cost of production of the first 40 articles is
$C(40) = 3(40)^2 + 300(40) + 800$ which gives C(40) = $ 17600.

7.113.

a. The total revenue of selling the first 100 devices is given by

$$M = \int_0^{100} M(p)\,dp = \int_0^{100} (-0.01p + 5)\,dp = \left(-\frac{0.01p^2}{2} + 5p\right)\Bigg|_0^{100}$$

which gives M = $ 450.

b. The total revenue for the second 100 units is given by

$$M = \int_{100}^{200} M(p)\,dp = \int_{100}^{200} (-0.01p + 5)\,dp = \left(-\frac{0.01p^2}{2} + 5p\right)\Bigg|_{100}^{200}$$

which gives M = $ 350.

7.115.

a. The cost of maintenance during the first 10 years is :

$$C = \int_0^{10} A(t)\,dt = \int_0^{10} (12t^2 + 100)\,dt = \left(4t^3 + 100t\right)\Big|_0^{10} = \$\,5000.$$

b. The cost of maintenance during the tenth year is :

$$C = \int_9^{10} A(t)\,dt = \int_9^{10} (12t^2 + 100)\,dt = \left(4t^3 + 100t\right)\Big|_9^{10} = \$\,1184.$$

7.117. The average number of sick people is given by

$$AV = \frac{1}{b-a}\int_0^8 D(x)\,dx = \frac{1}{8-0}\int_0^8 (45x^2 - x^3)\,dx =$$

$$\frac{1}{8}\left(15x^3 - \frac{1}{4}x^4\right)\Big|_0^8 \text{ persons.}$$

7.119.

a. m ' (n) = - 2 n + 2 + 3 ln (n + 1) + 3 = M(n) , then m (n) is an antiderivative of M(n).

b. The total cost is given by $C(n) = \int M(n)\,dn$ with C(0) = 2. Then

C(n) = m(n) + k , C(0) = 2 implies that k = 2.So

$C(n) = -n^2 + 2n + 3(n+1)\ln(n+1) + 2$.

The cost of 5000 books is C(5) = -25 + 10 + 18ln6 + 2

$= -13 + 18\ln 6 \approx 19.251$ thousands dollars which is $ 19251 .

c. The total cost C_1 , expressed in thousands of dollars of production of 5000 books the first day and 3000 books the second day is:

$C_1 = C(5) + C(3) = -13 + 18\ln 6 - 1 + 12\ln 4$ = $ 34887. The total cost for production of 4000 books per day is :

$C_2 = 2\,C(4) = 2\,[\,-16 + 8 + 15 \ln 5 + 2\,] = -12 + 30 \ln 5 \approx \$\ 36283.$

Since $C_1 < C_2$ we deduce that it is preferable to produce 5000 books the first day and 3000 books the second day.

7.121.

a. Note that d is expressed in hundreds of kilometers then, for a trip of 120 km, we have d = 1.2. $p_1 = D(1.2) = 5\,(1.2 + 2)\,e^{-1.2} \approx 4.82$.
For a trip of 120 km, a client is ready to pay \$ 4.82 for 100 km
$p_2 = S(1.2) = \dfrac{1.2+2}{5}\,e^{1.2} \approx 2.12$. For a trip of 120 km, the company is ready to offer a price of \$ 2.12 for 100 km .

b. At equilibrium, D(d) = S (d) . Its solution is ln 5 , then the demand equals the supply for $d_0 = \ln 5 \approx 1.61$ km . The distance d_0 that corresponds to the price of equilibrium is $100 \times \ln 5$ km ≈ 161 km.
d_0 is the solution of D(d) = S (d) so we can write $D(d_0) = S(d_0)$; to find p_0 we calculate $D(d_0)$ or $S(d_0)$.
$p_0 = S(d_0) = \dfrac{\ln 5+2}{5}\,e^{\ln 5} \approx 3.61$. The corresponding price of equilibrium is then \$ 3.61 for 100 km.

c. $F'(d) = -5\,[\,e^{-d} - e^{-d}(d+3)\,] = 5\,(d+2)\,e^{-d} = D(d).$

$$G = \int_0^{d_0} D(d)\,d_d - p_0 \times d_0 = G = \int_0^{\ln 5} D(d)\,d_d - (2+\ln 5) \times \ln 5$$

$$\text{But } \int_0^{\ln 5} D(d)\,d_d = F(d)\Big|_0^{\ln 5} = F(\ln 5) - F(0) = -\ln 5 - 3 + 15$$

Then $G = -\ln 5 + 12 - 2\ln 5 - \ln^2 5 = -\ln^2 5 - 3\ln 5 + 12 \approx \$\ 4.58$ for 100 km.

Appendix – A
Indeterminate Forms & Infinity

Let $l > 0$ and $k < 0$.

Convention of notation	Limit	Convention of notation	Limit
$+\infty+\infty$	$+\infty$	$\frac{0}{\infty}$	0
$-\infty-\infty$	$-\infty$	$\frac{0}{k}$	0
$(+\infty)(+\infty)$	$+\infty$	0^0	Indeterminate
$(-\infty)(-\infty)$	$+\infty$	$0\times\infty$	Indeterminate
$(+\infty)(-\infty)$	$-\infty$	$+\infty-\infty$	Indeterminate
$k(+\infty)$	$-\infty$	$\frac{\infty}{\infty}$	Indeterminate
$l(+\infty)$	$+\infty$	$\frac{l}{0^-}$	$-\infty$
$k(-\infty)$	$+\infty$	$\frac{k}{\infty}$	0
$l(-\infty)$	$-\infty$	$K+\infty$	$+\infty$
$\frac{+\infty}{0}$	$\pm\infty$	$K-\infty$	$-\infty$
$\frac{+\infty}{0^+}$	$+\infty$	$\frac{0}{0}$	Indeterminate
$\frac{+\infty}{0^-}$	$-\infty$	$\frac{k}{0}$	$\pm\infty$
$\frac{-\infty}{0}$	$\pm\infty$	$\frac{k}{0^+}$	$-\infty$
$\frac{-\infty}{0^-}$	$+\infty$	$\frac{k}{0^-}$	$+\infty$
$\frac{-\infty}{0^+}$	$-\infty$	$\frac{l}{0}$	$\pm\infty$
$\frac{0}{\pm\infty}$	0	$\frac{l}{0^+}$	$+\infty$

Appendix – B
Circular Functions of Remarkable Arcs

α	0 or 2π	$\frac{\pi}{6}$	$\frac{\pi}{4}$	$\frac{\pi}{3}$	$\frac{\pi}{2}$	π
Cos	1	$\frac{\sqrt{3}}{2}$	$\frac{\sqrt{2}}{2}$	$\frac{1}{2}$	0	- 1
Sin	0	$\frac{1}{2}$	$\frac{\sqrt{2}}{2}$	$\frac{\sqrt{3}}{2}$	1	0
Tan	0	$\frac{\sqrt{3}}{3}$	1	$\sqrt{3}$	undefined	0
Cot	undefined	$\sqrt{3}$	1	$\frac{\sqrt{3}}{3}$	0	undefined

Appendix – C
Sign of Trigonometric Functions

α	$(0, \frac{\pi}{2})$	$(\frac{\pi}{2}, \pi)$	$(\pi, \frac{3\pi}{2})$	$(\frac{3\pi}{2}, 2\pi)$
Cos	+	-	-	+
Sin	+	+	-	-
Tan	+	-	+	-
Cot	+	-	+	-

Appendix – D

Brief Trigonometric Formulas

1. $\cos^2\alpha + \sin^2\alpha = 1$
2. $\sec^2\alpha - \tan^2\alpha = 1$
3. $\csc^2\alpha - \cot^2\alpha = 1$
4. $\cos^2\alpha = \dfrac{1}{1+\tan^2\alpha} = \dfrac{\cot^2\alpha}{1+\cot^2\alpha}$
5. $\sin^2\alpha = \dfrac{1}{1+\cot^2\alpha} = \dfrac{\tan^2\alpha}{1+\tan^2\alpha}$
6. $\tan\alpha = \dfrac{\sin\alpha}{\cos\alpha}$
7. $\cot\alpha = \dfrac{\cos\alpha}{\sin\alpha}$
8. $\sec\alpha = \dfrac{1}{\cos\alpha}$
9. $\sec\alpha = \dfrac{1}{\cos\alpha}$
10. $\csc\alpha = \dfrac{1}{\sin\alpha}$
11. $\cos(\alpha-\beta) = \cos\alpha\cos\beta + \sin\alpha\sin\beta$
12. $\cos(\alpha+\beta) = \cos\alpha\cos\beta - \sin\alpha\sin\beta$
13. $\sin(\alpha-\beta) = \sin\alpha\cos\beta - \cos\alpha\sin\beta$
14. $\sin(\alpha+\beta) = \sin\alpha\cos\beta + \cos\alpha\sin\beta$
15. $\tan(\alpha+\beta) = \dfrac{\tan\alpha+\tan\beta}{1-\tan\alpha\tan\beta}$
16. $\tan(\alpha-\beta) = \dfrac{\tan\alpha-\tan\beta}{1+\tan\alpha\tan\beta}$
17. $\cos 2\alpha = \cos^2\alpha - \sin^2\alpha = 2\cos^2\alpha - 1 = 1 - 2\sin^2\alpha$
18. $\sin 2\alpha = 2\sin\alpha\cos\alpha$
19. $\tan 2\alpha = \dfrac{2\tan\alpha}{1-\tan^2\alpha}$
20. $\cos^2\alpha = \dfrac{1+\cos 2\alpha}{2}$
21. $\sin^2\alpha = \dfrac{1-\cos 2\alpha}{2}$
22. $\tan^2\alpha = \dfrac{1-\cos 2\alpha}{1+\cos 2\alpha}$
23. $\sin 3\alpha = 3\sin\alpha - 4\sin^3\alpha$
24. $\cos 3\alpha = 4\cos^3\alpha - 3\cos\alpha$

25. $\tan 3\alpha = \dfrac{3\tan\alpha - \tan^3\alpha}{1 - 3\tan^3\alpha}$

26. $\cos\alpha \cos\beta = \dfrac{1}{2}[\cos(\alpha+\beta) + \cos(\alpha-\beta)]$

27. $\sin\alpha \sin\beta = \dfrac{1}{2}[\cos(\alpha-\beta) - \cos(\alpha+\beta)]$

28. $\sin\alpha \cos\beta = \dfrac{1}{2}[\sin(\alpha+\beta) + \sin(\alpha-\beta)]$

29. $\cos\alpha + \cos\beta = 2\cos(\dfrac{\alpha+\beta}{2})\cos(\dfrac{\alpha-\beta}{2})$

30. $\cos\alpha - \cos\beta = -2\sin(\dfrac{\alpha+\beta}{2})\sin(\dfrac{\alpha-\beta}{2})$

31. $\sin\alpha + \sin\beta = 2\sin(\dfrac{\alpha+\beta}{2})\cos(\dfrac{\alpha-\beta}{2})$

32. $\sin\alpha - \sin\beta = 2\cos(\dfrac{\alpha+\beta}{2})\sin(\dfrac{\alpha-\beta}{2})$

33. $\tan\alpha + \tan\beta = \dfrac{\sin(\alpha+\beta)}{\cos\alpha\cos\beta}$

34. $\cot\alpha + \cot\beta = \dfrac{\sin(\alpha+\beta)}{\sin\alpha\sin\beta}$

35. $\tan\alpha - \tan\beta = \dfrac{\sin(\alpha-\beta)}{\cos\alpha\cos\beta}$

36. $\cot\alpha - \cot\beta = \dfrac{\sin(\beta-\alpha)}{\sin\alpha\sin\beta}$

37. $\sin\alpha + \cos\alpha = \sqrt{2}\cos(\dfrac{\pi}{4}-\alpha) = \sqrt{2}\sin(\dfrac{\pi}{4}+\alpha)$

38. $\cos\alpha - \sin\alpha = \sqrt{2}\cos(\dfrac{\pi}{4}+\alpha) = \sqrt{2}\sin(\dfrac{\pi}{4}-\alpha)$

39. $\cos\alpha = \dfrac{1-\tan^2\dfrac{\alpha}{2}}{1+\tan^2\dfrac{\alpha}{2}}$

40. $\sin\alpha = \dfrac{2\tan\dfrac{\alpha}{2}}{1+\tan^2\dfrac{\alpha}{2}}$

41. $\tan\alpha = \dfrac{2\tan\dfrac{\alpha}{2}}{1-\tan^2\dfrac{\alpha}{2}}$

42. $\cos(-\alpha) = \cos\alpha$
43. $\sin(-\alpha) = -\sin\alpha$
44. $\tan(-\alpha) = -\tan\alpha$
45. $\cot(-\alpha) = -\cot\alpha$

47. $\csc(-\alpha) = -\csc\alpha$
48. $\cos(\pi - \alpha) = -\cos\alpha$
49. $\sin(\pi - \alpha) = \sin\alpha$
50. $\tan(\pi - \alpha) = -\tan\alpha$
51. $\cot(\pi - \alpha) = -\cot\alpha$
52. $\cos(\pi + \alpha) = -\cos\alpha$
53. $\sin(\pi + \alpha) = -\sin\alpha$
54. $\tan(\pi + \alpha) = \tan\alpha$
55. $\cot(\pi + \alpha) = \cot\alpha$
56. $\cos(\frac{\pi}{2} + \alpha) = -\sin\alpha$
57. $\sin(\frac{\pi}{2} + \alpha) = \cos\alpha$
58. $\tan(\frac{\pi}{2} + \alpha) = -\cot\alpha$
59. $\cot(\frac{\pi}{2} + \alpha) = -\tan\alpha$
60. $\cos(\frac{\pi}{2} - \alpha) = \sin\alpha$
61. $\sin(\frac{\pi}{2} - \alpha) = \cos\alpha$
62. $\tan(\frac{\pi}{2} - \alpha) = \cot\alpha$
63. $\cot(\frac{\pi}{2} - \alpha) = \tan\alpha$

Appendix – E

Brief Economical Formulas

1. Total costs = C (q) = Fixed costs + Variable Costs
2. Average cost per unit = $\overline{C}\,(q) = \dfrac{C(q)}{q}$ for all $q > 0$.
3. Marginal cost = $M_c(q) = \dfrac{dC}{dq}$
4. Depreciable value = C - S
5. Annual depreciation = $\dfrac{C-S}{n}$
6. Total depreciation = $D(t) = \dfrac{C-S}{n}\,t$ for $t \in [0, n]$
7. Depreciated value = $V(t) = C - \dfrac{C-S}{n}\,t$ for $t \in [0, n]$
8. Revenue = (Price per Unit) × (Quantity Demanded)
9. Average revenue per unit = $\overline{R}\,(q) = \dfrac{R(q)}{q}$ for all $q > 0$.
10. Marginal revenue = $M_R(q) = \dfrac{dR}{dq}$
11. Profit = $P(q) = R(q) - C(q)$
12. $P(q) > 0 \Leftrightarrow$ the company is operating at a profit
13. $P(q) < 0 \Leftrightarrow$ the company is operating at a loss
14. $P(q) = 0 \Leftrightarrow$ the company is breaking – even
15. Average profit per unit = $\overline{P}\,(q) = \dfrac{P(q)}{q}$ for all $q > 0$.
16. Marginal profit = $M_P(q) = \dfrac{dP}{dq}$
17. $M_R(q) = M_c(q) \Leftrightarrow P(x)$ is maximum
18. If p is the price function and D is the demand function , then

    Elasticity of demand at price p is $E(p) = -p\,\dfrac{D'(p)}{D(p)}$
19. Demand is elastic at p_0 if $E(p_0) > 1$
20. Demand is inelastic at p_0 if $E(p_0) < 1$
21. Demand has unit elasticity if $E(p_0) = 1$
22. $E(p_0) > 1 \Leftrightarrow R(p)$ decreases as the price p increases
23. $E(p_0) < 1 \Leftrightarrow R(p)$ increases as the price p increases
24. $E(p_0) = 1 \Leftrightarrow R(p)$ is a maximum at p
25. If U (W) is the utility of wealth function , then :

    a. absolute risk aversion A R A = $A(W) = \dfrac{-U''(W)}{U'(W)}$

    b. relative risk aversion R R A = $R(W) = W\,A(W) = \dfrac{-W\,U''(W)}{U'(W)}$
26. $\min \sigma_p^2 = w'\Sigma_x w = w_1^2\sigma_1^2 + w_2^2\sigma_2^2 + 2w_1 w_2\,\rho\,\sigma_1\sigma_2$

27. Expected return on the optimum portfolio : $E(R_p) = w' E(R)$
$= w_1 \mu_1 + w_2 \mu_2$.

28. $V = \int_a^b \pi y^2 \, dx$.

29. Total revenue $r = \int \frac{dr}{dq} \, dq$

30. Total cost $\mathbf{C} = \int \frac{dc}{dq} \, dq$

31. $PV = \int_a^b Se^{-it} \, dt$

32. $FV = \int_a^b Se^{rt} \, dt$

33. $FV = PV(1+i)^n$

34. Annuity $A = R \times \frac{1-(1+i)^{-n}}{i}$.

www.ingramcontent.com/pod-product-compliance
Lightning Source LLC
Chambersburg PA
CBHW031819170526
45157CB00001B/118
* 9 7 8 1 4 6 2 8 8 7 8 9 7 *